Advanced Functional Piezoelectric Materials and Applications

Edited by

Inamuddin[1], Tariq Altalhi[2], Mohammad Luqman[3], Hamida-Tun-Nisa Chisti[4]

[1]Department of Applied Chemistry, Zakir Husain College of Engineering and Technology, Faculty of Engineering and Technology, Aligarh Muslim University, Aligarh-202002, India

[2]Department of Chemistry, College of Science, Taif University, 21944 Taif, Saudi Arabia

[3]Department of Chemical Engineering, College of Engineering, Taibah University, Yanbu, Saudi Arabia

[4]Department of Chemistry, National Institute of Technology, Srinagar, Jammu and Kashmir 190006, India

Published by **Materials Research Forum LLC**
Millersville, PA 17551, USA

Published as part of the book series
Materials Research Foundations
Volume 131 (2022)
ISSN 2471-8890 (Print)
ISSN 2471-8904 (Online)

Print ISBN 978-1-64490-208-0
eBook ISBN 978-1-64490-209-7

Distributed worldwide by

Materials Research Forum LLC
105 Springdale Lane
Millersville, PA 17551
USA
https://www.mrforum.com

Manufactured in the United States of America
10 9 8 7 6 5 4 3 2 1

Table of Contents

Preface

Piezoelectric materials are solid substances with the capacity to produce an internal electric charge in response to an external mechanical stress, notably, electricity created by pressure and latent heat. The primary subgroups of piezoelectric materials include piezo-crystals, piezo-ceramics, piezo-semiconductors, piezo-biological matter, and piezopolymers, among others. Piezoelectric materials possess a plethora of uses in sensors, actuators, power sources, motors, environmental, and biomedical domains, depending on their selected attributes. In light of this, piezoelectric materials and their utilization are a crucial component of sustainable future energy.

This book offers a descant of functional piezoelectric material critique techniques and applications. All aspects of piezoelectric materials are investigated, including their history, developments, properties, process design, and uses. Essential topics are addressed by prominent experts. Hence, this book is significant for researchers, students, educators, and engineers in the field with a multitude of basic specialties, including physics, chemistry, electronics, material science, biochemical engineering, and so on.

Chapter 1 wields several aspects and typical characteristic constants that are necessary for selection of suitable piezoelectric composition. With regard to piezoelectric materials that also exhibit ferroelectric activity, fundamental features derived from various characterization approaches were succinctly discussed.

Chapter 2 delineates how piezoelectric materials generate electrical energy as a result of mechanical stress. Piezoelectric materials have recently gained prominence as one of the key functional materials in precision and acoustic engineering sectors. This study describes the production of several piezoelectric materials (ceramic and biomaterials), as well as their applications and challenges.

Chapter 3 describes the feasibilities of different piezoelectric materials for execution as piezoelectric nanogenerators. The relation between piezoelectricity and crystal structure of the materials are described. Additionally, the major bottlenecks about the synthesis of materials and nanogenerator fabrication as well as performance with future possibilities are reflected in the literature.

Chapter 4 discusses in greater depth piezoelectric materials-based phototronics. The chapter summarizes this concept, from the fundamental of piezoelectricity to the advanced effects and applications of piezophototronics. The roles of piezoelectricity, semiconductors, and photoexcitation as building blocks are also explored to help grasp the notion of phototronics. The main detail focuses on the effects and applications of piezophototronics and gives an outlook on the future of these phototronics.

Chapter 5 considers the current research and advancements on piezoelectric composites and their creative applications. Additionally, this chapter focuses on the properties of composites made of piezoelectric materials. The benefits of these composites such as high connecting factors, low acoustic impedance, good fixation with human tissue, mechanical modulus, and so on are also discussed.

Chapter 6 deliberates the development of new materials for energy production and biomedical applications as scientists explore the potential uses of piezoelectric materials. With their accessibility and usefulness, these materials have demonstrated their revolutionary potential in the modern era. This chapter has provided an overview of the types of piezoelectric materials and their applications as sensors, energy harvesting generator cell and tissue regeneration.

Chapter 7 deals with piezoelectric thin films and discusses different lead free piezoelectric thin films, their synthesis method and application. Piezoceramic industry has been ruled with lead based ceramics for decades, but now lead free materials are in study. Lead free piezoelectric thin films such as in oxide, aluminum nitride and potassium sodium niobate based thin films are briefly deliberated.

Chapter 8 discusses about bulk lead-free piezoelectric materials and their application as actuators, resonators and sensors. After the shift to lead free based piezoceramics a lot of lead alternatives have come up in the horizon of piezoceramic industry, some of them such as sodium potassium niobate based.

Chapter 9 discusses about piezoelectric mechanism, different types of piezoelectric materials, and their fabrication methods. Additionally, utilization of these piezoelectric materials in different fields like wearable and implanted biomedical, energy, tissue engineering, and other fields are briefly discussed. The conclusion and future scope are also presented.

Advanced Functional Piezoelectric Materials and Applications   Materials Research Forum LLC
Materials Research Foundations 131 (2022) 1-36   https://doi.org/10.21741/978164490209-1

Chapter 1

Types, Properties and Characteristics of Piezoelectric Materials

Y. Kalyana Lakshmi[1,a], K.V. Siva Kumar[2,b] and S. Bharadwaj[3,c*]

[1]Department of Physics, University College of Science, Osmania University, Hyderabad, Telangana, 500007, India

[2]Ceramic Composite Materials Laboratory, Department of Physics, Sri Krishnadevaraya University, Anantapur, Andhra Pradesh, 515055, India

[3]Department of Physics, GSS, GITAM (Deemed to be University), Hyderabad, Telangana, 502329, India

[a] kalyaniyanapu@gmail.com, [b] sivakumar.sivani@gmail.com, [c] ba626k@gmail.com

Abstract

Piezoelectric materials are an important class of materials that find application in science and technology, in engineering and in modern warfare as pressure transducers, sensors and energy harvesting devices. Selection of suitable piezoelectric materials relies on several factors such as the type, Curie temperature, environmental stability, different physical properties and their characterization depending different measurement techniques. Several aspects and typical characteristic constants that are necessary for selection of suitable piezoelectric composition are very broadly highlighted in this chapter. Fundamental properties obtained from different characterization tools were concisely discussed with respect to piezoelectric materials which also show ferroelectric behavior.

Contents

1.   Introduction

The piezoelectricity is the ability of certain materials to develop electric charges proportional to the external mechanical stresses. Under stress, in the deformed materials, electrical charges with different polarity are produced on opposite faces, resulting in piezo-potential. These materials can operate in wide temperature (<200ºC) and frequency ranges, have resistance to minor shock, vibration, chemical atmosphere, etc. It further depends on design, material selection from different space groups and operating frequency, generally from 1 Hz to 1 MHz. Bulk piezoelectric materials from synthetic ferroelectric ceramics can be prefabricated into desired shape, area with high strength.

Piezoelectric materials with high dielectric constant have several advantages but also have certain disadvantages [1]. Piezoelectric crystals are often single crystalline materials, which have asymmetrical lattice and exhibit properties even without poling. These materials contain irregular domain structure that align in specific direction, under the process called 'poling'. During the process of poling the sample is subjected to high electric field gradient under elevated temperatures beyond the Curie temperature of the sample and then further cooled to room temperature. Common characteristics of piezoelectric ceramics materials show high dielectric constant or permittivity (ε'), moderate Curie temperature (T_c) and high piezoelectric constants. These properties make them useful in energy harvesting transducers or harvesters but low mechanical properties make them inappropriate for high frequency applications. Preparation of these materials require high degree of perfection in synthesis with high purity and fabrication of crystals with controlled precision. In order to improve, use of polymers enables higher flexibility, low density and resistance with higher piezoelectric voltages have attracted the attention but they possess lower d_{33} values [2-4].

Piezoelectric phenomenon can be found naturally occurring materials such as Quartz, topaz and other group minerals [5]. Piezoelectric materials are found naturally and can also be synthetically produced. Synthetically produced piezoelectric materials are similar to natural ones and possess at-least 32 crystal classes out of which 20 crystal classes exhibit piezoelectric properties. Table 1 shows the list of some materials which show piezoelectric phenomenon.

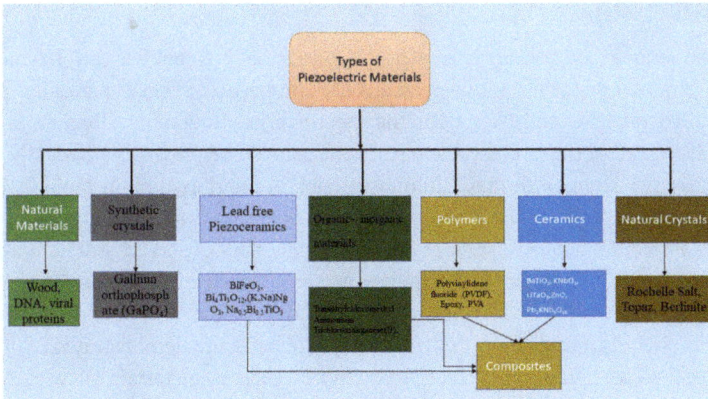

Fig. 1 Different types of piezoelectric materials

The electromechanical effects produced in any piezoelectrical material depends on piezoelectric effects, electrostriction, ferroelectricity (domain switching) and induced phase changes [6]. While selecting a material, one should be aware that each material with certain compositional aspects that show different properties and when doped, strain will vary under the influence of electric field or electrostriction. While selection of any material, one should properly synthesize material with optimum properties. Fig.2 clearly show selection of piezoelectric material should be based on the application, properties and material compatibility.

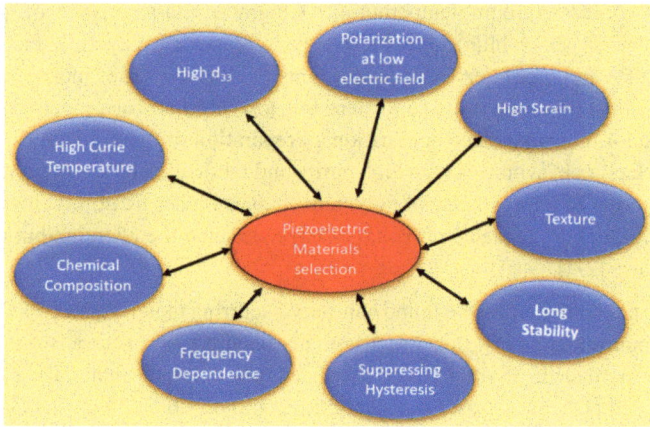

Fig. 2 Selection of piezoelectric materials based on the properties

1.1   Single crystals

Among the known piezoelectric materials, quartz and Rochelle salt have naturally occurring single crystals exhibit piezoelectric properties. These naturally occurring materials have to be carefully cut along the specific crystallographic direction to optimize the piezoelectric responses. Quartz and Rochelle salt can be grown synthetically but these have limited use. Along with these lithium sulfate, tourmaline, lithium niobate ($LiNbO_3$), $Sr_2Nb_2O_7$, $La_2Ti_2O_7$ and lithium tantalate ($LiTaO_3$) are useful in shock wave measurements and high temperature acoustic sensors [7-8]. It is generally observed that naturally grown crystals are more suitable for hydrophones in sonar application for prolonged periods. Although quartz provides good resistivity but its $d_{11} \sim 2.3$ pC/N whereas tourmaline has pyroelectric effects around 4×10^{-6} $C/(m^2.K)$ [9]. Rochelle salt demonstrated 2700 pC/N at $20°C$ but has disadvantage of hydroscopic nature whereas materials such as potassium dihydrogen phosphate (KDP) or ammonium dihydrogen phosphate (ADP) have piezoelectric activity around 23 and 49 pC/N [9] only.

1.2 Ceramics

Among ceramics, popularly known ferroelectric materials such as barium titanate($BaTiO_3$) and lead zirconate titanate (PZT) are widely used as piezoelectric materials The main issue of these materials is low operating temperature or Curie temperature and wear-tear effects. Several efforts have been made to improve their properties and enhance Curie temperature by doping different elements. Strontium or calcium mixed with lead titanate gives optimum properties useful for hydrophone applications while $Na_{0.5}Bi_{4.5}Ti_4O_{15}$ can be used for accelerometers up to a temperature range of 400°C [8]. Lead metaniobate ($PbNb_2O_6$) can be used for submergence hydrophone or non-destructive tool of testing [9]. Sodium potassium niobate ($NaK)Nb_2O_6$ and antimony sulfur iodide (SbSi) can also be used for high frequency transducer applications [10] but have limitation with Curie temperature. $PbTiO_3$ have been used to exhibit high piezoelectric coefficients which could be used for different acoustic applications. [11]. Maximum strain performance can be observed near morphotropic phase boundaries (MPB) with multiphase structure comprising of R(R3M) and T(P4mm) phase zones and much literature can be found [6,12]. Lot of work is done on improvement of piezoelectric coefficients for MPB based ceramics adjusting Curie temperature (T_c) near to room temperature.

1.3 Composites

Composite materials provide improvement in fabrication, designing and study of different mechanisms, especially when particle or grain size is lowered. Nano or sub micro-grain size of composite material leads to low yield loss in thin thickness layered materials, improving reliability and mechanical strength. Composites in piezoelectric materials research can be understood on the basis of three stages:

- First stage is to identify material which show highest piezoelectric coefficient, electromechanical coupling, large bandwidth with high sensitivity and source level.

- Second stage comprises of improving Curie temperature, lowering heat shunts, sustaining electric field and mechanical strength.

- Third stage comprises of improving quality factor (Q) and electromechanical coefficients for high frequency and low frequency applications.

Among popular piezoelectric materials, $Pb\left(Mg_{1/3}Nb_{2/3}\right)O_3 - PbTiO_3$ also known as PMT-PT has highest electro-strictive strain than $BaTiO_3$ [13]. $\left(K_{1-x}Na_x\right)NbO_3$ [14], $\left(K_{1-x}Na_x\right)NbO_3 - LiNbO_3$ [15]. PZT with PDMS show high d_{33} component nearing 78.33

pC/N with improvement in polarization in shorter cycles [16]. Some leadfree composites such as $Bi_{1/2}Na_{1/2}TiO_3$ and $BaTiO_3$ are primarily used to drive intended operation at particular frequency based on resonance and anti-resonance [17-18]. In $BaTiO_3$, grain oriented sintering in the direction of [100] was achieved 788 pC/N for d_{33} [19] whereas when mixed with pseudo-binary systems such as $Ba(Ti_{0.2}Zr_{0.2})O_3 - (Ba_{0.7}Ca_{0.3})TiO_3$ reported 620~755 pC/N [20-21]. A magnitude of 416 pC/N was recorded for (K,Na)NbO₃ when combined with lithium (Li), tantalum (Ta) or tin (Sb) with improved Curie temperature [22]. Adjustment of Nb concentration in $0.96[Bi_{0.5}(Na_{0.84}K_{0.16})_{0.5}(Ti_{1-x}Nb_x)O_3]-0.04SrTiO_3$ results in high strain around 0.438% at 50 kV/cm with d33 around 876 pm/V [23].

1.4   Polymers

Piezoelectric polymer materials have received lot of interest as they possess an ability to transform the vibrational energy into electrical in controllable environment. Polymer piezoelectric materials provide high flexibility, good mechanical strength, low processing cost, high biocompatibility and chemical resistance [24]. Among them polyvinylidene fluoride (PVDF) along with its copolymer possess highest piezoelectric coefficients showing a promising result when mixed with different inorganic fillers [25]. Polymers possess additional antioxidation along with mechanical, chemical resistance and hydrolytic properties [26]. High thermal, electrical actuation and modification in surface charge due to poling, altering responses has developed interest in polymer based piezoelectric materials [27]. PVDF mainly comprises non-polar (α), polar (β,γ,σ) and ε-phases [28] and among them β-phase can be optimized for high piezo-coefficient values [29]. Other polymers used along polyvinylidene fluoride (PVDF) such as trifluoroethylene (TrFE) have improved electromechanical responses due to β-phase and chain mobility [30]. The PVDF-TrFE have shown 40 times greater sensitivity and strength for pressure sensors [31]. Polymer based piezoelectric materials can provide energy harvesting concepts without external power from different ambient sources. They can be wearable and flexible for which then can be used in human gesture recognition techniques [32], real time monitoring devices [33], health assessment [34], electronic textile cardiogram [35] etc.

1.5   Sensor configuration based on shape and size

Sensitivity of any piezoelectric material is generally noted in open circuit voltage when applied stress is acted upon the material with thickness (t). Some sensors, especially in underwater hydrostatic acoustic environment have to work below <200 kHz frequency range for which sensitivity is very low. In hydrostatic modes of operation, $d_n = d_{31} + d_{32} + d_{33}$. For some ceramics, $d_{31} \approx d_{32}$ resulting $d_{33} \approx -2d_{31}$ and this cause d_n very small and insignificant. Moreover, $g_n \approx \dfrac{d_n}{\varepsilon \times \varepsilon_0}$ where ε as dielectric constant and for most of the piezoelectric materials, high dielectric constant is observed, for which g_n is small. Due to low g_n, pressure wave on the sample is very low for which low voltage is developed across.

There are number of ways to reduce the hydrostatic pressure effects on the piezoelectric ceramics and one of the ways is to encapsulate the material in polymer with air backing from one side. Different shapes, sizes and air backing are used for fabrication of piezoelectric transducers and among them materials that range from bars with different length or shapes, disk with different dimensions, cylinders with different diameter and lengths, spherical or flexors/benders etc. Preparation of cylinders or bars depends on the operating frequency or resonating frequency which further depends on the dimension of the sample. Furthermore, some transducers which operate under hydrophobic conditions have to be shielded from different pressure wave effects and they mostly operate under lower fundamental resonance conditions. For omni directional reception, spherical designed transducer is designed using two equally radially poled hemispheres [6]. Generally, spherical transducers have large $d_h.g_h$ product and stability of properties are found to be up to 7 MPa pressure.

1.6 Classification based on dimension

Based on the dimension of piezoelectric materials, it is further classified into

a) Zero-dimensional (0-D)

b) One-dimensional(1-D)

c) Two-dimensional(2-D)

d) three dimensional (3-D).

Zero-dimensional (0-D) have materials in nano clustered with high dispersion and surface area with easy preparation techniques. Choice of dimension play vital role in improving several properties. 1-D piezoelectric material provide high electro-mechanical capability where 3-D devices is crucial role for acoustic impedance. Based on the classification of size, shape, material or mixing with another materials, it can be said that piezoelectric materials can be used for different applications as shown Table 1 and additional literature can be understood from some review papers listed in Table 2.

Table 1: List of Piezoelectric materials used for different applications.

Sl.No	Material	Reference	Applications/Materials
	Table 1: List of Piezoelectric materials used for different applications.		
1.	BaTiO3	36	Flexible Materials
2.	BaTiO$_{3+}$ PMMA	37	Improved Bridges and dendritic structures
3.	BaTiO3@ PVDF-TrFE	38	Composites
4.	BiFeO3-PbTiO3-BaTiO3	39	high Curie temperature and large piezoelectric properties
5.	PZT+ thermoplastic polyurethane elastomer	40	Composite/ acoustic absorption coefficient
6.	CuO-doped PZT-PZNN	41	Composite/ energy harvester
7.	PZT/epoxy	42	Composite/ Ultrasonics
8.	NBT-BT type	43	Composites
9.	NBT-BT/PVDF composite	44	energy harvesting
10.	NBT–BZT	45	Energy storage and energy harvesting application
11.	ZnO+PDMS	46	flexible nanogenerator
12.	ZnO/CdS nanofibers	47	degradation of bisphenol A
13.	ZnO+KNN	48	Composites
14.	BCT-BS	49	Composites
15.	KNN-CT	50	Composites

Table 2: List of some review papers publish in piezoelectric materials

Sl.No	Title of the Review Paper	References
	Table 2: List of some review papers publish in piezoelectric materials	
1.	High voltage coefficient piezoelectric materials and their applications	51
2.	A critical review on lead-free hybrid materials for next generation piezoelectric energy harvesting and conversion	52
3.	Flexible nanogenerators for wearable electronic applications based on piezoelectric materials	53
4.	Piezoelectric materials for sustainable building structures: Fundamentals and applications	54
5.	Multicomponent nanostructured materials and interfaces for efficient piezoelectricity	55
6.	Recent development in lead-free perovskite piezoelectric bulk materials."	56
7.	Factors Affecting the Piezoelectric Performance of Ceramic-Polymer Composites: A Comprehensive Review	57
8.	Stretchable piezoelectric energy harvesters and self-powered sensors for wearable and implantable devices	58

9.	Technology transfer of lead-free (K, Na) NbO3-based piezoelectric ceramics.	59
10.	A review on vibration-based piezoelectric energy harvesting from the aspect of compliant mechanisms	60
11.	Development and outlook of high output piezoelectric nanogenerators	61
12.	Piezoelectric MEMS based acoustic sensors: A review	62
13.	Performance enhancements in poly (vinylidene fluoride)-based piezoelectric nanogenerators for efficient energy harvesting	63
14.	Flexible PVDF based piezoelectric nanogenerators	64
15.	A comprehensive review of flexible piezoelectric generators based on organic-inorganic metal halide perovskites	65
16.	Piezoelectric energy harvesters for biomedical applications	66
17.	Electrical stimulation and piezoelectric biomaterials for bone tissue engineering applications	67
18.	A review on ZnO-based piezoelectric nanogenerators: Synthesis, characterization techniques, performance enhancement and applications	68
19.	Progress in high-strain perovskite piezoelectric ceramics	69
20.	Frequency tunable, flexible and lowcost piezoelectric micro-generator for energy harvesting.	70
21.	Strategies to achieve high performance piezoelectric nanogenerators	71
22.	Piezoelectric sensors and sensor materials	72
23.	Piezoelectric materials for high temperature sensors	73
24.	Review on engineering structural designs for efficient piezoelectric energy harvesting to obtain high power output	74
25.	A review of power harvesting from vibration using piezoelectric materials	75

2.   Properties of piezoelectric materials

Piezoelectric materials have found its utility in sensors, actuators, transducers, transformers etc. Majority of piezoelectric materials have different orientations in crystal structure resulting deviation in center of symmetry, thus positive and negative charges do not cooccur unless external applied field and possess property of spontaneous polarization (P_s). Currently, a requirement for piezoelectric materials with treble performance in wide temperature and frequency range with improved response and high stability is desired, especially in attainable electric fields. Many materials have to encounter phase instability at different temperature [76-79], frequency and such materials results in different properties. When piezoelectric materials are operated at low frequencies, the conductivity induces charge drift which inhibits piezoelectric induced charges. While selection of piezoelectric materials, listed below properties have to be considered while designing for any application.

2.1 Basic equations

The piezoelectric materials convert mechanical energy into electrical energy following the relations:

$$d_{ijk} = \left(\frac{\partial D_i}{\partial T_{ijk}} \right)_E \qquad g_{ijk} = \left(\frac{\partial E_i}{\partial T_{ijk}} \right)_D \ \dots \ \ (1)$$

Where D_i is the electric displacement, E_i is the electric field, T_{jk} is the stress component, d_{ijk} is piezoelectric charge and g_{ijk} is the piezo voltage coefficients. The linear relationship which exists between electric displacement (D_i) and electric field (E_i) can be written as

$$D_i = \sum_{j=1}^{3} \varepsilon_{ij} E_j \qquad \dots\dots \ \ (2)$$

where ε_{ij} is the permittivity is a second order tensor. The interrelationship between stress (T) and strain (S) for mechanical properties can be viewed through the following relations:

$$T_{ij} = \sum_{i,j,k,l=-1}^{3} c_{ijkl} \cdot S_{kl}$$
$$S_{ij} = \sum_{i,j,k,l=-1}^{3} s_{ijkl} \cdot T_{kl} \qquad \dots\dots\dots \ \ (3)$$

Where c_{ijkl} is elastic stiffness constant and s_{ijkl} is the elastic compliance parameters of the transducer materials. Since the stress (T) and strain (S) are second rank tensors the elastic stiffness constant, c_{ijkl} and elastic compliance s_{ijkl} turns out to be fourth order tensors. Using Eq.(2) and Eq.(3), we can write

$$D_i = \varepsilon_{ij}^T E_j + d_{ij} T_j$$
$$S_i = \varepsilon_{ij}^T E_j + S_{ij}^E T_j \qquad \dots\dots \ \ (4)$$

Here, d_{ij} is strain or charge constant and superscripts (T and E) in some terms are coefficients at constant stress and electric fields. Based on the Eq.(3) and Eq.(4), further one-dimensional (1-D), two-dimensional (2-D) and three-dimensional (3-D) equations can be developed [80].

The entire set of piezoelectric coefficients can be broadly written as

- Stress (T)
- Strain (S)

- Electric Field (E)

- Electric displacement (D)

- Piezoelectric charge or strain coefficient (d)

- Piezoelectric stress coefficient (e)

- Piezoelectric voltage coefficient (g)

- Piezoelectric stiffness coefficient (h).

- Elastic stiffness coefficient (c)

- Elastic compliance coefficient (s)

Here, eq.(1) show piezoelectric charge or strain coefficient (d) and often referred as direct piezoelectric effect. Here, d_{ij} is intrinsic piezoelectric parameter and can be measured by quasi static d_{33}meter. The term d_{33} represents longitudinal piezoelectric coefficient, d_{15}represents shear piezoelectric coefficient and $d_{31 \text{ or }} d_{24}$ represent transverse piezoelectric coefficients. Piezoelectric stress coefficient (e) can be written as

$$e = \left(\frac{\partial D}{\partial S}\right) E = \left(\frac{\partial T}{\partial E}\right) S \dots \dots \quad (5)$$

The stiffness constant coefficient (h) can be written as

$$h = -\left(\frac{\partial E}{\partial S}\right) D = -\left(\frac{\partial T}{\partial D}\right) S \dots \dots \dots \quad (6)$$

The constants d, e, g and h can be further linked as

$$
\begin{aligned}
&d = \varepsilon^T g = S^E g \\
&e = \varepsilon^S h = c^E d \\
&G = \left(\frac{d}{\varepsilon T}\right) = S^D h \dots \dots \dots \quad (7) \\
&h = \left(\frac{e}{\varepsilon S}\right) = C^D g \\
&\varepsilon^T = \varepsilon^S + d.e
\end{aligned}
$$

2.2   Curie temperature

Basically piezoelectric materials are essentially ferroelectric in nature which exhibit Curie temperature (T_c) [76-79]. At Curie temperature, and beyond Curie temperature, the material reverts back to paraelectric phase, from ferroelectric phase, and all the piezoelectric parameters of the materials disappear. Even though poling may provide thermal stability to dipoles but as temperature increases, the random orientation increases dampening the piezoelectric properties, which is termed as "thermally activated aging". If the material is operated below Curie temperature, then not only it minimizes the aging effect. Prolonged usage of piezoelectric material may soften the crystal lattice and show strong polarization for small variation in electric or temperature change, causing a change in electromechanical properties [55,81].

2.3   Phase transition

Some materials such as quartz or gallium orthophosphate, have α-β transition as part of their symmetry. As temperature rises beyond this transition temperature the piezoelectric properties of these materials disappear.

2.4   High dielectric constant

Dielectric constant is a value of stored charge present in any material during an applied field. It is useful in overcoming the losses in any material. Electric displacement for any material under applied electric field is given by:

$$D_i = \varepsilon_{ij} E_j \ldots\ldots(8)$$

Where 'D' is electric displacement (C/m^2), ε_{ij} is measured as permittivity (F/m), E is the electric field (V/m). It should be known that when ε_{ij} is measured at lower frequency, it is referred as dielectric constant, mostly below 1 MHz and above 1 MHz, it is termed as relative permittivity. There are four types of polarization which contributes to dielectric value and they are

- Ionic polarization
- Electronic polarization
- Space charge polarization
- Oriental polarization.

Out of four, ionic and electronic polarization play curial role in determining the dielectric constant or permittivity value at different frequency zones. Apart from these, random dipole rotation due to thermal excitation and domain wall motion [82] due to applied field or impurities play vital role in actual value of dielectric constant.

In piezoelectric material, moderate ε_{ij} is essential to overcome the losses. High ε_{ij} may cause over damping and lower the piezoelectric voltage coefficient [83]. It is should be observed that ferroelectric properties are often dependent on charge carriers, mobility of charged electrons and domain walls, phase transition and polarization rotation. Some additional non ferroelectric materials dependent on Clausius- Mossotti relation [84].

2.5   Sensitivity

Sensitivity is one the important parameter of piezoelectric studies which should be as high as possible after removal of background noise at different temperature. Piezoelectric charge coefficient (d) can be interrelated with piezoelectric voltage coefficient (g) using an equation:

$$g = \left(\frac{d}{\varepsilon_o \varepsilon_{ij}} \right) \dots \dots \tag{9}$$

Here, piezoelectric charge coefficient (d) is the ratio of the short circuit charge per electrode with area to an applied stress whereas piezoelectric voltage coefficient (g) gives the ratio of field developed to the stress applied [55,81-83]. Strain developed in material to an applied electric field at electrodes or actuators is also referred as piezoelectric coefficient (d) whereas strain developed for applied charge is piezoelectric voltage coefficient (g) [55,81-83]. Typical values of 'd' should be in the range of >2000 pC/N for synthetic ferroelectric and 'g' values should be in the range of 0.02 to 0.2 Vm/N. Piezoelectric charge coefficients (d) should be generally high and dependent on permittivity whereas piezoelectric voltage coefficient (g) are dependent on temperature variation.

2.6   Electromechanical Coupling Factor (k)

In piezoelectric materials, the electromechanical coupling factor indicates the portion of mechanical energy which is transformed into electrical energy and vice versa. It is given by the relation:

$$k^2 = \frac{E_m}{E_o + E_m} \dots \dots \tag{10}$$

Where E_o is the portion of energy remained after not being converted into electrical energy and E_m is the fraction of energy which is being exchanged as electrical energy. It can be directly calculated using resonance frequency (f_r) and anti-resonance frequency (f_a) with relation:

$$k^2 = \left(\frac{\pi}{2} \right) \left(\frac{f_r}{f_a} \right) \cot \left(\frac{\pi}{2} \frac{f_r}{f_a} \right) \dots \dots \tag{11}$$

Advanced Functional Piezoelectric Materials and Applications   Materials Research Forum LLC
Materials Research Foundations 131 (2022) 1-36   https://doi.org/10.21741/978164490209-1

Since complete conversion of energy is not possible experimentally, hence k^2 is generally less than 1. Based on measurement of electromechanical coupling factors depending on direction, one can calculate planar (k_p), lateral (k_{31}), longitudinal (k_{33}), thickness shear (k_{15}) and thickness extension (k_t). The effective coupling factor for resonator [6-7] at any harmonic is written as

$$k_{eff}^2 = 1 - \left(\frac{f_r}{f_s}\right)^2 \dots\dots \tag{12}$$

Where f_r is the resonance frequency and f_a is anti-resonance frequency of piezoelectric material. It is observed that electromechanical coupling factor should be greater than 90% in relaxor crystals and impervious to temperature variation.

2.7   Resistivity (R) and time constant (RC)

In piezoelectric material, moderate to high resistance is desired such that large electric field during excessive charge leakage or breakdown during poling procedure. Resistance is required such that when a charge is developed, it should sustain for a longer period of time and thus get detected by electronic measurable systems. Thus, when the charge is developed in the material it is related to capacitance (C) in the material with resistance (R), duration of time is thus referred as RC time. Using this time constant, lower limiting frequency can be calculated given by the relation:

$$f_{LL} = \left(\frac{1}{2}\right) \times \pi \times R \times C \dots\dots \tag{13}$$

Below this frequency, the amount of charge developed will be drained off before it gets detected [85-87]. In many applications involving low frequency sensors, large time constant is required. It should be remembered that resistance depends on many factors such as frequency, temperature and material characteristics.

Electromechanical coupling factor (k) can be measured as the square root of the fraction of mechanical energy which is converted in electrical energy in each cycle. The value of k is amount of energy stored during each cycle or dissipation. Furthermore, the interrelation dependence of d_{ij} with k_{ij} can be written:

$$k_{15} = \left[\frac{d_{15}}{\sqrt{s_{44}^E \varepsilon_1^T}} \right]$$

$$k_{33} = \left[\frac{d_{33}}{\sqrt{s_{33}^E \varepsilon_3^T}} \right]$$

$$k_{31} = \left[\frac{d_{31}}{\sqrt{s_{11}^E \varepsilon_3^T}} \right] \qquad \cdots \cdots \qquad (14)$$

$$k_p = \left[\frac{\dfrac{d_{31}}{2}}{\left(S_{11}^E + S_{12}^E\right)\varepsilon_3^T} \right] = \left[\frac{k_{31}}{\sqrt{2(1-\sigma)}} \right] \quad \text{where } \sigma = \left(-\frac{S_{12}^E}{S_{11}^E} \right)$$

Here, σ is Poisson ratio, k_p is the planar coupling factor which arises from the application of stress from 1 & 2 axis. Generally, k_p should be as large as possible and k_{33} with k_{15} should be greater than 0.7. k_t is also important coupling factor where the subscript indicates the longitudinal thickness vibration and is given by the relation:

$$k_t = \left(\frac{e_{33}}{\sqrt{\varepsilon_3^S c_{33}^D}} \right) \cdots \cdots \qquad (15)$$

Where c_{33} is the stiffness constant gives the ratio of stress to strain in the selected direction 3. Generally, the $k_{33}^2 \sim \left(k_t^2 + k_p^2 - k_t^2 k_p^2 \right)$ and the other coupling constants in open and short circuit are given by interrelation:

$$S_{44}^D = S_{44}^E \left(1 - k_{15}^2 \right)$$

$$S_{33}^D = S_{33}^E \left(1 - k_{33}^2 \right)$$

$$S_{11}^D = S_{11}^E \left(1 - k_{31}^2 \right) \quad \cdots \cdots \cdots \qquad (16)$$

$$S_{12}^D = S_{12}^E - k_{31}^2 S_{11}^E$$

$$c_{33}^E = c_{33}^D \left(1 - k_t^2 \right)$$

Further interrelation using dielectric constant and electromechanical coupling can be written as

$$k_1^S = \left(1 - k_{15}^2 \right) k_1^T$$

$$k_3^S = \left(1 - k_p^2 \right)\left(1 - k_t^2 \right) k_3^T \cong \left(1 - k_{33}^2 \right) k_3^T \qquad \cdots \cdots \cdots \qquad (17)$$

2.7   Quality factors (mechanical and electrical)

Quality factors is one of important parameter in reshaping the power handling ability and bandwidth control of piezoelectric devices [85-86]. Quality factor (Q) is associated with both mechanical and electrical energy, is related to operational aspects of the application. When heat is generated, dielectric loss will increase and reciprocal of dielectric loss is quality factor, Q_E which is an indicator of off resonance aspects of the material whereas Q_M indicate the narrow resonance peak. As the temperature arises, damping or phonon scattering in material increases, thus varying quality factor. Moderate to high quality factor is essential for resonant type applications where high Q_M means narrow resonance. This further leads to determination of greater accuracy of resonance frequency or sensing precision. Mechanical quality factor (Q_M) can be estimated using the relation:

$$Q_M = \left(\frac{f_r}{f_2 - f_1} \right) \dots\dots\dots \tag{18}$$

Where f_r is the resonance frequency, $f_1 \& f_2$ are sideband frequencies at -3dB of the maximum admittance. Q_M is interrelated with electromagnetic coupling factor (k), free capacitance (C_f) and impedance (Z) at resonant frequency (f_r) using the relation

$$Q_M = \left(\frac{1}{2\pi f_r C_f Z_{min}} \right) \times \left(\frac{1}{k^2} \right) \dots\dots\dots \tag{19}$$

Quality factor (Q_M) should be high in theoretical nature for any sensors but often it is compromised for coupling factors and piezoelectric coefficients.

2.8   Figure of Merit (FOM) and strain coefficient

Figure of Merit (FOM) and strain coefficients are two important factors which are two be considered. FOM is the product of piezoelectric coefficient and mechanical quality factor [88] and can be written as product of $d_h.g_h$. FOM is used to ascertain which type of amplifier that can be used to overcome the noise and measure the quality of sensing capacity of any material.

2.9   Piezoelectric resonance frequency

In piezoelectric material, under electric field, many factors such as density (ρ), elastic constants (S^E), geometry of the sample [shape, thickness (t), length (l), size, area, weight] affects the resonant frequency and anti-resonant frequency [81-82]. The relation of resonant and anti-resonant frequency generally depends on the vibration mode, are given by the relations:

$$f_r = \left(\frac{1}{2l\sqrt{\rho S^E}} \right)$$
$$\ldots\ldots\ldots\ldots \qquad (20)$$
$$f_a = \left(\frac{1}{2t} \right)\left(\sqrt{\frac{C^D}{\rho}} \right)$$

At resonant frequency, impedance is minimum for which maximum current is obtained at low voltage while at anti-resonant frequencies, admittance is minimum recorded at high voltage with minimum current [89]. It is important to study the temperature dependence of material for wide frequency range.

2.10 Thermal expansion

In piezoelectric material, thermal expansion coefficient (α) is important factor and should have range from 1~1000 ppm/K at room temperature. Generally, thermal expansion of any material depends on the material composition, ionic or chemical strength, dopants, exchange interaction among dopants, anisotropic nature of the dopants and compounds, etc., [82]. Thermal expansion studies are important to understand the variation of material under different temperature condition and its interface with packaging device. Piezoelectric materials such as Barium Sodium Niobate (BNN), Ethylene Diamine Tartrate (EDT), Ammonium Dihydrogen Phosphate, Lithium Sulfate, Potassium Dihydrogen Phosphate (KDP) and Lead Niobate was briefly discussed with point group and its property elsewhere [90].

2.11 Ageing

While designing any transducer or actuator for any application, ageing effect have to be considered. As time progress, the thermal change, mechanical stress or transducing strong electrical signal will vary, affecting the linearity response of the material.

3. Characterization of piezoelectric materials

Piezoelectric materials have good actuators and performance parameters like dielectric constant, Curie temperature (T_c), piezoelectric coefficient (d) and electromechanical coupling parameter (k), reflecting different responses at different conditions such as pressure, stress, temperature and nature of material. Characterization of piezoelectric material is generally done by obtaining different parameters discussed in the previous section.

3.1   Measurement of piezoelectric coefficient

Calculating piezoelectric constant is important aspect in characterizing the piezoelectric materials. There are three commonly known methods and they are

- Frequency method
- Laser interferometry method
- Quasi-static method.

Frequency method is commonly used method but needs prior knowledge of some additional material coefficients is important in determining the piezoelectric coefficients [91-92]. Impedance analyzers can detect and measure high sensitivity in diverse frequency range. However, the main disadvantage of such techniques is to fabricate sample holder in specific dimension along with sample to suit the piezoelectric material. Furthermore, the calculation and measurement have to be satisfied on par with international testing agencies standards [93-94]. Comparison of all three method is elaborately discussed and can be found in the literature [95]. Piezoelectric materials with elastic nature show resonant (f_r) and anti-resonant frequency (f_a) using electric field as shown in Fig.3. A simple circuit comprising a LCR for small resistance (R), gives resonant frequency such that impedance $2\pi f_r L$ and $-\left(\dfrac{1}{2\pi f_r C_1}\right)$ are opposite in nature. The mechanically obtained resonant frequency (f_r) is in parallel with capacitance C_o. Thus (f_r is standing wave under zero field. Above f_r, the circuit becomes inductive in nature and circuit impedance becomes opposite, reaches maximum, referred as anti-resonance (f_a), is also a standing wave, but under open circuit.

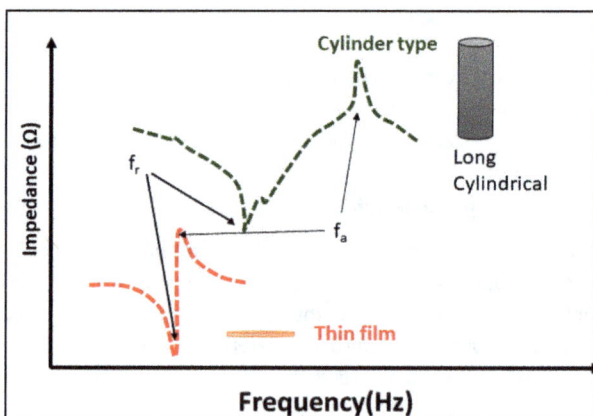

Fig. 3 Impedance versus frequency in piezoelectric materials.

The relation of resonant frequency and anti-resonant frequency with elastic parameters is given through relation:

$$2 \times f_r \times (Thickness\,/\,length\,/\,Area) = \frac{1}{\sqrt{\rho S^E}}$$

$$2 \times f_a \times (Thickness\,/\,length\,/\,Area) = \frac{1}{\sqrt{\rho S^D}}$$

....   (21)

Where ρ is the density of the sample. Using the resonant and anti-resonant frequency, it can be written as

$$\frac{f_a^2 - f_r^2}{f_a^2} = P.\left[\frac{S^E - S^D}{S^E}\right] = P.k^2 = \frac{C_1}{C_o + C_1}$$ ....   (22)

Where 'P' is the dimensionless shape factor which has a value of 0.80, depending on the sample. Furthermore,

$$f_r.(diameter) = \left[2.048 + 0.62(\sigma - 0.30)\right] \times \left[\frac{V}{\pi\sqrt{1-\sigma^2}}\right]$$

where   ....   (23)

$$V = \frac{1}{\sqrt{\rho.S_{11}^E}}$$

Here, V is the velocity of the compression wave in slim bar normal to the poling axis. Resonant frequency and anti-frequency can be clearly identified from the schematic Fig.3 which is further dependent on shape or volume. The quality factor (Q) is reciprocal of dissipation (D) can be measured directly from the impedance analyzer or spectrum analyzer. The quality factor is expressed as the ratio of strain (in phase) with stress to strain which is out of phase, in vibrating body. Generally, high quality factor is desirable in piezoelectric material and indicate the energy lost to mechanical damping when body is vibrating. The d_{33} and d_{31} coefficient [95] can be directly calculated using the equation:

$$d_{33} = k_{33} \left(\varepsilon_{33}^T S_{33}^K \right)^{1/2}$$

$$d_{33} = \sqrt{\left[\varepsilon_{33}^T S_{33}^D \frac{k_{33}^2}{1-k_{33}^2} \right]}$$

$$d_{33} = \sqrt{ \left(C_T \cdot \frac{t}{\frac{\pi d^2}{4}} \frac{1}{4\rho f_a^2 t^2} \left(\frac{\frac{\pi f_r}{2 f_a} \tan\left(\frac{\pi f_a - f_r}{2 f_a} \right)}{1 - \frac{\pi f_r}{2 f_a} \tan\left(\frac{\pi f_a - f_r}{2 f_a} \right)} \right) \right) } \cdot \quad \dots\dots\dots \quad (24)$$

$$d_{31} = k_{31} \sqrt{\varepsilon_{33}^T S_{11}^K}$$

$$d_{31} = \sqrt{ \left(C_T \cdot \frac{\pi f_a}{2 f_r} \frac{t}{4\rho l^2} \left(\frac{1}{\frac{\pi f_a}{2 f_r} - \tan\left(\frac{\pi f_a}{2 f_a} \right)} \right) \right) }$$

Where C^T is the measured capacity, t is the thickness, d is the diameter, ρ is the density and other terms were discussed in the previous sections accordingly.

In laser interferometry, piezoelectric charge constant can be measured by knowing the deflection caused due to elongation or displacement after voltage is applied on the sample. Laser interferometry mainly used to measure charge coefficients d_{31} and d_{33} in high accuracy but its high sensitivity is the main concern. Any small irregularities or abnormal vibration will lead to inaccurate values [96-97]. Piezoelectric coefficient can be measured only one time using this method where the whole optical set up is placed in a base for eliminate external vibration effects. Instruments operate in selected voltage range of mm.s/V to μm.s/V at selected frequency, as shown in Fig.4(b). To evaluate output and input, oscilloscope is used and any deviation is analyzed from ΔV on the applied sample. Measurement of voltage in open circuit and short circuit gives a resulting value of charge coefficient and this is equal to dimensional change of the sample. Piezoelectric coefficient (d_{33} or d_{31}) can be calculated using relation of change in length (Δl) to applied voltage (V_{in}) and given as:

$$d_{33} = \frac{\Delta l}{V_{in}}$$
$$\dots\dots\dots\dots\dots \quad (25)$$
$$d_{31} = \frac{\Delta l}{V_{in}}$$

In quasi-static method, an electric charge on a sample can be measured using amplifiers or voltmeter or d_{31}/d_{33} meter which is directly available in the market [98]. The piezoelectric charge constant (d_{ij}) is measured as electric charge (Q) value obtained during mechanically applied force(F) which results in deformation in desired direction and can be given as

$$Q = d_{ij} \times F \ \dots\dots\dots\dots \tag{26}$$

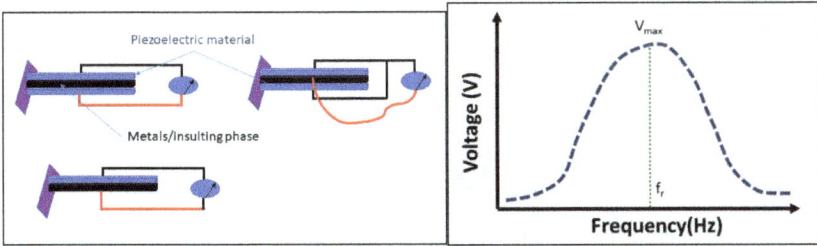

Fig. 4 a) Measurement of voltage or current in piezoelectric sample
b) voltage measurement at different frequency.

The sample can be transverse mode, radial mode, thickness extension mode, cylindrical mode. The charge which is generated on the sample when force develops on the electrode in the piezoelectric sample. Changes can be observed for applied field or positioning of weight on the lever as shown in Fig.4(a). Changes in force can be recorded in terms of voltage pulse in any oscilloscope and piezoelectric coefficient can be calculated using the relation:

$$d_{33} = \frac{V_{out}}{F_3 k_q}$$
$$\dots\dots\dots\dots\dots \tag{27}$$
$$d_{31} = \frac{V_{out}}{F_3 k_q}$$

Where F is the stress applied in the direction 3, k_q is charge sensitivity and V_{out} is the output voltage on the oscilloscope.

3.2   Measurement of dielectric constant

Measurement of dielectric constant in wide frequency range can be calculated using impedance analyzer. High dielectric constant ensures generation of high electrostatic forces which are responsible in actuation stress and strain. Material should have elastic ability defines material strength to defines electrostatic forces. As external electric field (E) is applied, an electric displacement is created towards and can be understood on the basis of $D = \varepsilon \times E$. Piezoelectric material produces strain (S) in electric field (E) and can be written as $S = d \times E$, where 'd' is displacement or elongation. Since, this strain can be either positive or negative and this depending on piezoelectric origin. The free charge density (σ) which appears at electrode can be $\sigma = \varepsilon \times E$. Under the applied field, the movement of free charges gives rise to polarization, increasing charge density by an amount and this can be understood by an equation: $P_p = e \times S = e \times d \times E$, then electrical displacement

$$D = \varepsilon \times E + P$$
$$D = \varepsilon \times E + e \times d \times E \ldots\ldots\ldots\ldots \qquad (28)$$
$$D = \overline{\varepsilon} \times E$$

Here $\overline{\varepsilon}$ is effective dielectric constant.

3.3   Measurement of Curie temperature

Improvement of Curie temperature is essential for any application in operating temperature conditions. Knowledge of Curie temperature gives a tentative idea of transition temperature or phase transition where ferro/piezoelectric materials changes into paraelectric materials. Curie temperature can be found through different measurement techniques such differential scanning calorimetry (DSc), dielectric constant versus temperature, polarization versus temperature etc. Dielectric constant and ferroelectric properties perform at optimum level at vicinity of Curie temperature, as shown in Fig.5.

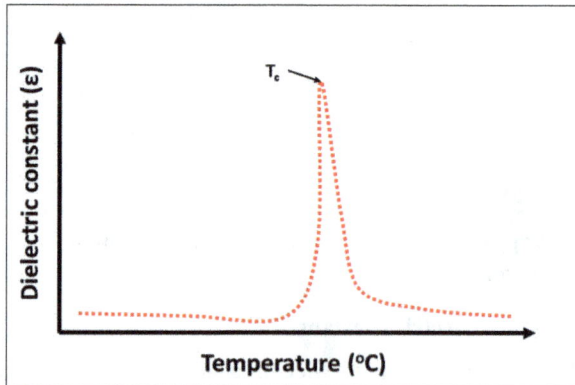

Fig. 5 Dielectric constant versus temperature.

Advanced Functional Piezoelectric Materials and Applications   Materials Research Forum LLC
Materials Research Foundations 131 (2022) 1-36   https://doi.org/10.21741/978164490209-1

3.4   Etching and poling

Some piezoelectric materials do not saturate even after applying high electric static field. For such materials, electric field is applied larger than saturation field at elevated temperature but below the Curie temperature and breakdown field of the material. Such process is called as 'Poling'. After poling, the crystal or material imparts symmetry (α m), orienting the domains in specific direction of the poling field, either by changing the angles or by reversal and thus enhances the polarization in the direction of poling field. Here, domains are small microscopic regions with atomic dipoles aligned along a particular direction. When the poling field is removed, majority of the domains remain aligned in the direction of the field but some may revert back to their position or direction under strained conditions, lowering the net polarization. Thus, both intrinsic and extrinsic conditions play vital role in determining the piezoelectric coefficients. For poled coefficients, d_{31}, d_{33} and d_{15} are main tensor coefficients to measure. Etching is a process to clean the surface of surface using strong acid or chemicals or any grated surface. Etching depends on either wet or dry etching state and further depends on rate of etching. Etching process influences on different piezoelectric characteristics [99].

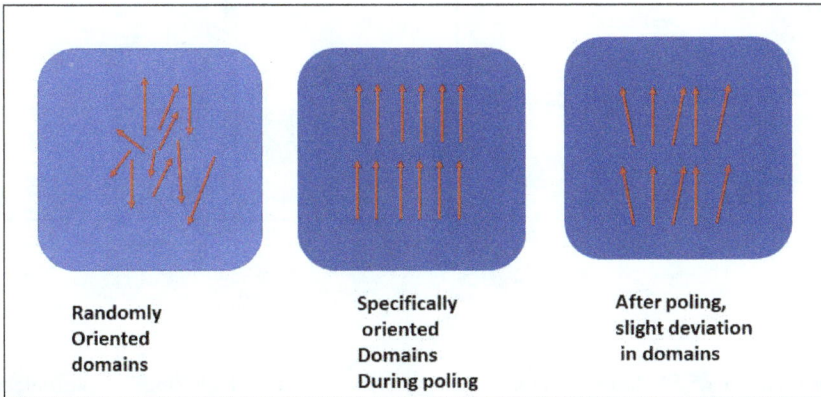

Fig. 6 Domain orientation during poling or etching and after poling.

By measuring grain size, we can actually measure grain size and grain boundaries formation which gives tentative idea on domain structure, as shown in Fig.6. Deformation may occurs inform of twinning or grain boundary sliding which occurs due to shear forces on comparison of normal forces, thus may cause grain roll over [100]. Dielectric constant and d_{33} increases with decrease in grain size and found to be maximum at 0.94 μm [101]. Moreover, the domain structure depends on the either X-cut or Y-cut or Z-cut and direction of domain at angle may provide instability on different ferroelectric properties.

3.5   Measurement of hysteresis (PE/SE) loops

Polarization (P) versus electric field (E) loop indicates the charge developed in terms of polarization (P) which is developed against the field (E) at a given frequency. For ideal capacitor which is linear in nature, a straight line is observed as current lead the voltage by 90°. However, for nonlinear ferroelectric materials, P-E loop is observed, as observed in Fig.7(a). Measuring the PE is important in piezoelectric materials as it indicates change of polarization with coupled states along with strain. PE loop show coercive field (E_c), remanent polarization (P_r) and saturation polarization (P_s) gives the account of extent of ferro/piezoelectricity in the samples. Polarization is interrelated to strain and small variation in current can describe displacement behavior in the sample.

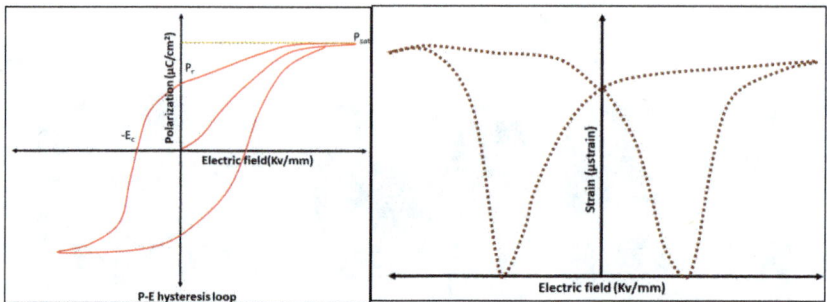

Fig. 7 a) PE hysteresis loop b) SE butterfly loop.

Measurement of SE loop is obtained as butterfly loop in which slope is piezoelectric coefficient (d_{33}) and is important for actuator application, as shown in Fig.7(b). SE loop show strain output is important as some piezoelectric materials show non-linear ferroelectric properties. Measurement of SE loop will quantify the dielectric loss as mechanical loss.

Table 3: List of some materials with some piezoelectric for comparison.

Material	d_{33} (pm/ V)	Strain (%)	E (kV/cm)	k_p	Ref.
Pb $(Zr_{0.7}Sn_{0.3})_x$ $[Ti_{1-x}]O_3$	680	0.34	50	-	[102]
$(Pb_{0.92}La_{0.08})[Zr_{0.65}Ti_{0.35}]O_3$	2000	0.2]	10	-	[103]
$(Pb_{0.90}Sr_{0.10})[Zr_{0.855}Ti_{0.145}]O_3$	1242	0.82	66	-	[104]
KNLNT-CaZrO3	1030	0.36	3.5	-	[105]
KNNS-BAH	507	0.152	3	-	[106]
KNN-LT-LS(LF4T)	750	0.15	2	-	[22]
$0.94(Bi_{0.5}Na_{0.5}TiO_3)-0.06[BaTiO_3]$	155	-	-	0.367	[107]
$0.94(Bi_{0.5}Na_{0.5}TiO_3)-0.04[\{Ba_{0.98}Ca_{0.02}\}(Ti_{0.94}Sn_{0.06})O_3]$	170	-	-	0.33	[108]
$\{[Bi0.5(Na0.84K0.16)0.5]0.96Sr0.04\}Ti0.975Nb0.025O3$	1400	0.70	-	-	[109]
(Ba0.85Ca0.15)(Ti0.9Zr0.1)O3	572	-	-	0.57	[110]
(Ba0.95Ca0.05)(Ti1−xSnx)O3 (x=0.09)	670	-	-	0.50	[111]
PVDF fibres with BT particles	50~ 130	-	-	-	[112]

Conclusions

Selection of a suitable piezoelectric material for its use as transducer, sensor or energy harvester depends on various parameters as mentioned in the above text. This will give the reader a bird's eye view of the field of piezoelectric materials, their types, characteristic coefficients and important properties. While selection of any piezoelectric material for sensing or transducer application, a detailed understanding of coefficients for different material should be understood. Hence, in table 3, an exploratory comparison was made for readers to understand the changes based on d_{33} coefficient, strain (%) and coupling coefficient (k_p).

References

[1] P. Dineva, D. Gross, R. Müller, T. Rangelov, Piezoelectric materials: In Dynamic fracture of piezoelectric materials, Springer, Cham, 2014, pp. 7-32. https://doi.org/10.1007/978-3-319-03961-9_2

[2] W. Li, Q. Meng, Y. Zheng, Z. Zhang, W. Xia, Z. Xu, Electric energy storage properties of poly(vinylidene fluoride), Appl. Phys. Lett. 96 (2010) 192905. https://doi.org/10.1063/1.3428656

[3] G.M. Sessler, Piezoelectricity in polyvinylidene fluoride, J. Acoust. Soc. Am. 70 (1981) 1596-1608. https://doi.org/10.1121/1.387225

[4] A. Vinogradov, F. Holloway, Electro-mechanical properties of the piezoelectric polymer PVDF, Ferroelectrics. 226 (1999) 169-181. https://doi.org/10.1080/00150199908230298

[5] M. de Jong, W. Chen, H. Geerlings, M. Asta, K.A. Persson, A database to enable discovery and design of piezoelectric materials, Sci. Data. 2 (2015) 150053. https://doi.org/10.1038/sdata.2015.53

[6] J. Bernard, Antiferroelectric ceramics with field-enforced transitions: a new nonlinear circuit element, Proceedings of the IRE. 49 (1961) 1264-1267. https://doi.org/10.1109/JRPROC.1961.287917

[7] O.B. Wilson, Introduction to the Theory and Design of Sonar Transducers, Peninsula Publishing, Los Altos, CA, 1988, 65-124.

[8] R.C. Turner, Pa A. Fuierer, R. E. Newnham, R. Thomas Shrout, Materials for high temperature acoustic and vibration sensors: A review, Appl. Acoust. 41 (1994) 299-324. https://doi.org/10.1016/0003-682X(94)90091-4

[9] S. Zhang,Y. Fapeng, Piezoelectric materials for high temperature sensors, J. Am. Ceram. Soc. 94 (2011) 3153-3170. https://doi.org/10.1111/j.1551-2916.2011.04792.x

[10] J.A. Gallego-Juarez, Piezoelectric ceramics and ultrasonic transducers, J Phys E Sci Instrum. 22 (1989) 804. https://doi.org/10.1088/0022-3735/22/10/001

[11] S.Zhang, F.Li, X.Jiang, J.Kim, J.Luo, X.Geng, Advantages and challenges of relaxor-PbTiO3 ferroelectric crystals for electroacoustic transducers-A review, Prog. Mater. Sci. 68 (2015) 1-66. https://doi.org/10.1016/j.pmatsci.2014.10.002

[12] H.G. Haertling, Ferroelectric ceramics: history and technology, J. Am. Ceram. Soc. 82 (1999) 797-818. https://doi.org/10.1111/j.1551-2916.1999.tb01840.x

[13] X.Jiang, F.Tang, J.T.Wang, T.P. Chen, Growth and properties of PMN-PT single crystals, Physica C Supercond. 364 (2001) 678-683. https://doi.org/10.1016/S0921-4534(01)00878-4

[14] J.Ma, S.Xue, X.Zhao, F.Wang, Y.Tang, Z.Duan, T.Wang, W.Shi, Q.Yue, H.Zhou, H.Luo, High frequency transducer for vessel imaging based on lead-free Mn-doped (K0. 44Na0. 56) NbO3 single crystal, Appl. Phys. Lett.111 (2017) 092903. https://doi.org/10.1063/1.4990072

[15] H.Du, F.Tang, D.Liu, D.Zhu, W.Zhou, S.Qu, The microstructure and ferroelectric properties of (K0. 5Na0. 5) NbO3-LiNbO3 lead-free piezoelectric ceramics, Mater. Sci. Eng. B: Solid-State Mater. Adv. Technol. 136 (2007) 165-169. https://doi.org/10.1016/j.mseb.2006.09.031

[16] K.K.Sappati, B.Sharmistha, Flexible piezoelectric 0-3 PZT-PDMS thin film for tactile sensing, IEEE Sens. J. 20 (2020) 4610-4617. https://doi.org/10.1109/JSEN.2020.2965083

[17] J. Rödel, K.G. Webber, R. Dittmer, W. Jo, M.Kimura, D. Damjanovic, Transferring lead-free piezoelectric ceramics into application, J. Eur. Ceram. Soc. 35 (2015) 1659-81. https://doi.org/10.1016/j.jeurceramsoc.2014.12.013

[18] C.Hong, H.Kim, B.Choi, H.Han, J.Son, et al., Lead-free piezoceramics-where to move on?, J. Materiom. 2 (2016) 1-24. https://doi.org/10.1016/j.jmat.2015.12.002

[19] S.Wada, K.Takeda, T.Muraishi, H.Kakemoto, T.Tsurumi, T.Kimura, Preparation of [110] grain oriented barium titanate ceramics by templated grain growth method and their piezoelectric properties, Jpn. J. Appl. Phys. 46 (2007) 7039-43. https://doi.org/10.1143/JJAP.46.7039

[20] W.F.Liu, X.B.Ren, Large piezoelectric effect in Pb-free ceramics, Phys. Rev. Lett. 103 (2009) 257602. https://doi.org/10.1103/PhysRevLett.103.257602

[21] Y.Liu, Y.Chang, F.Li, B.Yang, Y.Sun, et al., Exceptionally high piezoelectric coefficient and low strain hysteresis in grain-oriented (Ba,Ca)(Ti,Zr)O3 through integrating crystallographic texture and domain engineering, ACS Appl. Mater. Inter. 9 (2017) 29863-71. https://doi.org/10.1021/acsami.7b08160

[22] Y.Saito, H.Takao, T.Tani, T.Nonoyama, K.Takatori, Lead-free piezoceramics, Nature. 432 (2004) 84-87. https://doi.org/10.1038/nature03028

[23] R.Malik, J. Kang, A.Hussain, A.Ahn, H.Han, J.Lee, High strain in lead-free Nb-doped Bi1/2(Na0.84K0.16)1/2TiO3-SrTiO3 incipient piezoelectric ceramics, Appl. Phys. Exp. 7 (2014) 061502. https://doi.org/10.7567/APEX.7.061502

[24] D.Hu, M.Yao, Y.Fan, C.Ma, M.Fan, M.Liu, Strategies to achieve high performance piezoelectric nanogenerators, Nano Energy. 55 (2019) 288-304. https://doi.org/10.1016/j.nanoen.2018.10.053

[25] S. Gong, B. Zhang, J. Zhang, Z.L. Wang, K. Ren, Biocompatible poly(lactic acid)-based hybrid piezoelectric and electret nanogenerator for electronic skin applications, Adv. Funct. Mater. 30 (2020) 1908724. https://doi.org/10.1002/adfm.201908724

[26] S.A. Haddadi, S. Ghaderi, M. Amini, S.A.A. Ramazani, Mechanical and piezoelectric characterizations of electrospun PVDF-nanosilica fibrous scaffolds for biomedical applications, Mater. Today: SAVE Proc. 5 (2018) 15710-15716. https://doi.org/10.1016/j.matpr.2018.04.182

[27] P. Sengupta, A.Ghosh, N.Bose, S.Mukherjee, A.Roy Chowdhury, P.Datta, A comparative assessment of poly (vinylidene fluoride)/conducting polymer electrospun nanofiber membranes for biomedical applications, J. Appl. Polym. Sci. 137 (2020) 49115. https://doi.org/10.1002/app.49115

[28] K.K.Sappati, B.Sharmistha, Piezoelectric polymer and paper substrates: a review, J. Sens. 18 (2018) 3605. https://doi.org/10.3390/s18113605

[29] S.Gong, B.Zhang, J.Zhang, Z.L.Wang, K.Ren, Biocompatible Poly (lactic acid)-Based Hybrid Piezoelectric and Electret Nanogenerator for Electronic Skin Applications, Adv. Funct. Mater. 30 (2020) 1908724. https://doi.org/10.1002/adfm.201908724

[30] N.Meng, X.Zhu, R.Mao, M.J.Reece, E.Bilotti, Nanoscale interfacial electroactivity in PVDF/PVDF-TrFE blended films with enhanced dielectric and ferroelectric properties, J. Mater. Chem. C. 5 (2017) 3296-3305. https://doi.org/10.1039/C7TC00162B

[31] Ali, Faizan, Waseem Raza, Xilin Li, Hajera Gul, Ki-Hyun Kim., Piezoelectric energy harvesters for biomedical applications, Nano Energy. 57 (2019) 879-902. https://doi.org/10.1016/j.nanoen.2019.01.012

[32] K. Maity, S.Garain, K.Henkel, D.Schmeißer, D.Mandal, Self-powered human-health monitoring through aligned PVDF nanofibers interfaced skin-interactive piezoelectric sensor, ACS Appl. Polym. Mater. 2 (2020) 862-878. https://doi.org/10.1021/acsapm.9b00846

[33] D.Y.Park, D.J.Joe, D.H.Kim, H.Park, J.H.Han, C.K.Jeong, H.Park, J.G.Park, B.Joung, K.J.Lee, Self-powered real-time arterial pulse monitoring using ultrathin epidermal piezoelectric sensors, Adv. Mater.29 (2017) 1702308. https://doi.org/10.1002/adma.201702308

[34] M.T.Chorsi, E.J.Curry, H.T.Chorsi, R.Das, J.Baroody, P.K.Purohit, H.Ilies, T.D. Nguyen, Piezoelectric biomaterials for sensors and actuators, Adv. Mater.31 (2019) 1802084. https://doi.org/10.1002/adma.201802084

[35] M.N. Teferra, C.Kourbelis, P.Newman, J.S. Ramos, D.Hobbs, R.A. Clark,K.J. Reynolds, Electronic textile electrocardiogram monitoring in cardiac patients: a

scoping review protocol, JBI evid. synth. 17 (2019) 147-156.
https://doi.org/10.11124/JBISRIR-2017-003630

[36] J.Lim, H.Jung, C. Baek, G.T. Hwang, J. Ryu, D.Yoon, J. Yoo, K.I. Park, J.H. Kim,
All-inkjet-printed flexible piezoelectric generator made of solvent evaporation assisted
BaTiO3 hybrid material, Nano Energy. 41 (2017) 337-343.
https://doi.org/10.1016/j.nanoen.2017.09.046

[37] Y.Tang, L.Chen, Z.Duan, K.Zhao, Z.Wu, Enhanced compressive strengths and
induced cell growth of 1-3-type BaTiO3/PMMA bio-piezoelectric composites, Mater.
Sci. Eng. C .120 (2021) 111699. https://doi.org/10.1016/j.msec.2020.111699

[38] Y.Cho, J.Jeong , M.Choi , G.Baek , S.Park, H.Choi , S. Ahn , S.Cha , T. Kim , D.S.
Kang, J.Bae, BaTiO3@ PVDF-TrFE Nanocomposites with Efficient Orientation
Prepared via Phase Separation Nano-coating Method for Piezoelectric Performance
Improvement and Application to 3D-PENG, Chem. Eng. J. (2021) 131030.
https://doi.org/10.1016/j.cej.2021.131030

[39] H. Jia, J.Chen, Tailoring the tetragonal distortion to obtain high Curie temperature
and large piezoelectric properties in BiFeO3-PbTiO3-BaTiO3 solid solutions, J. Eur.
Ceram. Soc. 41 (2021) 2443-2449. https://doi.org/10.1016/j.jeurceramsoc.2020.11.053

[40] A.Yazdani, H.D. Manesh, S.M. Zebarjad, Piezoelectric properties and damping
behavior of highly loaded PZT/polyurethane particulate composites, Ceram. Int.
(2021) In-press July 15. https://doi.org/10.1016/j.ceramint.2021.07.126

[41] B.S.Kim, J.H. Ji, J.H. Koh, Improved strain and transduction values of low-
temperature sintered CuO-doped PZT-PZNN soft piezoelectric materials for energy
harvester applications, Ceram. Int. 47 (2021) 6683-6690.
https://doi.org/10.1016/j.ceramint.2020.11.008

[42] H.Jia, H.Li, B.Lin, Y.Hu, L.Peng, D.Xu, X.Cheng, Fine scale 2-2 connectivity
PZT/epoxy piezoelectric fiber composite for high frequency ultrasonic application,
Sens. Actuator A Phys. 324 (2021) 112672. https://doi.org/10.1016/j.sna.2021.112672

[43] R.McQuade, T.Rowe, A.Manjón-Sanz, L.De la Puente, M.R.Dolgos, An
investigation into group 13 (Al, Ga, In) substituted (Na0. 5Bi0. 5) TiO3-BaTiO3
(NBT-BT) lead-free piezoelectrics, J. Alloys Compd. 762 (2018) 378-388.
https://doi.org/10.1016/j.jallcom.2018.04.329

[44] M.V.Petrovic, F.Cordero, E.Mercadelli, E.Brunengo, N.Ilic, C.Galassi,
Z.Despotovic, J.Bobic, A.Dzunuzovic, P.Stagnaro, G.Canu, Flexible lead-free NBT-

BT/PVDF composite films by hot pressing for low-energy harvesting and storage, J. Alloys Compd. (2021) 161071. https://doi.org/10.1016/j.jallcom.2021.161071

[45] J.Xiao, J.Wang , S.Liu , Y.Wu , J.Xu , Z.Zhang , F.Wang , X.A.Wang, Y.Tang, H.Luo, Microstructure, electrical and optical properties of NBT-xBZT lead-free single crystals, J. Alloys Compd. 861 (2021) 157949. https://doi.org/10.1016/j.jallcom.2020.157949

[46] K.Batra, N.Sinha, B.Kumar, Ba-doped ZnO nanorods: Efficient piezoelectric filler material for PDMS based flexible nanogenerator, Vacuum. (2021) 110385. https://doi.org/10.1016/j.vacuum.2021.110385

[47] C.Zhang , N.Li , D.Chen , Q.Xu , H.Li , J.He , J.Lu ,The ultrasonic-induced-piezoelectric enhanced photocatalytic performance of ZnO/CdS nanofibers for degradation of bisphenol A, J. Alloys Compd.885 (2021) 160987. https://doi.org/10.1016/j.jallcom.2021.160987

[48] J.W.Li, Y.X.Liu, H.C.Thong, Z.Du, Z.Li , Z.X.Zhu, J.K.Nie, J.F.Geng, W.Gong, K.Wang, Effect of ZnO doping on (K, Na) NbO3-based lead-free piezoceramics: Enhanced ferroelectric and piezoelectric performance, J. Alloys Compd. 847 (2020) 155936. https://doi.org/10.1016/j.jallcom.2020.155936

[49] J.Mayamae, V.Wanwilai, S.Usa, Theerachai Bongkarn, Rangson Muanghlua, Naratip Vittayakorn, High piezoelectric response in lead free 0.9 BaTiO3-(0.1-x) CaTiO3-xBaSnO3 solid solution, Ceram. Int. 43 (2017) S121-S128. https://doi.org/10.1016/j.ceramint.2017.05.252

[50] R.C.Chang, S.Y.Chu, Y.F.Lin, C.S.Hong, P.C.Kao, C.H.Lu, The effects of sintering temperature on the properties of (Na0. 5K0. 5) NbO3-CaTiO3 based lead-free ceramics, Sens. Actuator A Phys. 138 (2007) 355-360. https://doi.org/10.1016/j.sna.2007.05.020

[51] T.E.Hooper, J.I.Roscow, A.Mathieson, H.Khanbareh, A.J.Goetzee-Barral, A.J.Bell, High voltage coefficient piezoelectric materials and their applications, J. Eur. Ceram. Soc. 41 (2021) 6115. https://doi.org/10.1016/j.jeurceramsoc.2021.06.022

[52] S. Banerjee, S.Bairagi, S.W.Ali, A critical review on lead-free hybrid materials for next generation piezoelectric energy harvesting and conversion, Ceram. Int.47 (2021)16402. https://doi.org/10.1016/j.ceramint.2021.03.054

[53] Z.Zhao, Y.Dai, S.X.Dou, J.Liang, Flexible nanogenerators for wearable electronic applications based on piezoelectric materials, Mater. Today Energy. (2021) 100690. https://doi.org/10.1016/j.mtener.2021.100690

[54] J.Chen, Q.Qiu, Y.Han, D.Lau, Piezoelectric materials for sustainable building structures: Fundamentals and applications, Renew. Sustain. Energy Rev.101 (2019) 14-25. https://doi.org/10.1016/j.rser.2018.09.038

[55] A.R.Chowdhury, J.Jaksik, I.Hussain, R.Longoria, O.Faruque, F.Cesano, D.Scarano, J.Parsons, M.J.Uddin, Multicomponent nanostructured materials and interfaces for efficient piezoelectricity, Nano-Struct. Nano-Objects. 17 (2019) 148-184. https://doi.org/10.1016/j.nanoso.2018.12.002

[56] T.Zheng, J.Wu, D.Xiao, J.Zhu, Recent development in lead-free perovskite piezoelectric bulk materials, Prog. Mater. Sci. 98 (2018) 552-624. https://doi.org/10.1016/j.pmatsci.2018.06.002

[57] P.Eltouby, I.Shyha, C.Li , J.Khaliq, Factors Affecting the Piezoelectric Performance of Ceramic-Polymer Composites: A Comprehensive Review, Ceram. Int.47 (2021)17813. https://doi.org/10.1016/j.ceramint.2021.03.126

[58] H.Zhou, Y.Zhang, Y.Qiu, H.Wu, W.Qin, Y.Liao, Q.Yu, H.Cheng, Stretchable piezoelectric energy harvesters and self-powered sensors for wearable and implantable devices, Biosens. Bioelectron. (2020) 112569. https://doi.org/10.1016/j.bios.2020.112569

[59] H.C.Thong, C.Zhao, Z.Zhou, C.F.Wu, Y.X.Liu, Z.Z. Du, J.F.Li, W.Gong, K.Wang, Technology transfer of lead-free (K, Na) NbO3-based piezoelectric ceramics, Mater. Today 29 (2019) 37-48. https://doi.org/10.1016/j.mattod.2019.04.016

[60] H.Liang, G.Hao, O.Z.Olszewski, A review on vibration-based piezoelectric energy harvesting from the aspect of compliant mechanisms, Sens. Actuator A Phys. 331 (2021) 112743. https://doi.org/10.1016/j.sna.2021.112743

[61] Q.Xu, J.Wen, Y.Qin, Development and outlook of high output piezoelectric nanogenerators, Nano Energy.86 (2021) 106080. https://doi.org/10.1016/j.nanoen.2021.106080

[62] W.R.Ali, M.Prasad, Piezoelectric MEMS based acoustic sensors: A review, Sens. Actuator A Phys. 301 (2020) 111756. https://doi.org/10.1016/j.sna.2019.111756

[63] J.Yan, M.Liu, Y.G. Jeong, W.Kang, L.Li, Y.Zhao, N.Deng, B.Cheng, G.Yang, Performance enhancements in poly (vinylidene fluoride)-based piezoelectric nanogenerators for efficient energy harvesting, Nano Energy.56 (2019) 662-692. https://doi.org/10.1016/j.nanoen.2018.12.010

[64] L.Lu, W.Ding, J.Liu, B.Yang, Flexible PVDF based piezoelectric nanogenerators, Nano Energy.78 (2020): 105251. https://doi.org/10.1016/j.nanoen.2020.105251

[65] V.Jella, S.Ippili, J.H.Eom, S.V.Pammi, J.S.Jung, V.D.Tran, V.H.Nguyen, A.Kirakosyan, S.Yun, D.Kim, M.R.Sihn, A comprehensive review of flexible piezoelectric generators based on organic-inorganic metal halide perovskites, Nano Energy.57 (2019) 74-93. https://doi.org/10.1016/j.nanoen.2018.12.038

[66] F.Ali , W.Raza , X.Li , H.Gul , K.H.Kim, Piezoelectric energy harvesters for biomedical applications, Nano Energy.57 (2019) 879-902. https://doi.org/10.1016/j.nanoen.2019.01.012

[67] D.Khare, B.Basu, A.K.Dubey, Electrical stimulation and piezoelectric biomaterials for bone tissue engineering applications, Biomaterials. 258 (2020) 120280.

[68] A.T.Le,M.Ahmadipour, S.Y.Pung, A review on ZnO-based piezoelectric nanogenerators: Synthesis, characterization techniques, performance enhancement and applications." J. Alloys Compd. 844 (2020): 156172. https://doi.org/10.1016/j.jallcom.2020.156172

[69] J.Hao, W.Li, J.Zhai, H.Chen, Progress in high-strain perovskite piezoelectric ceramics, Mater. Sci. Eng. R Rep. 135 (2019) 1-57. https://doi.org/10.1016/j.mser.2018.08.001

[70] J.Le Scornec, B.Guiffard, R.Seveno, V.Le Cam, Frequency tunable, flexible and low cost piezoelectric micro-generator for energy harvesting, Sens. Actuator A Phys. 312 (2020) 112148. https://doi.org/10.1016/j.sna.2020.112148

[71] D.Hu, M.Yao, Y.Fan, C.Ma, M.Fan, M.Liu, Strategies to achieve high performance piezoelectric nanogenerators, Nano Energy.55 (2019) 288-304. https://doi.org/10.1016/j.nanoen.2018.10.053

[72] J.F.Tressler, S.Alkoy, R.E.Newnham , Piezoelectric sensors and sensor materials, J. Electroceramics. 2 (1998) 257-272. https://doi.org/10.1023/A:1009926623551

[73] S.Zhang, Y.Fapeng, Piezoelectric materials for high temperature sensors, J. Am. Ceram. Soc. 94 (2011) 3153-3170. https://doi.org/10.1111/j.1551-2916.2011.04792.x

[74] N.Wu, B.Bao, Q.Wang, Review on engineering structural designs for efficient piezoelectric energy harvesting to obtain high power output, Eng. Struct. 235 (2021) 112068. https://doi.org/10.1016/j.engstruct.2021.112068

[75] H.A.Sodano, D.J.Inman, G.Park, A review of power harvesting from vibration using piezoelectric materials, Shock. Vib. 36 (2004) 197-206. https://doi.org/10.1177/0583102404043275

[76] T. R. Shrout, R. Eitel, C. Randall, High Performance, high Temperature Perovskite Piezoelectric Ceramics, in: N. Setter (Eds.), Piezoelectric Materials in Devices, Lausanne, Switzerland, 2002.

[77] R.C.Turner, P.A.Fuierer, R.E.Newnham, T.R.Shrout, Materials for High Temperature Acoustic and Vibration Sensors: A Review, Appl.Acoustics. 41(1994) 299-324. https://doi.org/10.1016/0003-682X(94)90091-4

[78] S. J. Zhang, J. Luo, D. W. Snyder, and T. R. Shrout, High Performance, High Temperature Piezoelectric Crystals: in Handbook of Advanced Dielectric, Piezoelectric and Ferroelectric Materials - Synthesis, Characterization and Applications, Edited by Z. G. Ye. Woodhead Publishing Ltd., Cambridge, England, 2008, pp. 130-57. https://doi.org/10.1533/9781845694005.1.130

[79] D. Damjanovic, Materials for High Temperature Piezoelectric Transducers, Curr. Opinion Solid State Mater. Sci. 3(1998) 469-73. https://doi.org/10.1016/S1359-0286(98)80009-0

[80] N. Setter, ABC of Piezoelectricity and Piezoelectric Materials: in Piezoelectric Materials in Devices, N. Setter(Eds.), Lausanne, Switzerland, 2002 , pp. 1-27.

[81] Y.Yan , J.E.Zhou, D.Maurya, Y.U.Wang, S.Priya, Giant piezoelectric voltage coefficient in grain-oriented modified PbTiO 3 material, Nat. Commun. 7 (2016) 13089. https://doi.org/10.1038/ncomms13089

[82] R. E. Newnham, Properties of Materials - Anisotropy, Symmetry, Structure. Oxford University Press, NY, 2005. https://doi.org/10.1093/oso/9780198520757.003.0005

[83] G. Gautschi, Piezoelectric Sensorices. Springer-Verlag, NY, 2002. https://doi.org/10.1007/978-3-662-04732-3

[84] P.J.Harrop, Temperature Coefficients of Capacitance of Solids, J. Mater. Sci.4 (1969). 370-4. https://doi.org/10.1007/BF00550407

[85] T. R. Shrout, R. Eitel, and C. Randall, High Performance, High Temperature Perovskite Piezoelectric Ceramics, : in Piezoelectric Materials in Devices, N. Setter (Eds.), Lausanne, Switzerland, 2002, pp. 413-32.

[86] R. C. Turner, P. A. Fuierer, R. E. Newnham, T. R. Shrout, Materials for High Temperature Acoustic and Vibration Sensors: A Review, Appl.Acoustics. 41 (1994) 299-324. https://doi.org/10.1016/0003-682X(94)90091-4

[87] S. J. Zhang, J. Luo, D. W. Snyder, T. R. Shrout, High Performance, High Temperature Piezoelectric Crystals: in Handbook of Advanced Dielectric, Piezoelectric and Ferroelectric Materials - Synthesis, Characterization and

Applications, Z. G. Ye (Eds). Woodhead Publishing Ltd., Cambridge, England, 2008, pp. 130-57. https://doi.org/10.1533/9781845694005.1.130

[88] S. Zhang, R. Xia, L. Lebrun, D. Anderson, T. Shrout, Piezoelectric Materials for High Power, High Temperature Applications, Mater. Lett. 59 (2005) 3471-5. https://doi.org/10.1016/j.matlet.2005.06.016

[89] K. Uchino, Ferroelectric Devices. CRC Press, NY, 2009.

[90] M.M. Choy, W.R. Cook, R.F.S. Hearmon, H. Jaffe, J. Jerphagnon, S.K. Kurtz, and S.T. Liu Landolt-BoÈrnstein Numerical Data and Functional Relationships in Science and Technology, edited by K.-H. Hellwege and A.M. Hellwege, (Springer-Verlag, Heidelberg, New York, 1979), Vol. 11, p. 328

[91] D.Berlincourt, T.Kinsley, T.M.Lambert, D.Schwartz, E.A. Gerber EA, I.E.Fair, IRE Standards on piezoelectric crystals: Measurements of piezoelectric ceramics, Proc. IRE. 149 (1961)1161. https://doi.org/10.1109/JRPROC.1961.287860

[92] B. Jaffe, W.R. Cook, H. Jaffe, Piezoelectric Ceramics, Academic Press - London and New York, 1971, p. 315. https://doi.org/10.1016/B978-0-12-379550-2.50016-8

[93] M.G.Cain, M.Stewart, Standards for piezoelectric and ferroelectric ceramics: in Characterisation of ferroelectric bulk materials and thin films, Springer, Dordrecht, 2014, pp. 267-275. https://doi.org/10.1007/978-1-4020-9311-1_12

[94] J.Fialka, P.Beneš, Measurement of piezoelectric ceramic parameters-A characterization of the elastic, dielectric and piezoelectric properties of NCE51 PZT, :In Proceedings of the 13th International Carpathian Control Conference (ICCC), 2012 , pp. 147-152. https://doi.org/10.1109/CarpathianCC.2012.6228632

[95] J.Fialka, P.Beneš, Comparison of methods of piezoelectric coefficient measurement, :In 2012 IEEE International Instrumentation and Measurement Technology Conference Proceedings, 2012 , pp. 37-42. https://doi.org/10.1109/I2MTC.2012.6229293

[96] L. Burianová, M. Šulc, M. Prokopová, Determination of the piezoelectric coefficients dij of PZT ceramics and composites by laser interferometry, J. Eur. Ceram. Soc. 21 (2001) 1387-1390. https://doi.org/10.1016/S0955-2219(01)00024-3

[97] J.W. Waanders, Piezoelectric Ceramics: Properties and Applications, Philips Components, Eindhoven - The Netherlands, 1991.

[98] J.Erhart, L. Burianová, What is really measured on a d -meter?, J. Eur. Ceram. Soc. 21 (2001) 1413-1415. https://doi.org/10.1016/S0955-2219(01)00030-9

[99] K.Nakamura, T.Tokiwa, Y.Kawamura, Domain structures in KNbO 3 crystals and their piezoelectric properties, Int. J. Appl. Phys. 91 (2002) 9272-9276. https://doi.org/10.1063/1.1476078

[100] W.Beere, Stresses and deformation at grain boundaries, Philos. Trans. Royal Soc. A PHILOS T R SOC A. 288 (1978) 177-196. https://doi.org/10.1098/rsta.1978.0012

[101] P.Zheng, J.L.Zhang, Y.Q.Tan, C.L.Wang, Grain-size effects on dielectric and piezoelectric properties of poled BaTiO3 ceramics, Acta Materialia. 60 (2012) 5022-5030. https://doi.org/10.1016/j.actamat.2012.06.015

[102] A.Pathak, C.Prakash, R.Chatterjee, Shape memory effect in PZST system at exact morphotropic phase boundary, Phys. Rev. B Condens. Matter. 404 (2009) 3457-3461. https://doi.org/10.1016/j.physb.2009.05.044

[103] L.X.Zhang, X.Ren, Y.Wang, X.Q.Ke, X.D.Ding, J.Sun, Novel electro-strain-effect in La-doped Pb (Zr, Ti) O 3 relaxor ferroelectrics, : In2009 18th IEEE International Symposium on the Applications of Ferroelectrics, 2009, pp. 1-4. https://doi.org/10.1109/ISAF.2009.5307586

[104] I.Dutta, R.N.Singh, Dynamic in situ x-ray diffraction study of antiferroelectric-ferroelectric phase transition in strontium-modified lead zirconate titanate ceramics, Integr. Ferroelectr. 131 (2011) 153-172. https://doi.org/10.1080/10584587.2011.616441

[105] S.Y.Choi, S.J.Jeong, D.S.Lee, M.S.Kim, J.S.Lee, J.H.Cho, B.I.Kim, Y.Ikuhara, Gigantic electrostrain in duplex structured alkaline niobates, Chem. Mater. 24 (2012) 3363-3369. https://doi.org/10.1021/cm301324h

[106] T.Zheng, W.Wu, J.Wu, J.Zhu, D.Xiao, Balanced development of piezoelectricity, Curie temperature, and temperature stability in potassium-sodium niobhrate lead-free ceramics, J. Mater. Chem. C. 4 (2016) 9779-9787. https://doi.org/10.1039/C6TC03389J

[107] C.Xu, D.Lin, K.W.Kwok, Structure, electrical properties and depolarization temperature of (Bi0. 5Na0. 5) TiO3-BaTiO3 lead-free piezoelectric ceramics, Solid State Sci. 10 (2008) 934-940. https://doi.org/10.1016/j.solidstatesciences.2007.11.003

[108] B.Wu, D.Xiao, W.Wu, J.Zhu, Q.Chen, J.Wu, Microstructure and electrical properties of (Ba0. 98Ca0. 02)(Ti0. 94Sn0. 06) O3-modified Bi0. 51Na0. 50TiO3 lead-free ceramics, Ceram. Int.38 (2012) 5677-5681. https://doi.org/10.1016/j.ceramint.2012.04.011

[109] X. Liu, X. Tan, Giant strains in non-textured (Bi1/2Na1/2) TiO3-based lead-free ceramics, Adv. Mater.28 (2016) 574-578. https://doi.org/10.1002/adma.201503768

[110] Y.Tian, L.Wei, X.Chao, Z.Liu, Z.Yang, Phase transition behavior and large piezoelectricity near the morphotropic phase boundary of lead-free (Ba 0.85 Ca 0.15)(Zr 0.1 Ti 0.9) O 3 ceramics, J. Am. Ceram. Soc. 96 (2013) 496-502. https://doi.org/10.1111/jace.12049

[111] L.F.Zhu, B.P.Zhang, X.K.Zhao, L.Zhao, F.Z.Yao, X.Han, P.F.Zhou, J.F.Li, Phase transition and high piezoelectricity in (Ba, Ca)(Ti1− x Sn x) O3 lead-free ceramics, Appl. Phys. Lett.103 (2013) 072905. https://doi.org/10.1063/1.4818732

[112] C.Mota, M.Labardi, L.Trombi, L.Astolfi, M.D'Acunto, D.Puppi, G.Gallone, F.Chiellini, S.Berrettini, L.Bruschini, S.Danti, Design, fabrication and characterization of composite piezoelectric ultrafine fibers for cochlear stimulation, Mater. Des. 122 (2017) 206-219. https://doi.org/10.1016/j.matdes.2017.03.013

Advanced Functional Piezoelectric Materials and Applications   Materials Research Forum LLC
Materials Research Foundations 131 (2022) 37-60   https://doi.org/10.21741/978164490209-2

Chapter 2

Fabrication Approaches for Piezoelectric Materials

Bhavya Padha, Sonali Verma, Sandeep Arya*

Department of Physics, University of Jammu, Jammu, J&K-180006, India

*snp09arya@gmail.com

Abstract

After decades of study and development, piezoelectric materials have been used in various applications. Piezoelectric material is highly acknowledged as one of the primary functional materials in precision and acoustic engineering fields. Researchers are being pushed to explore novel materials and device combinations for new applications due to increasing demand, notably from the electrical, energy, and biomedical sectors. On the other hand, engineers are always working to enhance existing technology. Since the field has such a broad reach, it is vital to present an overview of the many areas of piezoelectric materials. This chapter focuses on the fabrication of different piezoelectric materials, applications, and challenges.

Keywords

Piezoelectric Materials, Energy, Biomedical, Ceramics, Bio-Piezoelectric Materials

Contents

1.   Introduction

Ceramics are composed of electrically charged crystals in particular. Apart from ferroelectric ceramics, the electrical charges of the crystals balance out in many ceramics. Ferroelectric ceramics have piezoelectric capabilities because they are electrically polarised. Because of the random distribution of electrical charge-carrying crystals, does not display piezoelectric properties right after they are formed. Instead, a strong direct current (D.C.) voltage is used to align the crystals, polarising the ceramic and making it piezoelectric. When piezoelectric ceramics are polarised, they keep that polarisation even when the D.C. power is withdrawn. Barium titanate, potassium niobate, sodium tungstate, and lead zirconate titanate (PZT) are examples of piezoelectric ceramics. PZT, a mixture of lead zirconate and lead titanate, is the most extensively used. Some materials have lower piezoelectric sensitivity and are less stable at high temperatures than PZT. Furthermore, the hard or soft piezoelectric characteristics of PZT may be formed. Affected by environmental worries regarding the use of lead, these properties have resulted in the widespread use of PZT.

Sensors and actuators made of piezoelectric ceramics such as lead zirconate titanate (PZT) and barium titanate (BaTiO$_3$) have been widely employed. They have recently acquired appeal as a component in energy collecting systems. Their excellent electromechanical coupling qualities make them attractive for actuators, where relatively low electric field inputs are adequate to operate ferroelectric ceramics. On the other hand, Ceramics are limited in their uses due to their brittle nature. Several properties are essential in electromechanical devices, including lightweight, high electromechanical coupling constants, low thermal expansion and conductivity, mechanical flexibility and compliance, and so on. Electro-active composites with a variety of ingredients have been considered for this purpose. Newnham et al. [1] outlined the fundamental aspects of making active composites customized to the organization of the constituents to achieve desired qualities (connectivity). Composites having active piezoceramic inclusions of particles or long fibre forms spread in a continuous soft matrix, such as polymers, are the most frequent and useful kinds.

Quartz cannot be used at higher temperatures, so alternative piezoelectric ceramic materials with higher charge output are preferred. Synthetic lead titanate and lead zirconate ceramic material are reviewed as recent developments in sensors. Ceramic piezoelectric sensors have two advantages: controllable polarisation processes (i.e., polycrystalline components may be formed into various shapes/geometries) and a high-charge amplified output. During the polarisation process, the material is exposed to a high-intensity electric field, also known as poling. A piezoelectric material is formed when the electric dipoles are oriented. Even though piezoelectric ceramics perform the same fundamental function, different ceramic materials are employed in accelerometers based on the application's needs. Special ceramics for high-temperature operations, such as those encountered in turbine engine operation, have recently been discovered [2]. Ceramic-based sensing components find use in vibration measurement and low-cost automotive accelerometers, where low signal levels and extensive wire runs are the normal aerospace applications operating at high temperatures (900 °F).

Furthermore, in recent years, the introduction of bio-piezoelectric as well as ferroelectric materials, like $BaTiO_3$ and KNN [3, 4], results in the successful conversion of mechanical energy into electrical energy in biological systems [5], paving the way for the advancement of novel bio-piezoelectric materials in the field of biomedicine. For instance, piezoelectric materials can be converted into functionalized bio-piezoelectric materials [6] by using surface modifications and engineering techniques to connect with natural tissues (like muscles, skin, bones, cells, nerves). Furthermore, there have been various studies on the advancements of piezoelectric materials for sensing [7], energy harvesting [8], as well as in biocompatible devices [9], smart fabrics [10], tissue regeneration [11], and electronic skin [12]. Fig. 1 depicts the most recent biomedical applications of bio-piezoelectric nanostructures [13].

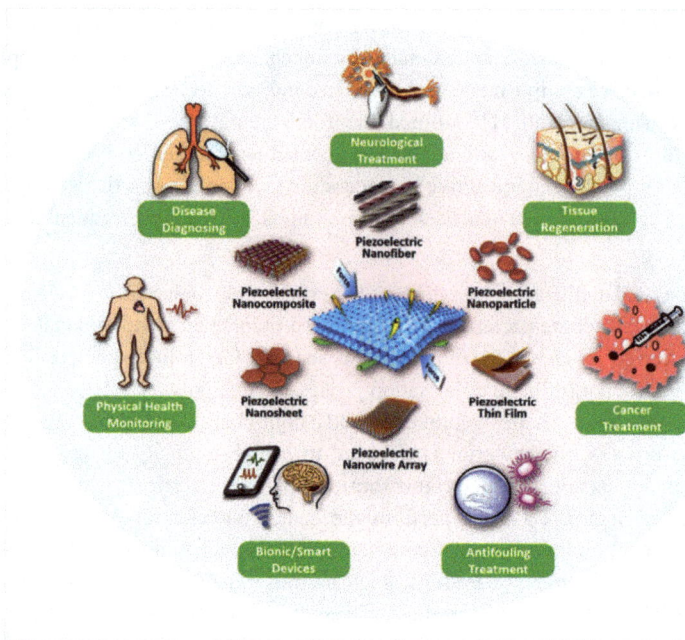

Fig. 1: Various biomedical applications of bio-piezoelectric platforms [13]. (Reproduced from [13] under Creative Commons Attribution 4.0 License. Copyright 2021, Wiley-VCH GmbH).

2. Preparation techniques for piezoelectric ceramics

Polycrystalline ceramics are commonly used to make ferroelectric devices. The two phases involved are synthesizing the ceramic material and sintering them. Crystalline materials are becoming more popular due to their improved performance over polycrystalline versions. In addition, template grain development procedures provide exhibit performance as that of a single crystal at a low cost.

2.1 Synthesis of ceramic powders

The essential elements to regulate in the raw powder are particle structure, size, and compositional consistency to achieve consistency in their piezoelectric properties. The most common approach is the oxide-mixing process, which involves burning unprocessed oxide particles and preparing their powdered form to get the appropriate chemical composition. However, wet chemical approaches (co-precipitation, alkoxide) have lately been used to manufacture ceramic devices since the oxide-mixing process makes it hard to achieve nanoscale compositional consistency [14]. Therefore, the fabrication of barium titanate (BT), lead zirconate titanate (PZT), and lead magnesium niobate (PMN) ceramics is discussed in this section.

2.1 Solid-state reaction

TiO_2, PbO, and ZrO_2 raw powders, in a suitable proportion, are combined and calcined for 1–2 hours at 800–900°C. The sample is next crushed and processed to produce fine particles. The disadvantages are that the milling method produces particles much smaller than 1 μm, and the milling medium contaminates the sample. In contrast, contaminated Fe has been used as a catalyst for the discovery of hard PZT unintentionally; before, PZT powders were ball-milled with steel (Fe) balls. BT is made using BaO and TiO_2 in equimolar amounts. $BaCO_3$ is preferred over BaO as the latter is costly and has low reactivity. $Pb[(Mg_{1/3}Nb_{2/3})_{1-x}Ti_x]O_3$ may be made using a similar calcination method that starts with PbO, MgO, Nb_2O_5, and TiO_2.

Additionally, this simple procedure produces a minor quantity of the second phase (pyrochlore). The additional amount of doped PbO in the sintering process is useful in suppressing this second phase [15]. Swartz et al. [16] presented a novel technique that considered the chemical reaction mechanism. They proved that a process using columbite $MgNb_2O_6$ and PbO might provide the ideal perovskite phase:

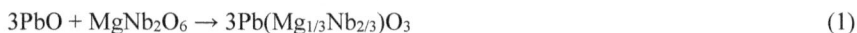

$$3PbO + MgNb_2O_6 \rightarrow 3Pb(Mg_{1/3}Nb_{2/3})O_3 \tag{1}$$

TiO_2, MgO, and Nb_2O_5 are combined and heated at 1000°C to make $Pb(Mg_{1/3}Nb_{2/3})O_3$-$PbTiO_3$ (PMN-PT). The columbite is then treated with PbO before calcining at 800–900°C. Excess MgO of several mol percent is especially efficient in attaining the ideal perovskite phase.

2.2 Co-precipitation

Because realistic piezoelectric ceramics are made using complicated solid-state devices, composition variation in ceramic molecules and purity are critical. This difficulty is inextricably linked to the strategy mentioned above, which combines solid-state reaction with mechanical mixing. To achieve increased homogeneity, the co-precipitation approach is used. First, a homogeneous precipitate is produced by introducing a precipitant to a mixed metal salts liquid solution. Then, the precipitate is transformed into uniform particles by thermal dissolving.

For instance, by dripping in oxalic acid, $BaTiO(C_2H_4)_2.4H_2O$ using an atomic scale Ba/Ti ratio of 1:1 was obtained as residue from $BaCl_2$ and $TiCl_4$ solution. This precipitate was

Advanced Functional Piezoelectric Materials and Applications   Materials Research Forum LLC
Materials Research Foundations 131 (2022) 37-60   https://doi.org/10.21741/978164490209-2

thermally disintegrated to produce fine stoichiometric $BaTiO_3$ powders having exceptional sintering properties. The basic components for $(Pb, La)(Zr, Ti)O_3$ (PLZT) are $Pb(NO_3)_2$, $ZrO(NO_3)_2.2H_2O$, $TiO(NO_3)_2$, and $La(NO_3)_3.6H_2O$ [17]. The solutions of all nitrates were combined proportionately in half litre ethanol. The PLZT oxalate was precipitated by gently dripping oxalic acid and ethanol in the nitric solution. At 800°C, thermal disintegration took place. The residue was dissolved by heating the solution to get the desired powder. In other situations, the precipitation process could create the oxide powder immediately.

2.3   Alkoxide hydrolysis

The hydrolysis creates metal oxide/hydrate by combining metal alkoxides $M(OR)_n$ (M: metal atom, R: alkyl) with alcohol, and water is introduced. This is also known as the sol-gel approach. The alkoxide technique can generate powders that are extremely fine and pure. Distillation is a simple way to purify metal alkoxides since they evaporate. Since no further ion doping is required during the hydraulic process, great purity may be maintained. The following is a summary of the hydrolysis and condensation processes:
(a) Hydrolysis

$$H-O-H + M-OR \rightarrow H-O-M + ROH \qquad (2)$$

(a) Alkoxylation

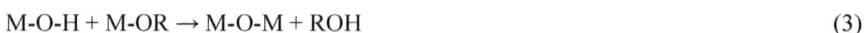

$$M-O-H + M-OR \rightarrow M-O-M + ROH \qquad (3)$$

(b) Oxidation

$$M-O-H + M - OH \rightarrow M-O-M + OH_2 \qquad (4)$$

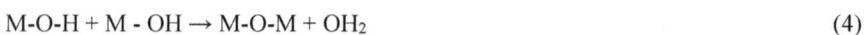

$Ba(OC_3H_7)_2$ and $Ti(OC_5H_{11})$ are diluted using isopropyl alcohol to make BT powders. Extremely fine $BaTiO_3$ powders having sufficient crystalline behaviour and a particle diameter of 10–100 Å may be achieved by selecting an adequate hydrolysis condition (pH). Compared to samples made by oxide-mixing, the hydrolytic technique produces a 99.98 percent pure sample, resulting in a significant improvement in the sintered ceramic's permittivity [18].

Pb alkoxide is more difficult to acquire for PZT production than Zr and Ti alkoxides. A two-step process is employed for simplicity: $(Zr, Ti)O_2$ is generated using an alkoxide and PbO. A sol-gel process paired using low-cost nanoparticles might be a viable path for lower expenditure. The alternative is to employ Pb alkoxide, which is made by mixing titanium isopropoxide $Ti[OCH(CH_3)_2]_4$, and zirconium n-butoxide $Zr(O(CH_2)_3CH_3]_4$ with lead acetylacetonate $Pb(CH_3COCHCOCH_3)$ for making a PZT precursor.

2.4   The sintering method

In this method, the agglomerated powder is burnt after being shaped into a suitable shape. The surface energy (surface tension) accelerates the diffusion of constituting atoms on surfaces of fine particles, promoting bonding at the interface of neighbouring molecules. Sintering is the term for the firing process that primarily removes pores and enhances the density of the ceramic. It's worth noting that the mechanical characteristics of the sintered body are influenced not only by the tiny crystalline particle's characteristics but also by the grain boundary. Mechanical strength is an instance of mechanical fracture in ceramic substances that rarely happens at the grain boundary.

On the other hand, the polycrystalline material has increased mechanical strength when the crystal has a significant cleavage feature. As a result, the grains develop, and the grain shape changes dramatically during sintering. However, it is well acknowledged that the qualities of the raw powder significantly impact the production environment and the features of the end product. Furthermore, the atoms' required diffusion length for sintering becomes shorter for fine powders, accelerating pore diffusion. Hence, high-density ceramics are produced. Much research on grain growth has been conducted. The common link between sintering period 't' and grain size 'D' is:

$$D^\beta - D_0^\beta \, \alpha t \tag{5}$$

$\beta = 2$ for normal growth, and $\beta = 3$ for aberrant growth of the grain. Doping should also be considered starting with the oxalic acid/ethanol technique. Dopants' primary function is to lower the sintering temperature, but they can also have secondary effects like suppressing or enhancing grain development. Excess PbO or Bi_2O_3 inhibits grain development in PZTs. The doping of Dysprosium (Dy) is particularly successful in reducing the particle size in $BaTiO_3$ [19]. Sintering of Hard PZTs was achieved at high temperatures, and inexpensive base metal electrodes cannot be used to manufacture multilayer parts. The influence of CuO and ZnO on the sintering process of a MnO_2 and Nb_2O_5 based PZT was also examined [20]. The inclusion of CuO created a liquid phase, which reduced the sintering temperature.

On the other hand, the piezoelectric characteristics of the sintering CuO-substituted ceramics at 950°C were reduced. The piezoelectric characteristics were significantly improved with the inclusion of ZnO. Creating a homogenous microstructure with big grains was credited with this improvement. The 0.2 weight percent CuO and 1.1 weight percent ZnO ceramics sintered at 950°C demonstrated outstanding dielectric and piezoelectric characteristics having values of $k_p = 0.532$, $Q_m = 750$, $d_{33} = 351$ pC N^{-1}, $\varepsilon_{33}/\varepsilon_0 = 1337$, and $T_C = 280$°C.

2.5   Templated grain growth

The piezoelectric response of polycrystalline ferroelectric ceramics is significantly improved by crystallographic texturing. Templated grain growth (TGG) is a technique for producing textured ceramics having single crystal-like characteristics. TGG produces an orientated material by nucleating and growing the required crystal on an oriented single crystal template and heating it. The template particles must be crystalline and stable for epitaxy and consequent orientated matrix formation sites. Messing et al. [21] observed that textured ceramic materials exhibit considerable improvements in piezoelectric characteristics compared to randomly oriented ceramics. Reactive-templated grain growth (RTGG) is a fundamental procedure that creates textured Pb-free piezoelectric ceramics. $Bi_4Ti_3O_{12}$ platelets were used to manufacture bismuth layered ferroelectric $ABi_4Ti_4O_{15}(A_{14}Na_{0.5}Bi_{0.5}$, Ca, Sr) ceramics with better piezoelectric characteristics. When compared to their arbitrarily oriented counterparts, textured ceramics were synthesized in $K_{0.5}Na_{0.5}NbO_3$ (KNN), $BaTiO_3$, and $Bi_{0.5}Na_{0.5}TiO_3$ (BNT)-based compositions having a Lotgering factor greater than 0.8 [22-24].

3.   Piezoelectric materials in device fabrication

A piezoelectric material may be used in various ways in the manufacturing of devices. For example, the piezoelectric material could be employed as a separator or an electrolyte to store energy. Alternative techniques have been used to prepare the piezoelectric material for fabricating devices, including vapour induced phase separation, mixed-oxide sintering, ultrasound irradiation, solvent film casting, and electrospinning [25-31]. PVDF is an important piezoelectric material. According to the literature review, PVDF film is commercially used as a separator for building supercapacitors. Song et al. [26] utilized PVDF film and an electrolyte of sulphuric acid (H_2SO_4)/polyvinyl alcohol (PVA). PVDF-based piezoelectric materials have also been documented to be synthesized using other solvents.

However, the literature reveals that piezoelectric materials are also used as electrolytes for supercapacitors. For example, Zhou et al. [32] produced an electrolyte by dissolving PVA, H_3PO_4 and KNN in deionized (DI) water. The resulting mixture was chilled before being poured onto a textured glass substrate after mixing for 2 hours. After drying, the resulting layer was scraped off the glass substrate, and $KNN/PVA/H_3PO_4$ electrolyte was produced. Considering the concept of piezo-electrolyte, Pazhamalai et al. [33] proposed a PVDF-based gel electrolyte by dissolving poly(vinylidene fluoride-co-hexafluoropropylene) (PVDF-co-HFP) in $CH_3C(O)N(CH_3)_2$ and $(CH_3)_2CO$. Then, after stirring and heating concurrently, 1 g tetraethylammonium tetrafluoroborate (TEABF$_4$) was added to the above solution, resulting in a translucent gel. Hence, piezoelectric materials can be used as electrolytes in different electronic devices.

4.   Bio-piezoelectric materials

4.1   Types bio-piezoelectric materials

Piezoelectric materials for biocompatible systems are generally categorized based on their chemical structure (Fig. 2):

- Bio-piezoelectric polymers
- Bio-piezoelectric ceramics
- Biomolecular piezoelectric materials
- Bio-piezoelectric nanomaterials

Perfect biocompatible piezoelectric materials are supposed to have great piezoelectric properties and surface characteristics modified according to the targeted application, such as hydrophilicity, roughness, and porosity, to meet the key needs of a biomedical application. Furthermore, bio-piezoelectric materials must be biocompatible and biodegradable to ensure biosafety and sufficient mechanical characteristics that offer high flexibility and robustness. Unfortunately, many piezoelectric materials cannot meet all of the aforementioned criteria. Thus, surface modifications of piezoelectric materials have become very significant in this area to achieve the necessary biological performance.

4.2   Synthesis strategies

To fulfil specific biomedical purposes, biocompatible piezoelectric materials are often formed into two kinds of nanostructures, i.e. thin films and nanoplatforms. The flexible thin films are attached to the skin, muscles, and other tissue surfaces for biosensing or disease therapy. More crucially, the large surface area provided by a biocompatible piezoelectric thin film allows for numerous bonding sites in electronic devices (such as inductors, capacitors, and resistors), allowing for the construction of smaller and transportable biomedical equipment. Magnetron sputtering, pulsed laser deposition, and solution casting are the most common synthetic procedures for creating bio-piezoelectric films. Bio-piezoelectric nanoplatforms have several distinct features [34], including extremely small size, good biocompatibility, large specific surface area, as well as outstanding piezoelectric properties, all of which considerably broaden their biomedical applications. Bio-piezoelectric nanoplatforms for biomedical applications have gained great popularity as research efforts on fabrication and surface modifications of these materials have increased in recent years [35]. Mechanical and chemical exfoliations, vapour phase deposition, hydrothermal, as well as sol-gel methods, have all been highly explored to synthesize bio-piezoelectric nanoplatforms.

4.2.1 Thin films

These biocompatible piezoelectric thin films play a significant role in the effective integration of various parts at a small scale due to the advent of micro-electric-mechanical devices in biomedicines. These thin films are mechanically flexible, simple to synthesize, inexpensive, and have good stability [36]. Furthermore, they can easily be coupled with semiconductors to react sensitively to micro-mechanical forces. A variety of methods for preparing high-quality biocompatible piezoelectric thin films have been examined, including pulsed laser deposition, magnetron sputtering, and solution casting. Magnetron sputtering is a well-established thin-film synthesis technique that may be utilized to synthesize films of various substrates, such as metals, ceramics, polymers and semiconductors. It has the advantages of faster and consistent film generation with high density [37]. Electrons are moved between the target and substrate with an electric field in magnetron sputtering, and a simultaneous magnetic field is also applied. Consequently, electrons interact with gas molecules, boosting plasma ionization [38]. The plasma interacts with the target due to a strong electric field, releasing target atoms that move to the substrate, hence creating a thin film [39].

By controlling deposition parameters like rate of gas flow, deposition time, substrate temperature, sputtering gas pressure, and annealing parameters, it is possible to make thin films with modified piezoelectric properties and conductivity using magnetron sputtering. For example, a series of ZnO thin films are biocompatible and exhibit piezoelectricity on a polyethylene terephthalate (PET) substrate by changing the oxygen flux at direct current magnetron sputtering reported in the literature [38]. The uniform thin film of ZnO produced had excellent mechanical strength and was well attached to the PET substrate. At the same time, the ZnO thin film deposition on the PET substrate improved its surface energy and hydrophobicity. Consequently, the thin film ZnO formed due to an oxygen flux had good piezoelectricity and strong substrate adhesion [39].

Pulsed laser deposition (PLD) is another technique for producing biocompatible piezoelectric thin films at low temperatures. In an ultra-high vacuum setup, high-voltage laser pulses are incident on the surface of the target. The target material is then vaporized quickly and deposited as a thin coating on the substrate [40, 41]. Non-contact laser heating successfully prevents sample contamination, resulting in highly pure films. Thus, PLD technique has been widely used to prepare biomedical micro-devices. For example, Scarisoreanu et al. [42] employed this approach to develop a high-quality lead-free biocompatible piezoelectric thin film of $(1-x)Ba(Ti_{0.8}Zr_{0.2})TiO_{3x}(Ba_{0.7}Ca_{0.3})TiO_3$, x = 0.45

Advanced Functional Piezoelectric Materials and Applications   Materials Research Forum LLC
Materials Research Foundations 131 (2022) 37-60   https://doi.org/10.21741/978164490209-2

(BCZT 45) using a Pt/Si substrate, and then deposited them on a Kapton polyimide polymer substrate. As a result of using compliant Kapton substrates, the obtained films exhibit outstanding piezoelectric properties and high biocompatibility and flexibility.

Furthermore, solution casting is also a facile as well as a widely employed method for synthesizing thin films. In this method, a powder specimen is mixed with an appropriate dispersant to make an evenly dispersed slurry, then cast into a thin film using a casting machine [43]. This approach has been extensively used to synthesize biocompatible piezoelectric thin films due to the equipment's simplicity with a good amount of production. By employing this approach, Hosseini et al. [44] created a free-standing biocompatible piezoelectric thin film using glycine and chitosan. This film was made by crystallographically orienting β-glycine crystals inside a chitosan matrix, making it exceptionally flexible. It will be used to make biocompatible pressure sensors in wearable electronics.

4.2.2 Nanoplatforms

Instead of the high potential of biocompatible piezoelectric nanoplatforms in biomedical applications, producing highly efficient nanoplatforms with controlled shapes and sizes (such as nanofibers, nanoparticles, nanowires, and nanoplatelets) and regulated crystal phases (such as cubic, quadrangle, and polyphase) remains a challenging issue. Several techniques have been devised to make high-quality bio-piezoelectric nanoplatforms, including hydrothermal, solvothermal, electrostatic spinning, and mechanical exfoliation. The hydrothermal and solvothermal technique effectively produces biocompatible piezoelectric nanoplatforms with different shapes and sizes under controlled conditions. Chemical processes are carried out at a high temperature in a sealed autoclave throughout their synthesis, resulting in high-quality nanocrystals via one-pot reactions [45]. Wang et al. [46] used hydrothermal reactions between $Ti(C_4H_9O)_4$ and $Ba(OH)_2$ to make ferroelectric tetragonal $BaTiO_3$ nanoparticles with an average size of 130 nm. The $BaTiO_3$ nanoparticles had a strong piezoelectric activity, which allowed them to generate radicals ($\cdot OH$ or $\cdot O^{2-}$) when ultrasonic vibrations were applied to help clean teeth. The hydrothermal approach may create hierarchical nanostructures with particular geometry due to its low reaction temperature and extensive application for manufacturing diverse morphologies, sizes, and dimensions [45]. Ha et al. [47] used a hydrothermal technique to create ZnO nanowires-decorated polydimethylsiloxane micropillar arrays. The nanowires of ZnO were formed on polydimethylsiloxane micropillars with controlled dimensions. The ZnO nanowires had a high bending property, superfast responsiveness, and minimal thermal expansion, and they may be used to make highly flexible electronic skins.

Advanced Functional Piezoelectric Materials and Applications   Materials Research Forum LLC
Materials Research Foundations 131 (2022) 37-60   https://doi.org/10.21741/978164490209-2

For the fabrication of nanofibers, electrostatic spinning is a typical process [48]. In a nutshell, the polymer-based electrolyte is sprayed constantly using a jet, followed by stretching in the presence of a strong electric field to generate electrospinning nanofibers [49]. The produced fibre membranes are elastomeric and have a high strain tolerance. Using a high electric field during the spinning process induces material polarisation, giving the nanofibers remarkable piezoelectric characteristics. Bairagi and Ali [50] used electrospinning to create KNN/ZnO integrated PVDF nanocomposites. The PVDF polymer was stretched mechanically as well as underwent in situ poling during the electrospinning process, transforming the non-polar α-phase into a highly polar β-phase to improve its piezoelectricity. By controlling the concentrations of the precursors as well as electrospinning parameters like flow rate, potential, and the separation between receiver and needle tip, biocompatible piezoelectric nanofibers with an analogous structure to that of the natural tissues can be created. By optimizing the spinning conditions, biocompatible piezoelectric nanofibers scaffold using poly-3-hydroxybutyrate-3-hydroxy valerate along with $BaTiO_3$ nanoparticles, shape and pore size analogous to real cartilage was also reported in the literature [51]. The inclusion of $BaTiO_3$ nanoparticles improved the mechanical characteristics and piezoelectricity and increased the dilapidation time. As a result, the scaffolds developed had outstanding mechanical characteristics as well as high piezoelectricity similar to real cartilage.

Under certain mechanical stresses, monolayer nanomaterials undergo exfoliation from their bulk counterparts, breaking the weak van der Waals' forces among the layers. Sometimes, mechanical exfoliation is also termed as Scotch tape technique [52]. Exfoliation is a simple and fast physical separation procedure to create biocompatible piezoelectric nanostructures. On the other hand, mechanical exfoliation has several drawbacks, including product inhomogeneity, limited stripping effectiveness, and poor control of nanomaterial morphology. As an extension of mechanical exfoliation, liquid exfoliation has recently been developed. In this situation, ultrasonic treatment in a suitable solution or surfactant breaks the weak van der Waal forces among neighbouring layers of bulk crystals [53]. Furthermore, by establishing a protective coating on the surface of the lamellar products, the solvent molecules can inhibit re-stacking and aggregation [54]. For example, Wu et al. [55] used an ultrasonic-assisted liquid-phase exfoliation approach to exfoliate monolayer nanosheets of MoS_2 from a bulk MoS_2 powder using NMP.

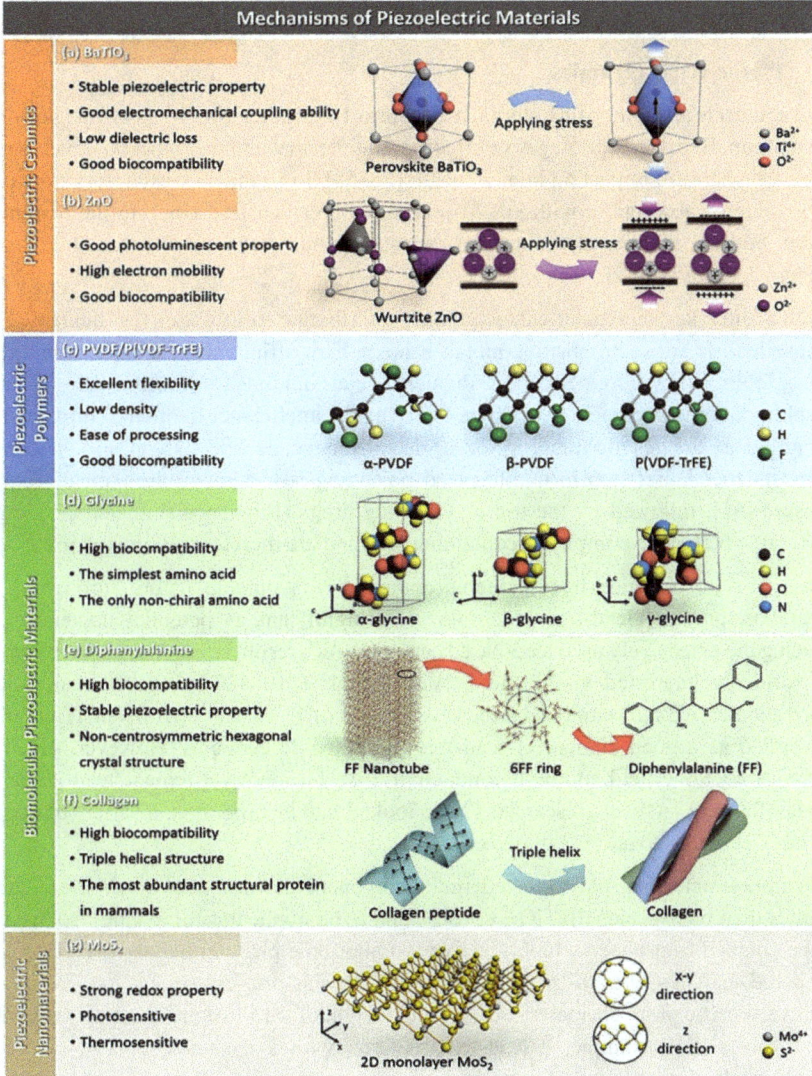

Fig. 2: Different bio-piezoelectric materials with their properties and structure [13]. (Reproduced from [13] under Creative Commons Attribution 4.0 License. Copyright 2021, Wiley-VCH GmbH).

5. Challenges

5.1 Piezoelectric ceramics

Fatigue, muscle and joint pains, stomach discomfort, and other symptoms of lead poisoning are common. Because of its negative effects on intellectual and neurological development, Pb leaves a harmful impact on human health [56–58]. In adults, the respiratory tract is the major route of absorption, with 30–70 percent of breathed lead entering the respiratory system. For a comparatively controlled exposure, blood lead levels vary in the range of 1.45 and 2.4 mol L^{-1} [59].

Lead has three key molecular characteristics contributing to its toxicity in humans. First, because lead is an electropositive metal, it has a high affinity for enzymes required for haemoglobin formation. Second, divalent lead inhibits mitochondrial oxidative phosphorylation comparable to calcium, lowering the intelligence quotient. Third, lead can also influence the genetic transcription of deoxyribonucleic acid (DNA) [60]. Therefore, the most crucial first step in treating lead poisoning is to remove the patient from the exposure [61], followed by the use of chelating drugs (Ethylenediaminetetraacetic acid (EDTA)), which form complexes containing lead and are therefore eliminated [62–65].

Because of its pollution-free and environmentally favourable nature throughout the preparation process, lead-free materials seem significant as potential contenders for replacing the extensively used lead-based ceramics. As a result, researchers from all around the world are interested in this field. With a high T_c of 320 °C, substantial residual polarization P_r of 38 μC cm^{-2}, and a coercive field E_c of 73 kV cm^{-1}, $(Bi_{1/2}Na_{1/2})TiO_3(BNT)$ is regarded as a good choice for lead-free piezoelectric ceramics. However, given the enormous coercive field and high conductivity, BNT cannot yet replace lead zirconium titanate (PZT). As a result, researchers have looked into incorporating a variety of dopants into BNT ceramics to address these issues.

Furthermore, scientists have created multicomponent devices with better piezoelectric qualities than binary ones. $BaTiO_3$, $KNbO_3$, and a rhombohedral compound, for instance, can be coupled to produce a highly active morphotropic phase boundary (MPB) system. Compared to PZTs, the Pb-free piezoelectric materials are less dense. As a result, the effective specific piezo characteristics per unit weight of lead-free materials are twice their actual values. For example, a lead-free material with a d_{33} value of 150 pC N^{-1} is approximately similar to a PZT material with a d_{33} value of 280–300 pC N^{-1}. In the 1950s, lead-free materials and others were produced concurrently with PZT. However, because of the outstanding piezoelectric capabilities of PZTs, researchers opted to focus on the PZT system, and hence less study on lead-free materials was done. Even though the original analysis of the PZT system indicated lower piezo characteristics, with the discovery of

MPB, inclusion of appropriate dopants, and other systematic studies, the values were much improved. Therefore, a systematic technique comparable to that used to improve the piezo characteristics of Pb-free materials may be necessary.

5.2   Bio-piezoelectric materials

These functional biomedical materials may be utilized as an instantaneous sensor to measure a range of crucial parameters in the human body like heart rate, respiration, and blood pressure. It may also be employed as an implantable medical device to attain a consistent energy supply. Advanced medical equipment, including cardiac pacemakers, artificial retinas, neurostimulators, as well as electronic skins, could all benefit from piezoelectric biosensors. While bio-piezoelectric platforms have traditionally been employed in bio-electronics, the current breakthroughs in these materials open new possibilities for biomedical bio-piezoelectric systems. Materials having piezoelectric catalytic properties can produce reactive oxygen species for treating cancer and antifouling due to their nanoscale size and capability for electromechanical conversion. Furthermore, bio-piezoelectric nanomaterials having high piezoelectric coefficients can effectively react to even mechanical strain of small magnitude. These intriguing materials can be employed as robust functional materials for various biomedical applications, including electronic skins, nerve stimulation, drug delivery, healing, as well as cancer therapy, owing to their capacity to convert mechanical energy into electrical one.

Despite the tremendous potential for piezoelectric materials in biomedical applications, there are still various difficulties in the advancement of biocompatible piezoelectric platforms:

- Prior to their practical deployment, bio-piezoelectric platforms must meet important biomedical requirements. Bio-piezoelectric materials are supposed to have great biocompatibility and outstanding mechanical strength, as well as these materials can efficiently convert mechanical energy into electrical energy for applications in biosensing and implant devices. In addition, immunogenicity, biodegradability, and tissue accumulation must all be taken into account when using bio-piezoelectric nanoplatforms to treat diseases. Although much research has been devoted to advancing bio-piezoelectric nanomaterials, the variety of materials available on current bio-piezoelectric platforms is still limited. Thus, there is a need to explore new piezoelectric and ferroelectric materials and composites and investigate their surface modifications. The introduction of lead-free ferroelectric polymers and ceramics would be very helpful in this area.

- The majority of bio-piezoelectric material research focuses on the shape, piezoelectric coefficient, and mechanical flexibility of the material. Their chemical characteristics, as well as their synergy with piezoelectricity, have received little consideration. Many bio-piezoelectric materials, like MoS_2 and black phosphorus, have remarkable chemical characteristics concerning photocatalysis and electrocatalysis. By combining the piezoelectricity of the material with its chemical characteristics, more innovative bio-piezoelectric platforms are probable to develop.

- The advancement of theoretical study has been considerably aided by the advancement of theoretical evaluations and simulations (density functional theory (DFT), molecular dynamics, and structural mechanics). Such methods can help improve materials design, which can help to speed up the research into bio-piezoelectric materials. The possibility of using high-throughput (HT) [66] as well as machine learning techniques for discovering new materials is particularly intriguing [67].

- One of the most common mechanical stimulations used to create a material's piezoelectricity is ultrasound. However, tissue damage can occur because of the ultrasonic cavitation effect when a high amount of ultrasound is used. As a result, more biocompatible ways of inducing material piezoelectricity should be developed. Furthermore, increasing the piezoelectric property of the existing bio-piezoelectric materials can reduce the amount of ultrasonic power supplied to a medically acceptable level [67].

Further, for better comparison, different synthesis techniques for piezoelectric materials and their permittivities as well as piezoelectric constants are discussed in Table 1.

Conclusion

Lead-free materials exhibit great potential for replacing the extensively used lead-based ceramics because of their eco-friendly nature throughout the preparation process, as a result of which researchers from all around the world show interest in this field. However, due to the exceptional piezoelectric capabilities of PZTs, researchers still opted to focus on the PZT system, as a result of which less study on lead-free materials was done. Even though the original analysis of the PZT system signifies lower piezoelectric characteristics, with the discovery of MPB, the inclusion of appropriate dopants, and other systematic studies, its value was much increased. Thus, a systematic technique comparable to that used to improve the piezo characteristics of Pb-free materials may be necessary. Likewise, the functional biomedical materials exhibiting piezoelectricity have the advantages of cost-effectiveness, easy fabrication, as well as consistent performance. Thus, these materials have gained great popularity in the field of biomedical applications. Because of the electromechanical properties of bio-piezoelectric materials, strain from biological activities (like muscle contractions, body movements, blood circulation, breathing, heartbeats, and so on) can be converted into electrical energy. Thus, both of these lead-free materials exhibits great potential to be employed as piezoelectric materials in various electronic devices.

Table 1: Synthesis techniques, specific permittivities, and piezoelectric constants for various piezoelectric materials.

Piezoelectric material	Synthesis Technique	Specific permittivities ε	Piezoelectric constants, d (pCN^{-1})	Reference
Quartz crystal (SiO_2)	Hydrothermal	$\varepsilon_{11}^{T}/\varepsilon_0 = 4.52$, $\varepsilon_{33}^{T}/\varepsilon_0 = 4.68$	$d_{11} = 2.31$, $d_{14} = 0.727$	[68, 69]
Lithium niobate $(LiNbO_3)$ crystal	Sol-gel	$\varepsilon_{11}^{T}/\varepsilon_0 = 84$, $\varepsilon_{33}^{T}/\varepsilon_0 = 30$	$d_{15} = 68$, $d_{22} = 21$, $d_{31} = -1$, $d_{33} = 6$	[69, 70]
Lithium tantalate $(LiTaO_3)$ crystal	Hydrothermal	$\varepsilon_{11}^{T}/\varepsilon_0 = 51$, $\varepsilon_{33}^{T}/\varepsilon_0 = 45$	$d_{15} = 26$, $d_{22} = 7$, $d_{15} = 68$, $d_{33} = 8$	[69, 71]
Lead zirconate titanate $(Pb(Zr,Ti)O_3)$ ceramics	Hydrothermal, sol-gel, solid-state reaction, pulsed laser deposition	$\varepsilon_{11}^{T}/\varepsilon_0 = $ from 1500 to 1700, $\varepsilon_{33}^{T}/\varepsilon_0 = $ from 1300 to 1700	$d_{15} = $ from 500 to 580, $d_{31} = $ from $-$170 to -125, $d_{33} = $ from 290 to 370	[69, 72-75]

References

[1] R.E. Newnham, D.P. Skinner, L.E. Cross, Connectivity and piezoelectric-pyroelectric composites, Mater. Res. Bull. 13 (1978) 525-536. https://doi.org/10.1016/0025-5408(78)90161-7

[2] A. Venkatanarayanan, E. Spain, Review of recent developments in sensing materials, in: S. Hashmi, G.F. Batalha, C.J. Van Tyne, B. Yilbas (Eds.), Comprehensive Materials Processing. Elsevier; Amsterdam, The Netherlands, 2014, pp. 47-101. https://doi.org/10.1016/B978-0-08-096532-1.01303-0

[3] M.Yuan, L. Cheng, Q. Xu, W. Wu, S. Bai, L. Gu, Z. Wang, J. Lu, H. Li, Y. Qin, T. Jing, Z.L. Wang, Biocompatible Nanogenerators through High Piezoelectric Coefficient 0.5 Ba (Zr0.2 Ti0.8) O3-0.5 (Ba0.7 Ca0.3) TiO3 Nanowires for In-Vivo Applications, Adv. Mater. 26 (2014) 7432-7437. https://doi.org/10.1002/adma.201402868

[4] C.K. Jeong, J.H. Han, H. Palneedi, H. Park, G.T. Hwang, B. Joung, S.G. Kim, H. J. Shin, I.S. Kang, J. Ryu, K.J. Lee, Comprehensive biocompatibility of nontoxic and high-output flexible energy harvester using lead-free piezoceramic thin film, APL Mater. 5 (2017) 074102. https://doi.org/10.1063/1.4976803

[5] L. Li, L. Miao, Z. Zhang, X. Pu, Q. Feng, K. Yanagisawa, Y. Fan, M. Fan, P. Wen, D. Hu, Recent progress in piezoelectric thin film fabrication via the solvothermal process, J. Mater. Chem. A 7 (2019) 16046-16067. https://doi.org/10.1039/C9TA04863D

[6] M.T. Chorsi, E.J. Curry, H.T. Chorsi, R. Das, J. Baroody, P.K. Purohit, H. Ilies, T.D. Nguyen, Piezoelectric biomaterials for sensors and actuators, Adv. Mater. 31 (2019) 1802084. https://doi.org/10.1002/adma.201802084

[7] G.T. Hwang, M. Byun, C.K. Jeong, K.J. Lee, Flexible piezoelectric thin-film energy harvesters and nanosensors for biomedical applications, Adv. Healthcare Mater. 4 (2015) 646-658. https://doi.org/10.1002/adhm.201400642

[8] N.A. Shepelin, A.M. Glushenkov, V.C. Lussini, P.J. Fox, G.W. Dicinoski, J.G. Shapter, A.V. Ellis, New developments in composites, copolymer technologies and processing techniques for flexible fluoropolymer piezoelectric generators for efficient energy harvesting, Energy Environ. Sci. 12 (2019) 1143-1176. https://doi.org/10.1039/C8EE03006E

[9] D. Jiang, B. Shi, H. Ouyang, Y. Fan, Z.L. Wang, Z. Li, Emerging implantable energy harvesters and self-powered implantable medical electronics, ACS Nano 14 (2020) 6436-6448. https://doi.org/10.1021/acsnano.9b08268

[10] F. Mokhtari, Z. Cheng, R. Raad, J. Xi, J. Foroughi, Piezofibers to smart textiles: A review on recent advances and future outlook for wearable technology, J. Mater. Chem. A 8 (2020) 9496-9522. https://doi.org/10.1039/D0TA00227E

[11] K. Kapat, Q.T.H. Shubhra, M. Zhou, S. Leeuwenburgh, Piezoelectric Nano-Biomaterials for Biomedicine and Tissue Regeneration, Adv. Funct. Mater. 30 (2020) 1909045. https://doi.org/10.1002/adfm.201909045

[12] H. Yuan, T. Lei, Y. Qin, R. Yang, Flexible electronic skins based on piezoelectric nanogenerators and piezotronics, Nano Energy 59 (2019) 84-90. https://doi.org/10.1016/j.nanoen.2019.01.072

[13] Q. Xu, X. Gao, S. Zhao, Y.N. Liu, D. Zhang, K. Zhou, H. Khanbareh, W. Chen, Y. Zhang, C. Bowen, Construction of Bio-Piezoelectric Platforms: From Structures and Synthesis to Applications, Adv. Mater. 33 (2021) 2008452. https://doi.org/10.1002/adma.202008452

[14] T. Kato, Fine ceramics technology, Fabrication technology of ceramic powder and its future, Industry Research Center, Japan, v3, 1983.

[15] M. Lejeune, J.P. Boilot, Ceramics of perovskite lead magnesium niobate, Ferroelectrics 54 (1984) 191-194. https://doi.org/10.1080/00150198408215848

[16] S.L. Swartz, T.R. Shrout, W.A. Schulze, L.E. Cross, Dielectric Properties of Lead-Magnesium Niobate Ceramics, J. Am. Ceram. Soc. 67 (1984) 311. https://doi.org/10.1111/j.1151-2916.1984.tb19528.x

[17] M. Tanada, H. Yamamura, S. Shirasaki, Abstract 22nd, Jpn. Ceram. Soc. Fundamental Div. 81 (1984).

[18] Y. Ozaki, Electron Ceram.,13 (1982) 26. https://doi.org/10.1002/chin.198241158

[19] A. Yamaji, Y. Enomoto, E. Kinoshita, T. Tanaka, Proc. 1st Mtg. Ferroelectric Mater. & Appl., Kyoto, 1977, p. 269.

[20] C.W. Ahn, H.C. Song, S. Nahm, S. Priya, S.H. Park, K. Uchino, H.G. Lee, H.J. Lee, Effect of ZnO and CuO on the sintering temperature and piezoelectric properties of a hard piezoelectric cermic, J. Am. Ceram. Soc. 89 (2006) 921-925. https://doi.org/10.1111/j.1551-2916.2005.00823.x

[21] G.L. Messing, S. Trolier-McKinstry, E.M. Sabolsky, C. Duran, S. Kwon, B. Brahmaroutu, P. Park, H. Yilmaz, P.W. Rehrig, K.B. Eitel, E. Suvaci, Templated Grain Growth of Textured Piezoelectric Ceramics, Crit. Rev. in Solid State Mater. Sci. 29 (2004) 45. https://doi.org/10.1080/10408430490490905

[22] T. Tani, T. Kimura, Reactive-templated grain growth processing for lead free piezoelectric ceramics, Adv. Appl. Ceram. 105 (2006) 55. https://doi.org/10.1179/174367606X81650

[23] Y. Saito, H. Takao, T. Tani, T.Nonoyama, K.Takatori, T. Homma, T. Nagaya. M. Nakamura, Lead-free piezoceramics, Nature 432 (2004) 84-87. https://doi.org/10.1038/nature03028

[24] K. Uchino, Manufacturing methods for piezoelectric ceramic materials, Advanced piezoelectric materials: Science and technology, Woodhead Publishing, United Kingdom, 2017, pp.385-393. https://doi.org/10.1016/B978-0-08-102135-4.00010-2

[25] Q. Meng, C. Du, Z. Xu, J. Nie, M. Hong, X. Zhang, J. Chen, Siloxene-Reduced graphene oxide composite hydrogel for supercapacitors, Chem. Engg. J. 393 (2020) 124684. https://doi.org/10.1016/j.cej.2020.124684

[26] R. Song, H. Jin, X. Li, L. Fei, Y. Zhao, H.Huang, H.L.W. Chan, Y. Wang, Y. Chai, A rectification-free piezo-supercapacitor with a polyvinylidene fluoride separator and functionalized carbon cloth electrodes, J. Mater. Chem. A 3 (2015) 14963-14970. https://doi.org/10.1039/C5TA03349G

[27] S. Sahoo, K. Krishnamoorthy, P. Pazhamalai, V.K. Mariappan, S. Manoharan, S.J. Kim, High-performance self-charging supercapacitors using a porous PVDF-ionic liquid electrolyte sandwiched between two-dimensional graphene electrodes, J. Mater. Chem. A 7 (2019) 21693-21703. https://doi.org/10.1039/C9TA06245A

[28] A. Maitra, S. Paria, S.K. Karan, R. Bera, A. Bera, A.K. Das, S.K. Si, L. Halder, A. De, B.B. Khatua, Triboelectric nanogenerator driven self-charging and self-healing flexible asymmetric supercapacitor power cell for direct power generation, ACS Appl. Mater.& Inter. 11 (2019) 5022-5036. https://doi.org/10.1021/acsami.8b19044

[29] K. Parida, V. Bhavanasi, V. Kumar, J. Wang, P.S. Lee, Fast charging self-powered electric double layer capacitor, J. Power Sources 342 (2017) 70-78. https://doi.org/10.1016/j.jpowsour.2016.11.083

[30] Y. Lu, Y. Jiang, Z. Lou, R. Shi, D. Chen, G. Shen, Wearable supercapacitor self-charged by P(VDF-TrFE) piezoelectric separator, Prog. Nat. Sci. 30 (2020) 174-179. https://doi.org/10.1016/j.pnsc.2020.01.023

[31] A. Ramadoss, B. Saravanakumar, S.W. Lee, Y.S. Kim, S.J. Kim, Z.L. Wang, Piezoelectric-driven self-charging supercapacitor power cell, ACS Nano 9 (2015) 4337- 4345. https://doi.org/10.1021/acsnano.5b00759

[32] D. Zhou, N. Wang, T. Yang, L. Wang, X. Cao, Z.L. Wang, A piezoelectric nanogenerator promotes highly stretchable and self-chargeable supercapacitors, Mater. Horiz. 7 (2020) 2158-2167. https://doi.org/10.1039/D0MH00610F

[33] P. Pazhamalai, K. Krishnamoorthy, V.K. Mariappan, S. Sahoo, S. Manoharan, S.J. Kim, A High Efficacy Self-Charging MoSe2 Solid-State Supercapacitor Using Electrospun Nanofibrous Piezoelectric Separator with Ionogel Electrolyte, Adv. Mater. Inter. 5 (2018) 1800055. https://doi.org/10.1002/admi.201800055

[34] C. Cui, F. Xue, W.J. Hu, L.J. Li, Two-dimensional materials with piezoelectric and ferroelectric functionalities, NPJ 2D Mater. Appl. 2 (2018) 1-14. https://doi.org/10.1038/s41699-017-0046-y

[35] F. Ali, W. Raza, X. Li, H. Gul, K.H. Kim, Piezoelectric energy harvesters for biomedical applications, Nano Energy 57 (2019) 879-902. https://doi.org/10.1016/j.nanoen.2019.01.012

[36] L.W. Martin, A.M. Rappe, Thin-film ferroelectric materials and their applications, Nat. Rev. Mater. 2 (2016) 1-14. https://doi.org/10.1038/natrevmats.2016.87

[37] G. Tan, K. Maruyama, Y. Kanamitsu, S. Nishioka, T. Ozaki, T. Umegaki, H. Hida, I. Kanno, Crystallographic contributions to piezoelectric properties in PZT thin films, Sci. Rep. 9 (2019) 1-6. https://doi.org/10.1038/s41598-018-37186-2

[38] J. Costa, T. Peixoto, A. Ferreira, F. Vaz, M.A. Lopes, Development and characterization of ZnO piezoelectric thin films on polymeric substrates for tissue repair, J. Biomed. Mater. Res., Part A 107 (2019) 2150-2159. https://doi.org/10.1002/jbm.a.36725

[39] N. Zhang, T. Zheng, J. Wu, Lead-free (K, Na) NbO3-based materials: Preparation techniques and piezoelectricity, ACS Omega 5 (2020) 3099-3107. https://doi.org/10.1021/acsomega.9b03658

[40] S.W. Zhang, Z. Zhou, J. Luo, J.F. Li, Potassium-Sodium-Niobate-Based Thin Films: Lead Free for Micro-Piezoelectrics, Ann. Phys. 531 (2019) 1800525. https://doi.org/10.1002/andp.201800525

[41] T.C. Kaspar, S. Hong, M.E. Bowden, T. Varga, P. Yan, C. Wang, S. R. Spurgeon, R.B. Comes, P. Ramuhalli, C.H. Henager, Tuning piezoelectric properties through epitaxy of La2Ti2O7 and related thin films, Sci. Rep. 8 (2018) 1-11. https://doi.org/10.1038/s41598-018-21009-5

[42] N.D. Scarisoreanu, F. Craciun, V. Ion, R. Birjega, A. Bercea, V. Dinca, M. Dinescu, L.E. Sima, A. Icriverzi, A. Roseanu, L.Gruionu, G. Gruionu, Lead-free piezoelectric (Ba, Ca)(Zr, Ti) O3 thin films for biocompatible and flexible devices, ACS Appl. Mater. Interfaces 9 (2017) 266-278. https://doi.org/10.1021/acsami.6b14774

[43] G. Ahn, S.R. Kim, Y.Y. Choi, H.W. Song, T.H. Sung, J. Hong, K. No, Facile preparation of ferroelectric poly (vinylidene fluoride-co-trifluoroethylene) thick films by solution casting, Polym. Eng. Sci. 54 (2014) 466-471. https://doi.org/10.1002/pen.23570

[44] E.S. Hosseini, L. Manjakkal, D. Shakthivel, R. Dahiya, Glycine-chitosan-based flexible biodegradable piezoelectric pressure sensor, ACS Appl. Mater. Interfaces 12 (2020) 9008-9016. https://doi.org/10.1021/acsami.9b21052

[45] B. Jiang, J. Iocozzia, L. Zhao, H. Zhang, Y.W. Harn, Y. Chen, Z. Lin, Barium titanate at the nanoscale: controlled synthesis and dielectric and ferroelectric properties, Chem. Soc. Rev. 48 (2019) 1194-1228. https://doi.org/10.1039/C8CS00583D

[46] Y. Wang, X. Wen, Y. Jia, M. Huang, F. Wang, X. Zhang, Y. Bai, G. Yuan, Y. Wang, Piezo-catalysis for nondestructive tooth whitening, Nat. Commun. 11 (2020) pp.1-11. https://doi.org/10.1038/s41467-020-15015-3

[47] M. Ha, S. Lim, J. Park, D.S. Um, Y. Lee, H. Ko, Bioinspired interlocked and hierarchical design of ZnO nanowire arrays for static and dynamic pressure-sensitive electronic skins, Adv. Funct. Mater. 25 (2015) 2841-2849. https://doi.org/10.1002/adfm.201500453

[48] B. Azimi, M. Milazzo, A. Lazzeri, S. Berrettini, M.J. Uddin, Z. Qin, M.J. Buehler, S. Danti, Electrospinning piezoelectric fibers for biocompatible devices, Adv. Healthcare Mater. 9 (2020) 1901287. https://doi.org/10.1002/adhm.201901287

[49] V. Aravindan, J. Sundaramurthy, P. Suresh Kumar, Y.S. Lee, S. Ramakrishna, S. Madhavi, Electrospun nanofibers: A prospective electro-active material for constructing high performance Li-ion batteries, Chem. Commun. 51 (2015) 2225-2234. https://doi.org/10.1039/C4CC07824A

[50] S. Bairagi, S.W. Ali, A hybrid piezoelectric nanogenerator comprising of KNN/ZnO nanorods incorporated PVDF electrospun nanocomposite webs, Int. J. Energy Res. 44 (2020) 5545-5563. https://doi.org/10.1002/er.5306

[51] J. Jacob, N. More, C. Mounika, P. Gondaliya, K. Kalia, G. Kapusetti, Smart piezoelectric nanohybrid of poly (3-hydroxybutyrate-co-3-hydroxyvalerate) and barium titanate for stimulated cartilage regeneration, ACS Appl. Bio Mater. 2 (2019) 4922-4931. https://doi.org/10.1021/acsabm.9b00667

[52] C. Cui, F. Xue, W.J. Hu, L.J. Li, Two-dimensional materials with piezoelectric and ferroelectric functionalities. npj 2D Materials and Applications, NPJ 2D Mater. Appl. 2 (2018) 1-1. https://doi.org/10.1038/s41699-017-0046-y

[53] J.H. Bang, K.S. Suslick, Applications of ultrasound to the synthesis of nanostructured materials, Adv. Mater. 22 (2010) 1039-1059. https://doi.org/10.1002/adma.200904093

[54] L. Cheng, X. Wang, F. Gong, T. Liu, Z. Liu, 2D nanomaterials for cancer theranostic Applications, Adv. Mater. 32 (2019) 1902333. https://doi.org/10.1002/adma.201902333

[55] C. Wu, T.W. Kim, J.H. Park, H. An, J. Shao, X. Chen, Z.L. Wang, Enhanced triboelectric nanogenerators based on MoS2 monolayer nanocomposites acting as electron-acceptor layers, ACS Nano 11 (2017) 8356-8363. https://doi.org/10.1021/acsnano.7b03657

[56] J.N. Gordon, A. Taylor, P.N. Bennette, Lead poisoning: case studies, Br. J. Clin. Pharmacol. 53 (2002) p.451. https://doi.org/10.1046/j.1365-2125.2002.01580.x

[57] D. Barltrop, A.M. Smith, Kinetics of lead interaction with human erythrocytes, Postgrad. Med. J. 51 (1985) 770-773. https://doi.org/10.1136/pgmj.51.601.770

[58] M.B. Rabinowitz, G.W. Wetherill, J.D. Kopple, Kinetic analysis of lead metabolism in healthy humans, J. Clin. Invest. 58 (1976) 260-270. https://doi.org/10.1172/JCI108467

[59] D. Courtney, S.R. Meekin, Changes in blood lead levels of solderers following the introduction of The Control of Lead at Work Regulations, Occup. Med. 35 (1985) 128-130. https://doi.org/10.1093/occmed/35.4.128

[60] P.L. Goering, Lead-protein interactions as a basis for lead toxicity, Neurotoxicology 14 (1993) 45-60.

[61] D.R. Baldwin, W.J. Marshall, Heavy metal poisoning and its laboratory investigation, Ann. Clin. Biochem. 36 (1999) 267-300. https://doi.org/10.1177/000456329903600301

[62] S.S. Kety, The lead citrate complex ion and its role in the physiology and therapy of lead poisoning, J. Biol. Chem. 142 (1942) 181-190. https://doi.org/10.1016/S0021-9258(18)72713-0

[63] W.J.H. Leckie, S.L. Tompsett, The diagnostic and therapeutic use of edathamil calcium disodium (EDTA, versene) in excessive inorganic lead absorption, Q. J. Med. 27 (1958) 65-82.

[64] H.V. Aposhian, DMSA and DMPS-water soluble antidotes for heavy metal poisoning, Ann. Rev. Pharmacol. Toxicol. 23 (1983) 193-215. https://doi.org/10.1146/annurev.pa.23.040183.001205

[65] J.H. Graziano, E.S. Siris, N.Lolacono, S.J.Silverberg, L.Turgeon, 2, 3-Dimercaptosuccinic acid as an antidote for lead intoxication, Clin. Pharmacol. Ther. 37 (1985) 431-438. https://doi.org/10.1038/clpt.1985.67

[66] M. de Jong, W. Chen, H. Geerlings, M. Asta, K.A. Persson, A database to enable discovery and design of piezoelectric materials, Sci. Data 2 (2015) 1-13. https://doi.org/10.1038/sdata.2015.53

[67] S. Chibani, F.X. Coudert, Machine learning approaches for the prediction of materials properties, APL Mater. 8 (2020) 80701. https://doi.org/10.1063/5.0018384

[68] A.C. Walker, Hydrothermal synthesis of quartz crystals, J. Am. Ceram. Soc. 36 (1953) 250. https://doi.org/10.1111/j.1151-2916.1953.tb12877.x

[69] Y. Saigusa, Quartz-based piezoelectric materials, in: Advanced Piezoelectric Materials, Woodhead Publishing, Nirasaki-city, Japan, 2017, pp. 197-233. https://doi.org/10.1016/B978-0-08-102135-4.00005-9

[70] M.A. Fakhri, E.T. Salim, M.H.A. Wahid, U. Hashim, Z.T. Salim, R.A. Ismail,. Synthesis and characterization of nanostructured LiNbO3 films with variation of stirring duration, J. Mater. Sci. Mater. Electron. 28(2017) 11813-11822. https://doi.org/10.1007/s10854-017-6989-0

[71] S. Takasugi, K. Tomita, M. Iwaoka, H. Kato,M. Kakihana, The hydrothermal and solvothermal synthesis of LiTaO3 photocatalyst: Suppressing the deterioration of the water-splitting activity without using a cocatalyst, Int. J. Hyd. Energy 40(2015)5638-5643. https://doi.org/10.1016/j.ijhydene.2015.02.121

[72] Z.C. Qiu, J.P. Zhou, G. Zhu, P. Liu and X.B. Bian, Hydrothermal synthesis of Pb (Zr0· 52Ti0· 48) O3 powders at low temperature and low alkaline concentration, Bull. Mater. Sci. 32 (2009) 193. https://doi.org/10.1007/s12034-009-0030-z

[73] P. Luginbuhl, G.A. Racine, P. Lerch, B. Romanowicz, K.G. Brooks, N.F. de Rooij, P. Renaud, N. Setter, Piezoelectric cantilever beams actuated by PZT sol-gel thin film, Sens. Actuat. A-Phys. 54 (1996) 530. https://doi.org/10.1016/S0924-4247(95)01196-X

[74] J. Zhao, L. Lu, C.V. Thompson, Y.F. Lu, W.D. Song, Preparation of (0 0 1)-oriented PZT thin films on silicon wafers using pulsed laser deposition, J. Cryst. Growth 225 (2001) 173. https://doi.org/10.1016/S0022-0248(01)00865-X

[75] M.I.S. Veríssimo, P.Q. Mantas, A.M.R. Senos, J.A.B.P. Oliveira, M.T.S.R. Gomes, Preparation of PZT discs for use in an acoustic wave sensor, Ceram. Int. 35 (2009) 617. https://doi.org/10.1016/j.ceramint.2008.01.016

Advanced Functional Piezoelectric Materials and Applications   Materials Research Forum LLC
Materials Research Foundations 131 (2022) 61-116   https://doi.org/10.21741/978164490209-3

Chapter 3

Piezoelectric Materials-based Nanogenerators

Ritamay Bhunia[1], Do Hwan Kim[1,2,*]

[1]Department of Chemical Engineering, Hanyang University, Seoul 04763, Republic of Korea

[2]Institute of Nano Science and Technology, Hanyang University, Seoul 04763, Republic of Korea

* dhkim76@hanyang.ac.kr

Abstract

With the progression of human civilization, the growing demands of smart electronics devices have compelled us to think about some effective alternative energy sources which can deliver the required power to these devices. Currently, flexible, lightweight, sustainable power sources can be alternatives to fulfill the demands. Piezoelectric nanogenerators are favorable candidates because they can be integrated with these portable personal electronic devices. A remarkable advancement in nanogenerators has been achieved in the synthesis process, energy conversion performance, environmental pollution due to conventional chemical batteries, and adaptability. This chapter presents the possibilities and implementation of piezoelectric materials for nanogenerator fabrication. This chapter would help the readers to get a clear perception of this topic.

Keywords

Piezoelectric, Nanogenerator, Sensor, Ceramic, ZnO, Polymer, Composite, PVDF, Cellulose

Contents

1.   Introduction

The energy was an indispensable factor from the beginning of our civilization and nowadays, to lead the modern lifestyle the consumption of energy has increased so much that we are facing a shortage of non-renewable sources of energy. Also, these non-renewable energy sources have large adverse effects on the environment. Keeping these in the background, renewable energy research has become the pathfinder to deal with the problems due to non-renewable sources [1-2]. Like solar energy, tidal energy, wind energy, hydropower, we also need to focus on the development of renewable energy sources which can provide low power for modern smart electronic devices. With the development of technology, we have reached a new society where the daily used electronics devices have been miniatured to a portable size, even smaller than conventional coin-battery sometimes. These smart electronics devices like mobiles, health monitoring sensors, smartwatches consume very low power (a few μW to mW). The necessity of self-powered devices drives us to develop a new system that can generate low electrical power (μW to mW) from other forms of energies of ambient. This leads to demonstrate first the revolutionary idea of nanogenerator (NG) utilizing the piezoelectricity of zinc oxide nanowires (ZnO NWs) by Z. L. Wang in the year 2006 [3]. Nanogenerators can be in various forms: piezoelectric [4-5], triboelectric [6-7], pyroelectric [8-9], electro-magnetic [10-11], even hybrid combining all effects [12-13]. The nanogenerator can be a compliment or even a replacement of batteries for the fabrication of self-chargeable smart electronics devices like biosensors [14], chemical sensors [15-16], photodetectors [17], wearable devices [18]. NGs are the future power sources for modern electronic devices and the succeeding technology of green renewable energy in the new epoch of the internet of things (IoT) [19].

However, this chapter will explore the possibilities and implementation of piezoelectric materials for nanogenerator fabrication. The piezoelectric effect is a phenomenon of polarization charge generation on the surface of specific types of dielectric materials for the effect of mechanical stress on it and the converse piezoelectric effect is the development of mechanical strain under applied external electric field. Riding on this property of the piezoelectric materials, researchers have developed a piezoelectric nanogenerator (commonly known as PENG) utilizing ubiquitous mechanical energies like walking, vibration, breezing, etc.

2.   Piezoelectricity and crystallography

The piezoelectric property of any material (single crystal, bulk, thin film of ceramic and polymers) are related to its crystal symmetry. After the discovery of piezoelectricity in 1880 by Jacques Curie and Pierre Curie in particular crystals (Rochelle salt, tourmaline, quartz, etc.), it was studied a lot and observed that this is strongly related to particular crystal symmetry [20]. However, during that time they could not predict and observe the converse piezoelectric effect where mechanical stress is observed upon implementation of electric field and it was mathematically predicted by Lippmann in 1881 [21]. After this, the converse piezoelectric effect was observed by Jacques and Pierre Curie [22]. The piezoelectric coupling can be linearly expressed through first rank tensor of electric displacement ($\vec{D}$ and $\vec{E}$) and the second rank tensor (stress tensor $\vec{S}$ and strain tensor $\vec{T}$). A third rank tensor is constituted by the piezoelectric coefficients. The piezoelectric linear constitutive relations are given by the following equations [23]:

$$S_{pq} = c_{pqrs}^E T_{rs} - e_{rpq} E_r \tag{1}$$

$$D_{p=} e_{prs} T_{rs} + K_{pr}^T E_r \tag{2}$$

Here, S_{pq} and T_{rs} are respectively the component of stress tensor and strain sensor. c_{pqrs}^E is elastic stiffness at the constant electric field, e_{rpq} is piezoelectric constant. These coupling coefficients (elastic stiffness and piezoelectric constant) have the following symmetries:

$$c_{pqrs}^E = c_{qprs}^E = c_{rspq}^E = c_{qpsr}^E \tag{3}$$

$$e_{rpq} = e_{rqp} \tag{4}$$

$$K_{pq}^T = K_{qp}^T \tag{5}$$

From these relations, we can have 27 piezoelectric stress constants, 81 elastic constants, and 9 dielectric constants. Generally, any third rank tensor has 27 independent components among which, in the case of piezoelectric tensor, the numbers of independent elements for elastic, piezoelectric, and dielectric constants are 21,18, and 6, respectively because of the symmetry relations of equations 3, 4, and 5. If any crystal has more symmetries, then the number of independent elements will be reduced more. As mentioned before, crystallographic symmetry plays an important role to introduce piezoelectricity in a material. From the definition of piezoelectricity, all the components of the piezoelectric tensor are zero in crystal with a center of symmetry. In this context, it should be remembered that 30% of all materials (nearly several million are discovered till now) statistically exhibits noncentrosymmetric crystal structure [24]. Moreover, only a few thousand have shown the piezoelectric property and among them, very few materials have enough piezoelectric polarization to be utilized for real energy harvesting applications.

3. Maxwell's equations and piezoelectric nanogenerator

Maxwell's equations are considered the most important formulas in science since it was established by Maxwell in 1861. We can see its application in every branch like static electricity, astrophysics, wireless communication, photonics, electromagnetic (EM) wave, and many fields dealing with electricity and magnetism. It is observed that the displacement current of Maxwell's equation plays a crucial role during the conversion of mechanical energies into electrical energies in nanogenerators (NGs) [25].

Let's focus on the four Maxwell's equations:

$$\vec{\nabla}.\vec{D} = \rho \text{ (Known as Gauss's law for electrostatic)} \tag{6}$$

$$\vec{\nabla}.\vec{B} = 0 \text{ (Known as Gauss's law for magnetism)} \tag{7}$$

$$\vec{\nabla} \times \vec{E} = -\frac{\partial \vec{B}}{\partial t} \text{ (Known as Faraday's law)} \tag{8}$$

$$\vec{\nabla} \times \vec{H} = \vec{J} + \frac{\partial \vec{D}}{\partial t} \text{ (Known as Ampere's circuital law with Maxwell's correction)} \tag{9}$$

Where, $\vec{D}, \vec{E}, \vec{B}, \vec{H}$ are the electric displacement field vector, electric field, magnetic field, and magnetizing field, respectively. The electric charge density is denoted by ρ. $\vec{J}$ is the electric current density.

Now, this electric displacement field vector can be expressed in terms of electric field ($\vec{E}$) and polarization field ($\vec{P}$) as following:

$$\vec{D} = \varepsilon_0 \vec{E} + \vec{P} \tag{10}$$

From, equations 9 and 10 it can be written that:

$$\vec{\nabla} \times \vec{H} = \vec{J} + \varepsilon_0 \frac{\partial \vec{E}}{\partial t} + \frac{\partial \vec{P}}{\partial t} \tag{11}$$

The second and third terms together are known as Maxwell's displacement current density (J_D).

$$J_D = \varepsilon_0 \frac{\partial \vec{E}}{\partial t} + \frac{\partial \vec{P}}{\partial t} \tag{12}$$

This displacement current is not like conventional electric current due to the flow of electrons, it consists of a time-varying electric field in any media that develops electromagnetic field and the contribution from the movement of polarized charges in materials. The first term of equation 12 i.e. $\varepsilon_0 \frac{\partial \vec{E}}{\partial t}$ develops the theory of electromagnetic wave and the second term ($\frac{\partial \vec{P}}{\partial t}$) i.e. the motion of dielectric polarization charges contributes to the output of nanogenerators.

Fig. 1 describes the working mechanism of a PENG. Mechanical stress deforms the piezoelectric insulator and develops the positive and negative polarization charges at the top and bottom sides of the piezoelectric layer, as demonstrated in Fig. 1b. The number of polarized charges is linearly dependent on the mechanical stress. That means an increase of applied force results development of more polarization charges (σ_p) up to a certain polarization saturation value. Due to the opposite polarization charges generation, a piezoelectric potential is developed between the two sides of the piezoelectric material. This piezopotential is balanced by the flow of electrons through externally connected load resistance between the top and bottom electrodes. If we assume the induced charge density on the electrode is σ then we can express the respective electric field (E) as follows:

$$E = \sigma - \sigma_p/\varepsilon \tag{13}$$

Now, if there is no external electric field applied and the polarization vector is along with the z-axis along the height of the piezoelectric insulator material then the equations 10 and 12 will be transformed to:

$$D(z) = P(z) = \sigma_p(z) \tag{14}$$

$$J_D(z) = \frac{\partial P(z)}{\partial t} = \frac{\partial \sigma_p(z)}{\partial t} \tag{15}$$

The equation 15 is indicating that the output current of PENG is the rate of change of polarization charges. The open-circuit voltage (V_{oc}) is given by the following equation:

$$V_{oc} = \frac{z\sigma_P(z)}{\varepsilon} \tag{16}$$

If the external load resistance (R) is connected between two electrodes of PENG, then the current transport relation of PENG will be:

$$RA\frac{d\sigma}{dt} = z\,[\sigma_P(z) - \sigma(z)]/\varepsilon \tag{17}$$

Where A is the area of electrodes, z in the thickness of the piezoelectric layer.

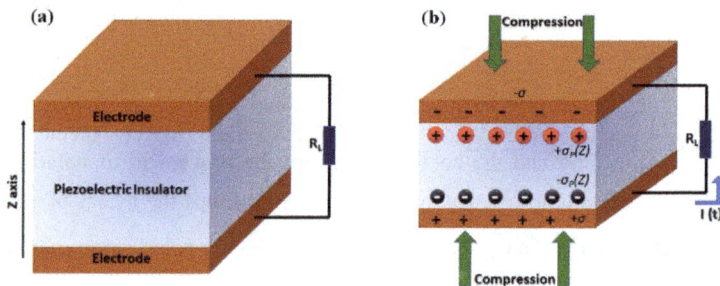

Figure 1. Working mechanism of PENG (a) when there is no stress on the material (b) under compressive stress

4.   Piezoelectric materials for nanogenerators

Piezoelectric nanogenerators with high-performance were manufactured using various piezoelectric materials including zinc oxide (ZnO) [3,26-27], lead zirconate titanate (PZT) [28-29], barium titanate [30-31], KNbO₃ [32]. We can broadly divide the piezoelectric materials used for fabrication of NGs in three types: ceramic, pure polymer and ceramic-polymer composite.

4.1   Ceramic

4.1.1   Zinc oxide

Nanogenerators (NGs) were first demonstrated by Wang and Song utilizing zinc oxide (ZnO) nanowires (NWs) and conductive tip of atomic force microscope (AFM), as shown in Fig. 2a [3]. ZnO has few advantages for the fabrication of NGs: it exhibits semiconducting and piezoelectric behaviors both, it is biocompatible materials and can be configured in various structures like nanowires [33], nanorings [34], nanobelts [35], nanosprings [36]. Highly aligned ZnO nanorod can be synthesized on the c-plane oriented α-Al_2O_3 substrate by vapor-liquid-solid (VLS) technique [3]. VLS is a mechanism to grow a one-dimensional nanostructure from chemical vapor deposition. However, at the first report of NGs the authors could not measure the current, they were able to measure the voltage around 9 mV. In this observation, the piezoelectric and semiconducting property of ZnO developed the strain field and the charge separation throughout the NWs because of compression by the AFM tip. the Schottky barrier between the semiconductor ZnO and metal tip of AFM generated the electrical voltage. It is noteworthy to mention that the AFM tip can only produce stress on a single NW which causes a less piezo-output voltage. This output can be enhanced by the actuation of all NWs simultaneously with the mechanical energy of ultrasonic waves [37]. Still, it did not enhance the output voltage significantly and these initial studies were not compatible with flexible polymer substrate for more realistic application. ZnO nanowires can be developed on a plastic substrate like polyimide (Kapton), polyethylene napthalate (PEN), poly(dimethylsiloxane) (PDMS) in a solution of zinc precursor using a synthetic chemistry process for demonstration of piezoelectric power generation from NWs on the flexible substrate [26,38-39]. Though, the output voltages and current of the studies were in the range of mV and nA, respectively. The Schottky junction between the ZnO NWs and the metal electrode is not very effective for good output voltage because of leakage current. This can be avoided by introducing a capping layer of polymer over NWs like Fig. 2b [40]. Also, the buffer layer helps to distribute mechanical stress homogeneously on nanowires, introduces robustness, and protects the NWs from their brittle nature due to pressure on them. We can grow the ZnO layer on metal or indium teen oxide (ITO), as the bottom electrode, deposited substrate (rigid [40] or flexible [41-42]) by RF sputtering or simple spin coating. This ZnO layer acts as seed layer, and then the ZnO NWs can be grown on this seed layer by hydrothermal. After the synthesis of NWs, one should anneal the ZnO for better crystalline and electrical property at nearly 200 °C. Polymer capping is done by simple spin coating of PMMA, PDMS on the ZnO NWs layer. The top electrode is deposited by thermal evaporation above this polymer layer. Enhanced rectified output (37 V open-circuit voltage, V_{oc}, and 12 μA short-circuit current, I_{sc}) by commercially available bridge rectifier under 1 MPa pressure has been reported from the capping-like structure of 12 µA height NWs over 1 x 1 cm^2 NG area by Zhu et al. [40]. A parallel connection between many nanogenerators can enhance the output current by linear superposition. Those types of 9 NGs in parallel increased V_{oc} to 58 V and I_{sc} to 134 µA height NWs with the maximum power density of 0.78 W/cm^3 by simple hand-palm pressing [40]. Pd-Au, silver, carbon-based electrodes (carbon nanotube or CNT, graphene) can be used for top electrodes [42]. However, carbon-based electrodes may reflect the lower conductivity and hence comparatively little lower performance of nanogenerator can be seen [43-44]. Single-walled CNT (SWCNT), and graphene have other superiority like transparency, high mechanical strength over metal electrodes. So, for the fabrication of flexible, rollable, transparent PENG SWCNT, graphene can be utilized as an electrode. They can show 70-80% transparency in the visible range [43-44]. Carbon based electrodes show better current density than other well-known transparent electrodes like ITO.

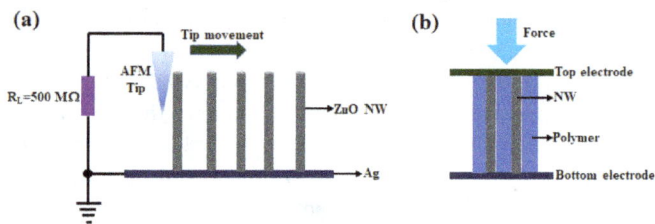

Figure 2. (a) Demonstration of first PENG based on the movement of AFM tip over vertically grown ZnO NWs, (a) Polymer capping of around NW to prevent fracture during force

Vertically aligned ZnO NWs for PENG have some bottleneck problems during practical application. You cannot have good power output during lateral bending because of its perpendicular presence on the substrate. Laterally grown ZnO NWs can show superior property over vertically grown NWs because there is more chance to good contact between all portions of nanowires and electrodes in lateral assembly [45-46]. Apart from nanowire structure different nanostructures like nanosheets, nanoparticles can be utilized for making PENG. One-dimensional (1D) ZnO nanowires, nanorods-based PENGs can only show a satisfactory performance only if when they show a high aspect ratio [47]. Because the enhanced piezoelectric co-efficient can be seen in 1D nanowires structure in case of high aspect ratio. nanowalls also have many attractive properties like high surface to volume ratio, nanometer range thickness, mechanical robustness which makes them potential candidates for nanogenerators, bio-sensors, chemical sensors, energy storage, etc. [48-49]. The buckling behavior of ZnO nanosheets makes them more stable under mechanical loads.

It has been found that the mechanical durability of the 2D ZnO nanosheet is more than 1D ZnO nanowires from nanoindentation analysis [50]. Vertically grown ZnO nanosheets of width 80 nm and height 3 μm over aluminum (Al) coated flexible polyethersulphone (PES) substrate provided the maximum DC output power density ~11.8 μW/cm² at 4 kgf force after passing through a conventional bridge rectifier [50]. It should be kept in mind that the output performance of PENG during open-circuit measurement value and DC output value after rectification are different. Because one can have a significant power loss across the bridge rectifier during rectification, shown in Fig. 3a.

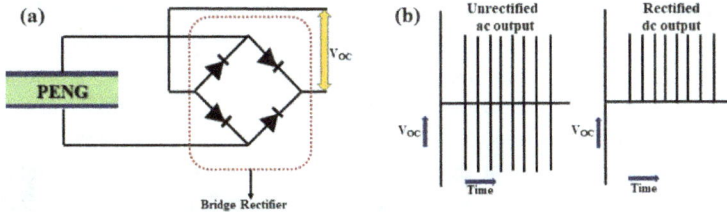

Figure 3. (a) Circuit diagram of rectification of PENG output through bridge rectifier, (b) Voltage profile of PENG before and after rectification.

However, rectification is an unavoidable step for NGs for fruitful utilization of PENG output. After rectification one can have a dc output profile like Fig. 3b which can be directly fitted to a capacitor for storing or any electronic device. People also tried ZnO nanosheet networks based flexible PENG on another substrate like ITO-coated PET sheet [51]. Despite the many efforts, still, the ZnO based PENG suffers from lower performance due to the poor piezoelectric coefficient and screening effect of piezo-potential for ZnO. The defects and surface states of ZnO act as free charge carriers which causes damping or screening over the developed piezoelectric potential [52-53]. Generally, hydrothermally grown ZnO nanostructures contains a numerous defect concentration. This reduces the crystal quality and this synthesis grown ZnO surface desorbs hydroxide. Also, in the open air, the oxygen molecules are adsorbed by pristine ZnO and these molecules are transferred to the O_2^- ions by capturing free electrons from the conduction band. This leads to lower conductivity of ZnO in air. UV band generates electron-hole pairs and these holes drift towards the surface of ZnO getting attracted by O_2^- present at the surface and holes neutralize these oxygen ions after recombination with these. Thus, the rest of the electrons remain unbound as free carriers and enhanced the conductivity [54]. This enhanced number of free carriers causes problems like the screening effect. The piezoelectric potential is strongly diminished due to the increment of free carrier density. Still, it is one of the most favored materials for the fabrication of PENG because of its easy synthesis process at a comparatively lower temperature. Surface passivation through thermal annealing, plasma treatment, organic hybridization enhances the output performance [55-57]. Through surface passivation the surface of the material is rendered inert. The material's properties do not change because of interaction with air, water molecules, etc. in contact with the material's surface. Plasma treatment in an oxygen chamber for a prolonged time (30-60 mins) and time-dependent thermal annealing at ~300-400 °C reduces the defect state, improves the crystal structure, and removes the hydroxide ions from the surface. Plasma treated, thermal annealed ZnO based PENG performance can reach up to 20 V V_{oc} and 6 μA I_{sc} with 0.2 W/cm^3 which is 20 folds superior to without treatment ZnO based PENG [55]. Organic hybridization can also be a fruitful way to enhance the ZnO based PENG performance. During hybridization, compatible p-type polymer (like Poly(3-hexylthiophene, commonly known as P3HT) is deposited on the surface of n-type ZnO NWs to build a p-n junction at their interface. Phenyl-C_{61}- butyric acid methyl ester (PCBM) can be added to P3HT to uplift the carrier transport. It is reported that ZnO/P3HT-PCBM assembly increased the output performance of PENG showing 18 times voltage and 3 times current improvement with a total output power density of 0.88 W/cm^3 [56]. The holes and electrons diffuse towards n-type ZnO and p-type P3HT from P3HT and ZnO side, respectively, and develops a depletion layer at the interface. As a result of this formation of a p-n junction, the free electrons are neutralized or passivated by the holes from the P3HT side and thus the piezopotential and performance are both enhanced. During forward and backward bending of ZnO cantilever PENG, the ZnO film experiences compressive and tensile strain. The negative and positive piezopotential are developed at the electrode/ZnO interface due to tensile and compressive strains, respectively [58]. Apart from this layer passivation technique, doping is also a pathway to improve PENG performance. The substitution of transition or alkali metal ions in Zn or O sites gives rise to interesting changes like ferroelectricity, piezoelectricity, ferromagnetism, conductivity [59-62]. Magnitude of the piezoelectric coefficient (d_{33}) directly affects the piezoelectric nanogenerator performance. Bulk ZnO exhibits 9.93 pm/V d_{33} value and one of the highest d_{33} values achieved from ZnO nanobelt is 26.7 pm/V [63]. Where, vanadium doped ZnO can show up to 110 pC/N value of d_{33} and improved NG performance is reported [27,64]. Other metals like cobalt (Co), sodium (Na), silver (Ag), lithium (Li), yttrium (Y) has been successfully utilized for doping in ZnO and several times greater piezoelectric coefficient and PENG output are observed [65-66]. ZnO based nanoparticles, rather than nanorod, nanosheets have exhibited their effective potential for PENG [67]. Yttrium doped ZnO nanoparticle shows a gigantic d_{33} value around 420 pm/V and 20V V_{oc} has been observed under gentle finger tapping (~0.01 kgf) [65]. Besides ZnO, many other ceramic piezoelectric materials were utilized for energy harvesting using their piezoelectric property.

4.1.2 Barium titanate

Barium titanate (BaTiO$_3$ or BT) is well known ferroelectric material. Ferroelectric materials possess two equilibrium spontaneous polarization orientations in absence of an external electric field. The polarization vector can be oriented according to the external electric field. Ferroelectric materials have the property to show piezoelectricity and pyroelectricity. BT is of perovskites material which general formula is ABO$_3$, where A and B are two metals. Fig. 4 expresses the structure of BT where we can see Ba^{2+} ions (A site in ABO$_3$) are located at the eight corner points of a cubic unit cell and Ti^{4+} ion (B site) occupies the center of the unit cell, while the six O^{2-} ions situate at the face-centers of the unit cell and constitute BO$_6$ octahedra [68]. When the crystal structure of BT shifts from cubic to tetragonal phase, the movement of Ti^{4+} (either towards up direction or down direction from its equilibrium, as shown in Fig. 4b) and O^{2-} ions relative to Ba^{2+} ions gives rise to spontaneous polarization in BT crystal. Being an m$\bar{3}$m centrosymmetry crystal, the cubic phase of BT is paraelectric and non-piezoelectric. The cubic m$\bar{3}$m phase of perovskite crystals exhibits seven proper ferroelectric phases in different temperature ranges: 1 tetragonal (4mm), 1 rhombohedral (3m), 1 orthorhombic (mm2), and 3 monoclinic (m) and 1 triclinic (1) symmetry [69]. At normal atmospheric pressure, the BT crystal undergoes a sequential first-order phase transition with temperature: cubic to tetragonal (at 131 °C), tetragonal to orthorhombic (at 0 °C), orthorhombic to rhombohedral (-90 °C) [70]. The direction of spontaneous dipole polarization along parallel to the edge, along the face diagonal, and along body diagonal in case of tetragonal, orthorhombic, rhombohedral, respectively. Each phase transition shows distinct thermal, mechanical, and piezoelectric properties. The first principle simulation study indicates that the spontaneous polarization of perovskite BT is greatly influenced by the covalent bonding between the Ti^{4+} and O^{2-} ions. BT is an ionic bonding crystal, but the simulation study reveals the existence of covalent bonding between Ti 3d and O 2p states [71]. It was predicted that this covalent bonding may distort the crystal structure like Fig. 4b and introduce a dipole moment inside the crystal. Thus, the covalent bonding between titanium and oxygen atom gives rise the spontaneous polarization in BT. Roughly, piezoelectric strain (S) can be expressed through the following equation [72]:

$$S = 2\varepsilon P_s QE$$

Where, ε, P_s, Q, and E are dielectric permittivity of the material, spontaneous polarization, electrostriction coefficient and induced electric field, respectively.

Like ZnO, we can also follow the same kinds of nanostructures with other piezoelectric ceramic materials. Vertically aligned BT NWs have been synthesized on rigid fluorine-doped tin oxide (FTO) glass substrate by two steps hydrothermal process and utilized as an energy harvester placing the NWs in between two electrodes [73]. BaTiO3 can show more enhanced performance than ZnO, because of its greater piezoelectric coefficient vale than ZnO. However, it is important to perform DC electric field poling of BT particles or NWs for enhancement of the polarization behavior [73]. For the increment of piezoelectric behavior, electric field poling is an important step. The magnitude of the DC electric field for poling of any material should be greater than the coercive field of that material [74]. Barium titanate horizontally aligned nanofibers based flexible PENG on PET substrate achieved high piezoelectric output performance of 0.1841 µW under 0.002 MPa with 2.67 V and 261.4 nA maximum voltage and current, respectively [75]. The dimension of the nanofibers was 1.7 mm x 0.7 mm x 0.65 mm (length x width x thickness). Vertically aligned BT nanotube arrays can be sandwiched between two ITO-PET flexible substrates for the fabrication of PENG. A maximum 10.6 V and 1.1 µA output voltage and current were recorded during repetitive bending at a frequency of 0.7 Hz [76]. The ceramic piezoelectric materials are brittle in nature. To increase robustness during strain, one can use a flexible polymer matrix like PDMS to contain the ceramic nanowires or tubes. We can also use BT nanoparticles instead of wires in polymer matrix for making flexible, robust PENG [77]. The possibilities of all ceramic polymer based composite materials will be discussed later.

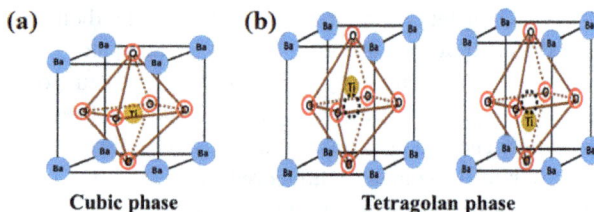

Figure 4. Crystal structures of BaTiO3 (a) cubic phase and (b) tetragolan phase

4.1.3 Lead zirconate titanate (PZT)

Lead zirconate titanate (PZT) is one of the most common in piezoelectric electronics devices because of its high ability of electro-mechanical coupling [78-80]. Due to this property one just needs a small amount of electric field to actuate the PZT. Various nanostructures have been investigated for the fabrication of PENGs. PZT is a mixture of two solid solutions of $PbZrO_3$ and $PbTiO_3$. This was first analyzed by G. Shirane et al. in 1952 and in 1953, E. Sawaguchi introduced its phase diagram [81-84]. The piezoelectricity of PZT was first demonstrated by B. Jaffe et al in 1955 [85]. After this discovery, PZT has been intensely developed and used by many research groups as piezoelectric materials for devices. PZT has a perovskite crystalline structure ABO_3. Perovskite structured materials exhibit interesting properties like ferroelectricity, piezoelectricity, optoelectronic, semi-conducting behavior. It has been observed that when A site is filled by Pb atom, PZT shows large piezoelectricity. Moreover, the piezoelectric property can be tuned by changing the compositional formula of PZT. Generally, the ABO_3 structure, as shown in Fig. 4, contains A-site cations at each corner of a cuboid and the oxygen octahedra remain inside the box. The B-site cation is placed into the center of the cuboid box. Lead (Pb) ions fill A-site and B-site is permeated randomly by Ti and Zr ions. In his theoretical study, R. E. Cohen reported that the bonding environment of $PbTiO_3$ is little different concerning to $BaTiO_3$, as described before. Pb 6s electrons contribute to the covalent bonding between Ti and O ions in $PbTiO_3$. Therefore, $PbTiO_3$ exhibits larger piezoelectricity than $BaTiO_3$ [86-87]. This concludes that the PZT perovskite structure performs well as a key material for a piezoelectric nanogenerator. PZT nano-powders can be used in rubber matrix for making stretchable piezoelectric nanogenerator which delivered an excellent output power ~81.25 $\mu W/cm^3$ nanofibers by electrospinning has shown a great potential for energy harvesting [88]. It is crucial to make electrodes for stretchable devices. One should take care of the fact that the resistance of the electrode should not be changed vigorously during stretching. Silver coated microsphere embedded in rubber matrix has been used as a stretched electrode in the case of PZT-based stretchable PENG by X. Niu et al. [80]. Electrospun PZT nanofibers are also extensively used for the fabrication of PENGs. 60 nm diameter and 500 μm long PZT fibers have been synthesized from PZT/PVP solution through the electrospinning method [29]. The pure perovskite phase of this fiber can be achieved after annealing at 650 °C for 30 mins. Embedding in the PDMS polymer matrix introduces flexibility. Interdigitated electrodes on Si substrate were used by this group, as shown in Fig. 5a. They achieved a maximum power of 0.03 μW across 6 MΩ resistance at 40 Hz vibration. However, due to the inflexibility of substrate this device cannot be useful for practical applications for flexible electronics. Other groups tried also with the PZT NWs, fibers for making piezoelectric nanogenerators [28,89-90].

Advanced Functional Piezoelectric Materials and Applications   Materials Research Forum LLC
Materials Research Foundations 131 (2022) 61-116   https://doi.org/10.21741/978164490209-3

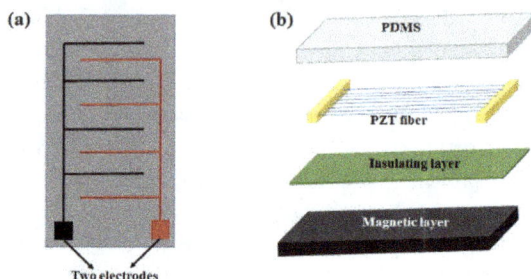

Figure 5. (a) Schematic interdigitated electrodes used on PENG in ref. [29], (b) Structure of magneto driven PENG used in ref. [89]

Contactless magneto-driven PENG has been demonstrated utilizing a magnetic layer with the piezoelectric layers introducing an insulating layer in between these [89]. A highly aligned PZT fibers have been synthesized over an insulated magnetic layer composed of Fe_3O_4 to make magneto-driven PENG. A d.c. electric field (4 MV/m at 130 °C about 10 min) is required to pole for better alignment of dipoles for this device. Two ends of this PENG are fixed so that when a bar magnet approaches the energy harvesting device, the whole device bends. The structure of this device has been depicted in Fig. 5b. This PDMS encapsulated magneto-driven PENG delivers a maximum of 170 µW/cm³ power across the 100 MΩ load resistance and open-circuit voltage 3.2 V from an active PZT area of 17. 5 mm² and 5 µm thickness [89]. Portable personal electronics has compelled us to fabricate wearable energy-harvesting device which can deliver power to the device. A wearable, flexible piezoelectric nanogenerator has been reported by PZT nanowires in parallel to each other which generates 6 V V_{oc} and 45 nA I_{sc} to power a liquid crystal display and UV sensor [28]. Tetrabutyl titanate, zirconium acetylacetonate, lead subacetate were added by this group according to the molar ratio of the composition formula of PZT in the solution of ethanol, acetylacetone, acetic acid for stirring more than 24 h. Mixing a very little amount of PVP (M.W.: 1300000) is very important to control the dimension of NWs. During electrospinning of the above solution, the distance between jet and collector was kept at about 25 cm. 20 °C temperature and 30% humidity were maintained. They applied 25 kV voltage to the jet which helped to develop parallel PZT NWs. Multipair of parallel electrodes were used to collect the aligned NWs. Coulomb force and the parallel electrodes take part an important role to align the NWs. As the electrodes are grounded, when the positively charged NWs fall on the electrodes after discharging from the jet, they just become electrically neutral after coming into contact with the grounded electrodes. So, on

Advanced Functional Piezoelectric Materials and Applications   Materials Research Forum LLC
Materials Research Foundations 131 (2022) 61-116   https://doi.org/10.21741/978164490209-3

the electrodes, the nanowires orient randomly. On the opposite side, the nanowires align parallel in-between space of two parallel electrodes due to Coulombic repulsive force. These nanowires remain electrically charged for some time as they can not meet grounded electrodes. After electrospinning, the NWs were dried and calcined at 650 °C for 3 h in a furnace. Before calcination the diameter was 530 nm, calcination removed the polymer layer coatings and decreased the diameter to 370 nm. PET substrate was chosen to fabricate this wearable device. An ultrathin film of PDMS was spin-coated on the substrate following by the placing of parallel aligned PZT NWs on the PDMS layer after cutting from the electrospinning NWs. It was further covered by the PDMS layer and silver electrodes were deposited on the two ends of NWs. Thus, it was demonstrated as flexible wearable PENG after performing electric field polarization of PZT at 4 kV/mm dc electric field at 130 °C for 15 mins. Ultralong vertically aligned PZT [$Pb(Zr_{0.52}Ti_{0.48})O_3$] NWs assembled PENG showed a dramatic performance [90]. The novelty of this design of PENG was to use ultralong 420 μm vertically aligned PZT on a single substrate instead of stacked layers of shorter wires. This strategy can introduce more flexibility of substrate removing the possibility of superfluous assembly of rigid substrates. To make this device, they first made parallel PZT NWs as stated above of reference 27 and stacked them layer by layer after cutting. The diluted PDMS was used to fill the interspace gaps between wires and layers. After solidification of PDMS at 80 °C for a couple of hours, the laterally aligned PZT layers were cut into many small cuboids and rotated to align them vertically. PDMS encapsulation was done for binding again. Metal electrodes were deposited on the top and bottom sides of these vertical aligned NWs and poled with 5 kV/mm dc electric field at 130 °C for 15 min to give the final structure of this PENG. This unique long vertical NWs structure PENG delivered a huge V_{oc} 209 V and I_{sc} 8.13 μA/cm^2. This large output was sufficient to stimulate the frog's nerve and to power a LED directly without storage, which is a great advancement of self-powered devices with PENG.

4.2   Polymer

Piezoelectric pure polymer and or its composites with ceramic piezoelectric have more advantages over the only ceramic because of their inherent flexibility and high breakdown field [91-92]. You can see the comparison between polymer and ceramic in Table 1. The rigidness, brittleness, and high processing temperature impede the ceramic piezoelectric for direct implementation in flexible energy harvesting devices. The piezoelectric polymers which can satisfy the requirements of soft, lightweight, flexible piezoelectric energy harvesting devices include poly(vinylidene fluoride) (PVDF) and its copolymer poly(vinylidene fluoride-co-trifluoroethylene) (PVDF-TrFE), poly(L-lactic acid) (PLLA) [91,93-95]. It is noteworthy to mention that polymer does not have good piezoelectric coefficient like inorganic piezoelectric materials, still polymer and its composites are most widely used for mechanical stability, robustness in all atmospheres. Piezoelectric polymer materials are classified according to their topology and composition: pure piezoelectric polymer, piezoelectric polymer-ceramic composite, and ferroelectret polymers. The piezoelectric property of the pristine polymer is controlled by the molecular structure mainly, whereas, in the case of void polymer the internal dipoles due to charged gas voids in between thin polymer layers. The combined piezoelectric effect of polymer and ceramic plays a great role to tune the piezoelectric property of the composites.

Table 1. Comparative view of piezoelectric properties between ceramics and polymer

	Piezoelectric Ceramics	**Piezoelectric Polymers**
Dielectric Constant	High (>500)	Low (<20)
Piezoelectricity	High	Low
Mechanical flexibility	Poor	Flexible

4.2.1 PVDF and its copolymer

The most explored pristine piezoelectric polymers are PVDF and its copolymer PVDF-TrFE. Kawai reported the piezoelectric property of polarized PVDF which was few folds superior to other polarized polymers like polycarbonate, polyethylene, and polytetrafluorethylene [96]. The flexible PVDF and its copolymer PVDF-TrFE reveals excellent dielectric properties, piezoelectricity, pyroelectricity, and ferroelectricity along with high mechanical strength and chemical resistance [91,97-99].

PVDF is a semi-crystalline polymer having four major different crystalline polymorphs: α, β, γ and δ as expressed in Fig. 6 [100]. Among all phases, α is non-polar form and the other three phases are polar because of their parallel packing of CH_2-CF_2 dipoles. The dipole moment originates from the CH_2-CF_2 unit of PVDF. This unit contains a positive H and negative F atom which develop a net dipole moment. The most common α phase contains trans-gauche-trans-minus gauche ($TGT\bar{G}$) packed in an anti-parallel way whereas, all-trans molecules ($TTTT$) oriented in parallelly in β polymorph. Parallel packing of ($T_3GT_3\bar{G}$) chains construct the γ conformation. A δ phase is parallel form of the non-polar α phase. It is mentioned that the above-stated parallel and antiparallel alignment depends on perpendicular dipole orientation to the polymer chain axis. The piezoelectric property of PDVF is determined by the presence of the β phase because the β crystalline phase has the highest spontaneous dipole moment due to all molecular diploes are aligned in a single direction [101-103]. The γ and δ polymorphs do not have well-understood properties because of the existence of various polymorphs and non-crystalline regions. Various techniques have been adopted to increase the β phase content as well as to improve the piezoelectric property such as high electric field poling [104-105], mechanical stretching [106-108], thermal annealing [109-110], introducing nano-filler in PVDF matrix [111-113]. It is noteworthy to mention that besides nanofiller polar solvent like dimethylformamide (DMF) also takes part to align the PVDF chain. The dipolar interaction and formation of hydrogen bond between the CH_2-CF_2 dipoles and the molecules of polar solvent lead to the all-trans ($TTTT$) planar configuration of CH_2-CF_2 dipoles during drying at room temperature. Moreover, if the thermal annealing is followed for the crystallization of PVDF then this large thermal external energy dominates this polar solvent interaction influence for $TTTT$ planar configuration [114-115]. PVDF and its copolymer does not exhibit a high magnitude of the piezoelectric coefficient, still, the efficient flexibility, toughness, chemical resistance, harmlessness to the bodies makes them a very interesting alternative for the fabrication of PENG. They show the highest piezoelectric response d_{31} ~20 to 25 pC/N and d_{33} ~ -30 to -40 pC/N among all other piezoelectric polymers [116-118]. Pure PDVF thick film-based transparent energy harvesting device has been demonstrated, which has been utilized for monitoring the muscular movement by fixing it on a finger. This PENG device is fabricated sandwiching 80 μm thick PVDF film (1.5 cm x 2.5 cm) by spin-casting of two silver NWs layers. Poly(2-hexyl-2,3-dihydrothieno[3,4-b][1,4]dioxine:dodecyl sulfate (PEDOT-C6:DS) solutions are coated on the AgNWs layers. After every step of spin-coating, the layer is annealed at 60 °C for 15 min. The combination of AgNWs and PEDOT-C6:DS act as top and bottom electrodes. PEDOT-C6:DS can help to improve the uniformity of NWs on the polymer and enhances the response during applied stress. To increase the β phase content in PVDF for improved piezoelectric performance thermal annealing of PVDF is done at 90 °C. This device is able to generate ~ 7 V V_{oc} and 1.11 μA I_{sc} at an optimized device vibration frequency of 8 Hz [119]. Enhancement of the piezoelectric performance of pure PVDF and its copolymers can also be achieved by high electric field poling. However, this process is complicated, time-consuming, risk of high voltage and also unidirectional polarization limits the flexibility of the device in any direction [120]. The PVDF fibers synthesized by the electrospinning process have a good piezoelectric behavior because of high β phase content. The high electric field and stretching force during electrospinning help -CH_2-CF_2-dipoles alignment leading to β phase induction. Self-powered aligned conducting PVDF nanofibers are utilized to fabricate piezoelectric sensors for health monitoring. 120 kV/m dc electric field was introduced to align PVDF fibers during this electrospinning. Conducting polyaniline (PANI) coated PVDF nanofibers have good conducting behavior due to enhanced interlinking of conducting paths between electrodes and the piezoelectric layer. This kind of electrode compatibility is good for fabrication of flexible piezoelectric energy harvesters [121]. The electrospinning technique can promote higher crystallinity (49%) and higher β phase content (76.2%) in PVDF nanofibers [122]. Like PVDF, its copolymer PVDF-TrFE is another largely accepted piezoelectric polymer for steric hindrance from the extra fluorine atoms of TrFE. The inclusion of TrFE induces an all-trans stereochemical configuration in PVDF-TrFE [123-124]. This is the reason that PVDF-TrFE contains highly the piezoelectric β-crystalline phase. Highly aligned PVDF-TrFE (PVDF:TrFE 70:30) nanofibers are synthesized through the electrospinning process for fabrication of flexible PENG [124-125].

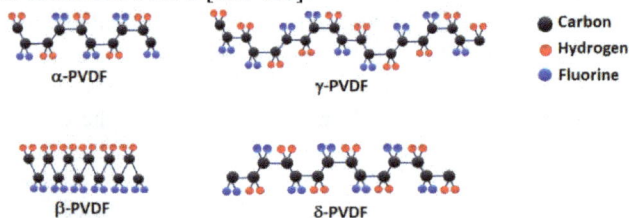

Figure 6. Different crystal structures of PVDF

4.2.2 Polylactic acid

Polylactic acid (PLA) based polymers are becoming an interesting candidate for piezo-nanogenerator for their piezoelectric properties, easy fabrication, biodegradability, and biocompatibility [126]. The great advantage is that PLA does not need any electrical poling to induce piezoelectricity [127-128]. When shear stress is applied on helical PLA molecules in the direction of the helix axis (Fig. 7), the carbonyl functional group rotates slightly and develops net polarization perpendicular to the plane shear strain [129]. Therefore, uniaxial stretching during the fabrication of PLA film induces piezoelectricity. Annealing at 140 °C from 3-4 h is needed after the uniaxial stretching of the film. But remember, PLA does not show any ferroelectricity and pyroelectricity as it exhibits only shear piezoelectricity. A biocompatible fiber mat has been synthesized by dispersing $BaTiO_3$ in PLA. This composite fiber mat is capable to show the d_{33} value of nearly 0.5 nC/N. The piezoelectric output is enough to power a small LCD by finger tapping on a PENG device made of 0.15 mm thick and 2 cm^2 area of this composite fiber mat [130]. PLA based thermally stable unimorph cantilever PENG is utilized for energy harvesting application. This cantilever is constituted of a double layer of PLA and thick PET substrate for the passive layer. The maximum output power and voltage are recorded as 14.17 μW and 9.05 V through the load resistance of 0.9 MΩ [131]. A wind-driven hybrid nanogenerator is reported where the PLA polymer layer was served as a piezoelectric energy harvester and the PET layer as a triboelectric energy harvester [132].

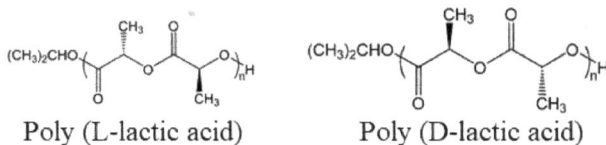

Figure 7. Two isomers of polylactic acid

4.2.3 Cellulose

Cellulose, the most abundant polymer on the Earth, is a polysaccharide consisting linear chain of many repetitive β linked D-glucose units. Every glucose unit has three hydroxyl groups and the unit are linked via glucosidic bonds. Cellulose is a highly crystalline polymer because the formation of strong hydrogen bonds between these hydroxyl groups [133-134]. Cellulose I and II are commonly available among the 4 crystal structures of cellulose, known as Cellulose I, II, III, and IV [135]. In 1955, the piezoelectricity of cellulose (wood) was discovered by Fukada [136]. The piezoelectric effects of wood corresponding to d_{14} and d_{25} have been reported in that article. The range of d_{14}= -d_{25} is about 10^{-9}. Due to the noncentrosymmetric arrangement of hydroxyl groups, development of a net dipole moment occurs, which leads to the piezoelectricity in cellulose. The piezoelectric coefficient d_{33} of natural cellulose is 0.4 pC/N [137]. Research is going on to improve the piezoelectric properties of cellulose by different processing techniques and also by inclusion of piezoelectric nanofillers in the polymer matrix. In one study, 27 wt% $MnFe_2O_4$ nanoparticles (~4 nm) are dispersed in bacterial cellulose and the composite piezoelectric film is synthesized by hot pressing at 60 °C for 8 h, this film exhibits a drastic change in the value of d_{33}. The d_{33} changes from ~5 pC/N (only bacterial cellulose) to ~23 pC/N [138]. Besides the fillers inclusion, another approach to improve the piezoelectricity of cellulose is changing the processing technique. Stretching during synthesis can arrange the dipole in an order which leads to enhancement of piezoelectricity [139]. A flexible piezoelectric nanogenerator is introduced utilizing the cellulose microfiber and PDMS blend with MWCNT as conducting filler. Cellulose microfiber powder, MWCNT, and PDMS are mixed in ratio 5:0.5:100 and poured into petridish and dried at 80 °C after removal of bubbles. After removing from petridish, the composite film is cut into pieces (length x width x thickness 40 mm x 28 mm x 2mm) to make PENG, and the Al foil is utilized as top and bottom electrodes. It delivers high electrical output V_{oc} ~30 V and power density ~9 $\mu W/cm^3$ through 30 MΩ load resistance under repeated hand punching. This power is enough to glow 22 green LEDs when the PENG is attached directly to the LEDs assembly. Also, a capacitor is charged to 5 V and a small LCD screen is powered up using this. PDMS produces Si-O-Si fracture and develops free radicals due to mechanical force. These free radicals and the crystallinity of cellulose together produce good piezoelectric performance. [140]. High-performance flexible PENG consisting of porous cellulose nanofiber (CNF) and PDMS is also demonstrated. High output signals: 60.2 V V_{oc}, 10.1 μA I_{sc}, and 6.3 mW/cm^3 power density are recorded from this PENG [141]. Under mechanical vibration, the -SiO⁻, =Si· free radicals are generated in the PDMS and thus more dipoles are developed to improve the PENG performance [142]. The composite, composed of barium titanate nanoparticles (~50 nm) and carboxymethylated nanocellulose, is used to demonstrate eco-friendly PENG. Before making composite, BT nanoparticles are annealed at 1000 °C for 10 h to induce more tetragonal phase of BT which improve the piezoelectricity. BT (30-60 wt%) nanoparticles are dispersed in an aqueous suspension of 1wt% nanocellulose and cast into a petridish and dried at 50 °C. For improvement of the piezoelectric property of BT nanoparticles, further d.c. electric field poling of the composite film is done at 150 kV/mm at 100 °C for 12 hr. Finally, the PENG performance has been evaluated by putting this poled film in between flexible ITO-coated PET substrates, which act as electrodes, under periodic compression of 5 kPa. Nanocomposite film with 40 wt% BT NPs exhibits the highest PENG outputs: voltage ~2.86 V, current 262.4 nA, and electrical power ~378.2 nW. This power is enough to charge the capacitors after rectification [143]. Cellulose/PDMS composite with Gold (Au) nanoparticles has been explored to demonstrate mechanical energy harvesting. The PENG device (3 x 8 cm^2) delivers V_{oc} 6V and power density 8.34 mW/m^2 when 3 N periodic force is applied on it. 1.8% energy conversion efficiency is calculated for this kind of PENG device. The PENG can light two commercial blue LEDs [144].

4.3   Ferroelectret

There is some other kind of material which is called ferroelectret or piezoelectret. The particular non-polar porous polymers showing ferroelectric-like properties under high electric field are known as ferroelectret materials [145]. During the corona poling process, electrical breakdown of gas happens inside the void after a threshold electric field of the gas (mainly air) remained inside the macro-sized void of a piezoelectrically active polymer foam. The microplasma, which is developed due to this breakdown, discharges and charges deposit on the top and bottom layers of void spaces, as shown in Fig. 8b. This atomic-sized dipole orientation resembles the dipoles of ferroelectric materials. The ferroelectret materials also exhibit dipole switching simultaneously with the reversal of the electric field. After the formation of polarized pores or voids, the ferroelectret with electrodes attracts the surface charges. The surfaces charges start to redistribute again when the polarization of pores is changed by the external mechanical stress or temperature. Because of the generation and redistribution of those charges, the electric current flows through an external circuit connected between two electrodes (Figs. 8c and 8d). Therefore, the ferroelectret materials show piezoelectricity and pyroelectricity. This unique property of this ferroelectret materials demonstrates their potential to be used for energy harvesting. Thermoplastic non-polar polymers like polyurethane (PU), polyolefin (PO), cyclo-olefin polymers are extensively used to synthesis ferroelectret. Polyvinyl chloride (PVC), poly(ethylene terephthalate) (PET), poly(ethylene napthalate) (PEN) have also been explored as ferroelectret energy harvesting materials. It should be kept in mind that, the dielectric and insulating property of the used polymer should be very good to maintain high voltages to create charges in voids. When a high electric field is applied across the void enriched polymer Townsend discharge begins. It is a gas ionization process where free electrons are drifted by this high electric field. If the electric field is very small, then these free electrons cannot acquire enough kinetic energy to reach the anode side. They then lose their whole energies during collisions. Thus, after a particularly high electric field, the charged voids create dipole structures as shown in Fig. 8a and the polymer with closed foam structure can be treated as ferroelectret or piezoelectret. The piezoelectric response of these materials is determined by the porosity, sizes of pores. Many research articles demonstrated energy harvesting using ferroelectret materials. You can see the details of those reports in Table 2. As this chapter deals with only pure piezoelectric materials, we will not go into details about these piezoelectret materials based nanogenerators. However, the general device structure of these nanogenerators is the same as PENG.

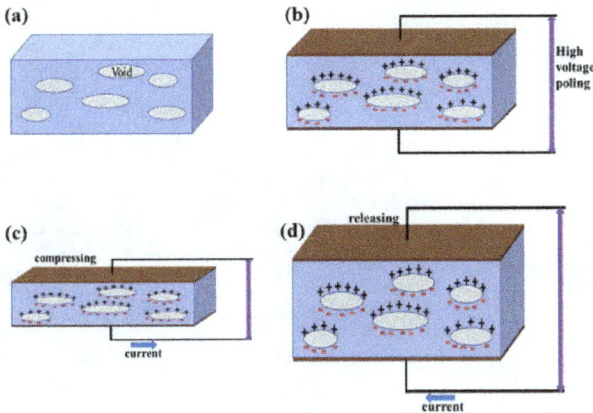

Figure 8. (a) General structure of ferroelectret materials, (b) Formation of charges inside voids under High voltage poling

Table 2. Summary of Ferroelectret/Piezoelectret Materials utilized as Nanogenerator

Materials	Manufact-uring Process	Pore Dimension	Poling Condition	Piezoelectric Coefficient	Device Dimen-sion	Harvested Output Power/Voltage	Ref. No.
Fluoroethylen epropylene (FEP)/PTFE	Stacking layers by hot pressing. Foam bought from company	3 [μm]	Corona Poling (20 [kV] for 5 [min])	N. R.	13 x 13 [mm^2]	32 [V] (p-p) & 0.56 [μA] (p-p) under 30 [N]	[206]
Polypropylne (PP)	Repeated folding	N. R.	Corona Poling (20 [kV] for 1 [min])	N. R.	30 x 10 [mm^2], 38 [μm]	641 [mW] (Seismic mass 4 [g], 36 [Hz] vibration)	[207]
Poyethylene (PE)	Blowing agent (ADZ powder) with heat treatment	90-340 [μm]	Corona Poling (35 [kV] for 1 [min])	d_{33}: 200 [pC/N]	600 [μm]	0.46 [V] and -1 [V] for 90 [μm] void (800 [N] force)	[208]
PVDF-HFP	Solution process with Mg-salt	4.6 [μm] diameter	Self-poled	N. R.	18 x 35 [mm^2], 0.3 [mm]	Stored power 60 [nW] in 4.7 [μF] capacitor	[209]
PVDF-HFP	Solution casting, ZnO NPs etching	0.9 [μm] diameter	Self-poled	d_{33}: -15.2 [pC/N]	3.6 [cm^2], 330 [μm]	V_{oc}: 9 [V] I_{sc}: 1.3 [mA/cm^2] under 0.36 [MPa]	[210]
Fluoroethylen epropylene (FEP)	Template-pattering and fusion-bonding	N. R.	N. R.	d_{33}: 1700 [pC/N]	7.1 [cm^2], 60 to 500 [μm]	73 [mW] (Seismic mass 60 [g], 130 [Hz] vibration)	[211]
Fluoroethylen epropylene (FEP)	Template-pattering and fusion-bonding	N. R.	Direct contact charging (1.4 to 2 [kV])	d_{33}: 340 [pC/N]	3.14 [cm^2], 300 [μm]	3.5 [μW/g] (Seismic mass 44.1 [g], 650 [Hz] vibration)	[212]
Crosslinked PP	Bought from Market	400 [μm] diameter	Corona Poling (25 [kV] for 1 [min])	N. R.	N. R.	80 [μW] from Seismic mass 27 [g]	[213]
PDMS	Molding	Cylindrical cavity of 40 [μm] height and 100 [μm] diameter	N. R.	N. R.	4 [cm^2], 250 [μm]	103 [nW] (load: 217 [MΩ])	[214]
EVA/BOPP	Hot Pressing	1.4 [mm] x 0.3 [mm]	Corona Poling (-18 [kV] for 5 [mins])	N. R.	5 x 5 [cm^2]	Rectified current 0.7 [mA] to 1 [mA]	[215]

Advanced Functional Piezoelectric Materials and Applications   Materials Research Forum LLC
Materials Research Foundations 131 (2022) 61-116   https://doi.org/10.21741/978164490209-3

4.4 PVDF based composite

Though the PVDF and its copolymer have exhibited great potential for flexible, stretchable piezoelectric nanogenerators, still they suffer from a low piezoelectric coefficient than inorganic piezoelectric materials like PZT, PMN-PT, ZnO, etc. Therefore, we can introduce the idea of superposition of the properties like flexibility, high electrical breakdown, the high mechanical strength of the polymer, and high piezoelectric and dielectric values of ceramic materials in ceramic/polymer composite. These types of composites have few positive aspects: the piezoelectric ceramic filler increases the piezoelectric performance overall, during the synthesis these nano-fillers also act as nucleation agents which increase the crystalline phase content of polymer by electrostatic interaction, the positions of fillers in the flexible composite act as stress concentration points those further generate larger deformation in composite and enhance the piezoelectric output, sometimes the conducting fillers improve output current from the piezoelectric layer during the performance of PENG. However, beyond a threshold concentration of conducting fillers in the piezoelectric polymer matrix degrade the performance due to the formation of conducting leakage paths between two electrodes leading to the early electrical breakdown, as demonstrated in Fig. 9. There are many hydroxyl (-OH) groups attached to the surface of the nanoparticles because of the absorption of water vapor from the atmosphere. During solution process, the strong hydrogen bonds between H atoms of OH functional groups and F atoms of CF_2 dipoles of PVDF (O-H$\cdots$F-C) are formed. In this way, they serve as nucleating agents for PVDF crystals. However, the nucleation phenomenon is strongly dependent on the specific area (surface area per unit volume) of the particles. As the particle's size increases, the specific surface area decreases. Therefore, the probability of forming a hydrogen bond (O-H$\cdots$F-C) will be reduced when the particle size increases, that will lead to poor crystallization towards *TTTT* planar configuration of electroactive β-phase PVDF. It is reported that the β-phase is decreased when the PZT particle of size is varied from nm to μm. Piezo-property also is found to decrease with the increment of particle size of PZT [146]. This same trend also can be seen in the case of other filler particles [147]. Like the size of the particle, volume concentration also plays a great role in the crystallization process. Beyond a threshold concentration, the polymer backbone movement is opposed by the agglomeration of nanoparticles in the polymer matrix [91,109]. Therefore, one should optimize the size and concentration of filler for polymer composite to have the best piezoelectric performance. Various kinds of nano-fillers including carbon-based materials [148-151], lead-based oxide perovskite materials [152-155], lead-free oxide perovskites [156-158], metal oxides [93,159], and others including halide perovskites [152-153,160-163] have been utilized fruitfully [164] in PVDF based composites for fabrication of flexible high performance based PENG.

Advanced Functional Piezoelectric Materials and Applications   Materials Research Forum LLC
Materials Research Foundations 131 (2022) 61-116   https://doi.org/10.21741/978164490209-3

Piezoelectric fillers not only perform as nucleating agents for crystallinity of PVDF but also contribute to uplift the piezoelectric property of the pristine polymer. When some piezoelectric fillers are included in the PVDF matrix, due to the interaction between PVDF and the filler, the PVDF backbone tends to align which imposes a synergetic effect on the crystallinity up to a threshold filler concentration. Beyond this threshold aggregation of fillers impedes the polymer chain mobility. In an equilibrium state, the electron cloud density is distributed asymmetrically on two sides of the PVDF chain for the variation of electron attraction by the polymer molecular group. Under some mechanical stress on the PENG device, this electron cloud equilibrium is perturbed, and the charges start to migrate through the polymer chain to establish this equilibrium again. This migration develops a potential, called piezo-potential [95]. Simultaneously, the piezoelectric filler also contributes to piezo-potential in each layer of the lamellar crystal structure. When mechanical stress is applied to the material, the piezoelectric dipoles in the ceramic material align like Fig. 10 and the opposite charges are induced at the interfacing polymer layer. Thus, the PDVF composite with piezoelectric fillers can perform at an enhanced level.

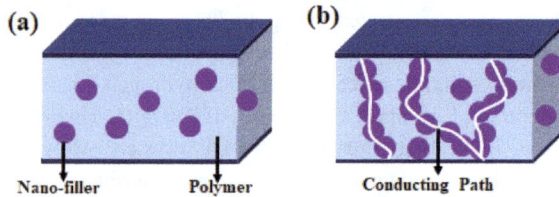

Figure 9. (a) Nano-fillers are distributed homogeneously in polymer matrix in lower concentration, (b) Higher concentration of nano-fillers forms conducting paths between top and bottom electrodes.

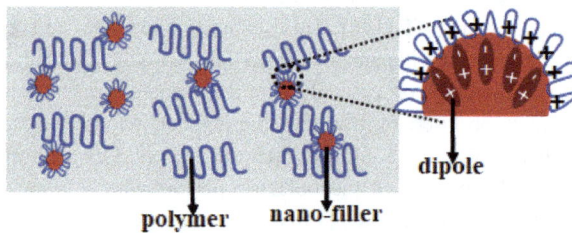

Figure 10. Wok mechanism of contribution of ceramic piezoelectric filler in polymer matrix

4.4.1 Ceramic filler

Lead based piezoelectric oxide perovskites like PZT, PMN-PT, PZN-PT pursue excellent piezoelectric coefficient, electromechanical coupling coefficient. So, these oxide materials can be fruitfully used as composite fillers with PVDF for improved piezoelectric performance. Being a high-performance-based piezoelectric inexpensive material, PZT is mostly explored as filler in PVDF-based polymer for PENG devices. Though piezo-filler can increase the β-phase content of PVDF, the size and concentration of filler have a strong influence on the crystallinity of PVDF, as discussed before. Flexible piezoelectric energy harvesting has been demonstrated by Tian et al. using the PZT nanoparticles embedded into the PDVF matrix. Their device can provide V_{oc} and I_{sc} 2.51 V and 78.43 nA, respectively, when an external pressure of magnitude 85.59 kPa is applied to it [165]. When surface-functionalized Nb-doped PZT [5 mol % Nb-doped MPB composition of $Pb(Zr_{0.52}Ti_{0.48}O_3)$] was introduced in PVDF based copolymer PVDF-TrFE thin film of thickness 5-6 μm fabricated through simple spin-coat technique, it exhibits an enhanced piezoelectric and ferroelectric property. Piezoelectric coefficient (d_{33}) and remnant polarization (P_r) increase up to -101 pC/N and 9.1 μC/cm², respectively, in case of 10 wt% NPZT content PVDF-TrFE film, whereas, in the case of pristine polymer the values are -45 pC/N and 7.5 μC/cm² [109]. Functionalization of nanoparticles surface helps homogeneous dispersion of the nanoparticles in the matrix. Flexible PENG is introduced by coating this composite on conductive PET substrate. 9.64 V V_{oc} is achieved by this flexible PENG of 0.6 cm² effective area only from simple bending with fingers [109]. PMN-PT, PZN-PT have higher piezoelectric properties than PZT. Tape-casted PMN-PT/PVDF based flexible PENG device delivered 10.3 V and 46 nA, which are 13 and 4.5 times larger than pure PVDF, respectively [166]. Electrospun PZN-PZT/PVDF-TrFE fiber can be developed on ITO-coated PET substrate to fabricate flexible PENG. The enhanced energy harvesting performance (~3.4 V output voltage and ~240 nA output current) is recorded for the 20 vol% nanoparticles-incorporated polymer matrix [167]. The output voltage and current are nearly 6 and 3 times greater than pure PVDF-TrFE nano-fibers. However, it should be noted that synthesis of PMN-PT and PZN-PZT with appropriate ratio and crystalline phase is difficult and not cost-effective. Apart from these lead-based ceramics, $BaTiO_3$, and its family BCZT, BZT based PVDF composites can be the eco-friendly materials for flexible PENG [158,168]. ZnO also has been explored as an effective filler for its cheap and easy synthesis and decent piezoelectric nature. ZnO/PVDF composite is explored as flexible PENG for charging capacitor just by tapping by Bhunia et al. [93]. The a.c. output (4 V) is utilized to light a red LED after charging a capacitor through a bridge rectifier. Surface modified ZnO has been utilized as a nanofiller in the

PVDF-TrFE matrix to enhance the performance of PENG [169]. We can summarize the performance of all types of fillers in the PVDF matrix in Table 3.

Table 3. PVDF Based Piezoelectric Nanogenerators

Materials	Processing Method	Poling Details	Device Size A: [cm²] t: [μm]	PENG Output			Demonstration of Application	Ref.
				Voltage [V]	Current [μA]	Power [μW/cm³]		
PT/rGO	Spin Coating	85 [MV/m]	A:4 t:6	89.7	0.91	340	Capacitor charging	[91]
PVDF/ZnO	Spin Coating	5 [MV/m] for 2 [h]	A:0.5 x 0.5 t:~10	4	-	-	Capacitor charging, LED glowing	[93]
PVDF-TrFE (PT)	Spin Coating	0.8 [MV/cm] at 100 [ºC], 30 [min]	A:0.09 t:6.5	7	58	-	-	[117]
PVDF	Purchased	No	A:1.5 x 2.5 t:80	7.02	1.11	1.18	Capacitor charging, LED glowing	[119]
PVDF	Electro-spinning	No	A:4x2 Fiber D: 0.8 [mm]	10	6.1	0.8	Human Health Monitoring	[121]
PVDF/GO	Spin Coating	5 [MV/m] for 2 [h]	A:1x1 t: 50	28	0.077	-	Capacitor Charging	[149]
PT-PMN-PT/PT-rGO/PT-PMN-PT	Spin Coating	No	A: 3x2.5 t:350	8.5	3	174.29	-	[155]
PVDF/ BaTiO₃ (BT)	Electro-spinning	No	-	6	-	-	-	[157]
PT/BZT-BCT	Electro-spinning	1.5 [kV/mm] for 30 [min]	A: 1 t:400	13.01	1.8	365	Power for force sensor	[168]
PVDF/AlO-rGO	Solution Casting	No	A:7.82 t: -	36	0.8	27.97	LED glowing, Powering Different Electronic Modules	[176]
PVDF/ZnO/ BaTiO₃	Electro-spinning	No	-	12	-	-	-	[177]
Porous PVDF	Solution Casting, Template etching	60 [MV/m] for 2 [h]	A:1 x 2 t:28	11	9.8	-	Capacitor charging	[216]
PVDF/ZnO	Spin Coating	1.2 [MV/cm]	A: N.R. t:4	0.4	0.029	-	-	[217]

PVDF/ZnO	Solution Casting	No	A: 0.8x0.8 t:20	24.5	1.7	32500	Capacitor charging, LED glowing	[189]
PVDF/ BaTiO₃/Ag-NWs	Solution Casting	No	A:2x2 t: 120	52	3.2	9860	Harvesting energy from car motion	[218]
PVDF-TrFE	Electro-spinning	No	A:4x2.5 t:3 [mm]	150	-	8.75	LED glowing	[219]
Polydopami ne@BaTiO₃ / PVDF-TrFE	Electro-spinning	50 [MV/m] for 2 [h]	A:2.5x2.5 t: 60	6	1.5	146.33	Capacitor charging	[220]
PT/BZT-BCT	Solution Casting	4 [kV/mm] for 40 [min]	A: 1.6x1.6 t:20	8.11	2.83	163.47	-	[221]

4.4.2 Carbon-based filler

Carbon based materials like carbon nano-tube (CNT), graphene oxide (GO), reduced graphene oxide (rGO) have attracted the researchers' interest as reinforcing filler in polymer matrix for their multifunctional properties like electrical conductivity, thermal conductivity, low density, mechanical property. These carbon-based fillers enhance the all-trans conformation of PVDF through the strong electrostatic interaction between -CF$_2$ dipoles and π electrons of these materials contained at their surfaces. Multiwalled CNT based PVDF nanocomposites (MWCNT/PVDF) fibers have been fabricated by electrospinning process and an improvement in the piezoelectric property is observed [150]. The high electric field poling during the electrospinning and the inclusion of MWCNT increase the β-phase quantity in these fibers. CNT can also enhance the PENG performance by introducing conductive paths which provide the paths between two electrodes to the separated piezoelectric charges developed during strain on the polymer matrix. A significant increment of PENG performance can be observed in potassium sodium niobite (KNN)/PDVF composite when 0.1% CNT is introduced in it. The PENG device containing 0.1%CNT/3%KNN/PVDF can deliver 23.24 V and 9 μA output voltage and current, respectively, where these values are 17.5 V and 0.522 μA in case of 3%KNN/PVDF [170]. Self-poled, recyclable polymer PENG with enhanced output performance has been reported introducing single-walled CNT (SWCNT) in PVDF-TrFE [171]. Besides these CNT structure, two-dimensional GO and rGO also contain many functional groups like OH⁻, COO⁻, -C=O at basal planes. The electrostatic interactions between the fluorine, hydrogen groups of PVDF, and oxygen functional groups of GO, rGO rotate the polymer chain which further form electroactive β-phase. However, it should be kept in mind that after a certain optimal concentration of the polymer, the chain rotation is resisted by the GO, rGO sheets and because of conducting nature, higher concentration leads to the possibility of increase in leakage current which degrade the PENG performance [91]. Due to the electrostatic interaction between rGO and PVDF, the piezoelectric coefficient and remnant polarization can be uplifted to a higher value. 78.6% and 69.3% increments than neat PVDF in the piezoelectric constant and remnant polarization have been demonstrated due to 0.1 wt% rGO inclusion in PVDF matrix [91]. A very few amounts of GO in the PVDF matrix leads to an increment the PENG output voltage up to 20 V, where the pristine PVDF PENG device of area 1 x 1 inch^2 delivers around 0.2 V [149]. Flexible PENG of area 2 x 2 cm^2 with 0.5 wt% rGO/PVDF-TrFE composite fabricated on conducting PET substrate delivers very high V_{oc} ~90 V and load current ~0.92 μA through 1 MΩ load resistance. The V_{oc} and load current are increased 20 and 25 times, respectively, than the pure PVDF-TrFE film due to more β-phase formation and electrical poling [91].

4.4.3 Metal based filler

From the past research articles, it can be concluded that a very controlled amount of metallic nanoparticle incorporation enhanced the piezoelectric performance of that polymer. Titanium di-oxide (TiO$_2$), iron oxide, zinc oxide can increase the piezoelectric performance of the PVDF composite. Iron oxide (Fe$_3$O$_4$)/PVDF has revealed a great enhanced piezoelectric response [172]. When TiO$_2$ nanoparticles have been used as fillers in PVDF for the synthesis of the composite film the same effect has been observed [173]. The metal-organic framework (MOF) can also be explored through Co$_3$[Co(CN)$_6$]$_2$ particles with PVDF. The superior specific surface area ables the strong interfacial interaction between metallic particles and PVDF and upgrades the piezoelectric properties. The coefficients d$_{31}$ and d$_{33}$ of these composites with 0.6 wt% of filler are -33 pC/N and 37 pC/N, respectively, which are almost double in comparison with pure PVDF.[174] In this context, it should be remembered that as the metal based particles are conducting, so higher concentration may cause an early electrical breakdown. Therefore, a very less and controlled amount is required to achieve the best property.

4.4.4 Other fillers

Hybrid fillers like two or more different materials with different properties can also be useful for upgrading the performance level of PVDF based composite PENG. The diversified effects of various fillers act as the leading path towards a better property. rGO/ZnO doped PVDF nanocomposite has been utilized with 83% β-phase crystallinity. However, the possibilities of these materials are not demonstrated for fabrication of PENG.[175] Aluminum oxide (AlO) decorated rGO has been mixed in PVDF for fabrication of hybrid PENG which exhibits trustable energy harvesting performance. This device is capable of electrical energy generation from various body movements and machine vibration. The interaction between the PVDF chain and functional groups of AlO-rGO results a more crystalline nature. Being lower resistive material, rGO also helps charges' movements and increases the PENG performance.[176] The output power density of this PENG is 27.97 $\mu W/cm^3$. ZnO and BaTiO$_3$ are used as a combined filler also in PVDF fiber for fabrication of piezoelectric energy harvesting device which is capable of delivering 12 V [177].

Biobased materials are always very attractive materials to researchers for their bio-compatibility. PVDF based biopolymer composites are reported by many scientific research groups for enhancing the device's performance and reliability in real-life applications.

5.   Applications of piezoelectric nanogenerator

Piezoelectric energy harvesting devices have revolutionary applications because of their multi-functionality, reliability, easy processing, cost-effectiveness. The first step for real use of the harvested energy is the convert it to unipolar voltage train. Because the output of PENG is very random with bipolar voltage peaks. The bridge rectifier is used for the transformation of voltage from bipolar to unipolar. After voltage rectification, we can store it in a capacitor or a battery for a long-time use. The total circuit diagram has been shown in Fig. 3a.

5.1   Power source of electronic devices

PENG with this circuit assembly demonstrates the potential application for lighting the LEDs after storing sufficient charge in a capacitor [16,46,90,93,178-182]. Glowing of LEDs is a good demonstration to utilize this very low power coming from the PENG. However, this very less power is not up to the required level for a small electronic device. Demonstration of powering LCD has been reported also using the power generated by PENG directly or after charging a capacitor [28,179,183]. Polymer dispersed liquid crystal device (PDLC) is driven by the NW based PENG device. The output of the PENG is able to keep the PDLC device "on state" when there is no external power backup attached to the device [184]. But it is noteworthy to mention that the cases of powering the display continued for a very short duration (few seconds). It will be interesting to compare the ratio between the applied input mechanical force on PENG and the duration of continuous glowing of LCD, LED, or powering any other electronic devices.

Advanced Functional Piezoelectric Materials and Applications   Materials Research Forum LLC
Materials Research Foundations 131 (2022) 61-116   https://doi.org/10.21741/978164490209-3

5.2   Sensing application

A portable, self-powered ultra-violet (UV) sensor is reported where the PENG device is connected directly to ZnO NWs based UV sensor. When there is no UV radiation on the sensor, no charge carriers into ZnO NWs are developed and the voltage drop across the sensor is high (around 0.2 V) due to the output of flexible PENG. This voltage drop significantly reduces after the exposure to UV radiation on the sensor. The radiation creates charge carriers in ZnO and the resistance of ZnO is reduced. Thus, the sensor works with UV radiation taking the power flexible PENG [28]. A self-powered electronic device that can transmit data through wireless way is demonstrated as another application of PENG [185]. A piezoelectric energy harvester and storage modules are together acted as power source of this above system. This total system is consisted of a sensor, a data processor, and a controller to analyze the information collected from the sensor, data transmitter, and receiver with the power module, depicted in Fig. 11. They have used the ZnO NW based PENG which can deliver 10 V peak to peak for each bending. The power, which is stored in the capacitor after rectification, is fed to a photosensor and a transmitter. The power requirement for the photosensor is high, it requires approximately 1000 bending of PENG to active. Whereas, the transmitter can be operated at low power. Therefore, for practical application, it is understood that the output of PENG should be increased and the power requirement for the photosensor should be decreased.

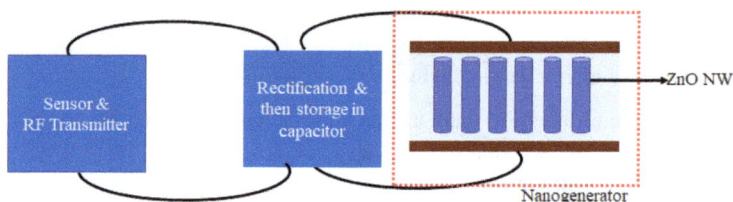

Figure 11. Schematic diagram of self-powered radio frequency (RF) transmitter as demonstrated in ref. [202]

Self-charging power cell (SCPC) can be explored using piezoelectric layer. The SCPC hybridizes the energy generation from PENG and energy storage [186-187]. So far, two types of PENG based SCPC are developed. One approach is the incorporation of piezoelectric film inside an electrochemical cell as a piezoelectric separator. When some mechanical stress is applied to this type of SCPC, the developed piezoelectric potential directly drives the motion of ions in the electrolyte towards the respective electrodes (anode and cathode). The other type is the direct connection through an external circuit between PENG and the electrochemical cell. The rectified DC output voltage of PENG provides the required charges to the energy storage cell. The prototype SCPC has been demonstrated by replacing the conventional polypropylene separator with piezoelectric polymer PVDF in a Li-ion coin cell [186]. The schematic of this coin cell is depicted in Fig. 12.

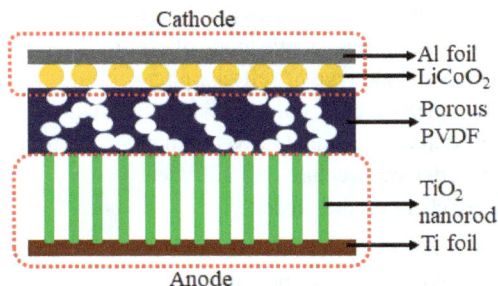

Figure 12. The structure of self-charging power cell demonstrated in ref. [203]

The compressive strain on this battery develops the piezoelectric potential in poled porous PVDF layer, this potential leads the Li^{2+} ions migrations from cathode to anode direction. On aluminum foil, the cathode is composed of a mixture of $LiCoO_2$, conductive carbon, and binder. The TiO_2 (anatase crystal phase) nanotube arrays on Ti substrate has been used as anode. During self-charging, this prototype device can raise the voltage level from 327 mV to 395 mV in 4 mins when the compressive force is applied on it at 2.3 Hz. Again, after the self-charging process, the device discharges and returns to its initial state of 327 mV through a discharge current of 1 mA. A flexible self-charging Li-ion battery, which is introduced using PVDF as a piezo-separator, can charge the battery from 500 mV to 832 mV within 500 sec by periodic stress of 34 N at 1 Hz [188]. This flexible battery has a discharge capacity of 0.266 mAh. Due to the flexibility of substrate, the strain is large compared to the above mentioned Li-ion coin cell covered with the hard stainless-steel case. Efforts can be given to increase the self-charging performance of SCPC by incorporation of high piezoelectric coefficient based inorganic piezoelectric materials in PVDF like ZnO [189-190], PZT [191], $NaNbO_3$ [192], $0.5(Ba_{0.7}Ca_{0.3})TiO_3$-$0.5Ba(Zr_{0.2}Ti_{0.8})O_3$ [193] and also piezoelectric biomaterials like fish swim bladder [194]. This piezoelectric based electrochemical process is also explored in the case of self-powered pseudocapacitive supercapacitor [195-197], Na-ion battery [198], electric double-layered supercapacitor [199-202]. Solid piezo-electrolyte based composite can also be introduced in SCPC where this layer simultaneously acts a piezo potential source and electrolyte [187]. The advantage of this approach is the simplicity in fabrication and compact assembly into a single device. However, the voltage enhancement of the electrochemical cell is entirely controlled by the piezoelectric potential and we only can increase the voltage to a certain mV which is not sufficient for practical application. Therefore, in another way scientists have tried to fabricate the total system of self-charging power cell developing the two systems (PENG and storage capacitor) separately and have connected these two parts through an external rectification circuit. After charging this capacitor we can use this power for an active mode of electronic devices [46,185]. But, the major disadvantage of this second type of self-charging system is low discharging ability even after having high piezoelectric output because of input and output impedance mismatch between the external circuit and PENG output port, respectively. Another major issue is the complicated device structure and sufficient power drop across the external circuit component before charging the capacitor.

Besides these types of applications, PENG can also be used to monitor various health indexes like pulse rate, blood pressure, breathing condition, muscular movements, etc. [203-204].

6. Challenges and future scopes

Nevertheless, the feasibility of the practical application of a piezoelectric nanogenerator is still at the very teenage stage. From the discovery of piezoelectric nanogenerator numerous efforts are implemented about the cost-effective easy fabrication, performance enhancement, and real-life application of PENGs. However, improvement is required for few crucial drawbacks.

Materials and Processing Issues:

With the developments of the PENG, various methods have been pursued to synthesize the different kinds of piezoelectric materials. Among the ceramic piezoelectric materials lead zirconate titanate (PZT) and PZT based other ceramics like PLZT, PMN-PT have been explored as the top-rated materials for the high d33 values, superior dielectric properties, and steadiness in different atmospheric conditions. However, lead (Pb) is a highly toxic material and causes serious health issues after a certain limit. Researchers have put attempts at other piezoelectric materials like $(K, Na)NbO_3$, BCT, BCZT, ZnO those are the most promising and environment friendly materials for the absence of lead. But during the synthesis, the conventional solid-state method faces some critical problems. Sometimes it needs high-temperature sintering to achieve the exact crystalline phase. Exact molar ratio mixing is the most concerning thing to avoid the existence of an impure phase in the material. The impure crystalline phase can decrease the piezoelectric property significantly. Another main issue is these materials are not capable of fulfilling the requirement of having flexibility because of their brittleness. Therefore, the upgradation of performance on a flexible substrate is very demanding in the present scenario. Piezoelectric composites consist of ceramic and polymer have shown their potential as the most favored materials for the fabrication of flexible piezoelectric nanogenerators. This composite can have flexibility like polymer and exhibits high piezo coefficients for the presence of ceramic. Although, there are still several challenges that are needed to overcome. It is very tough to synthesis the nanomaterials having a specific structure, size and crystal phase during industry oriented large-scale production. The final product of standard nanomaterials' synthesis techniques like hydrothermal and sol-gel is highly influenced by pH of the solution, temperature and time, volume and concentration of reactant. Other major concern is the high surface energy of nanomaterials which causes significant agglomeration. The aggregation leads to a reduction in the electrical property. Surface functionalization of nanomaterials with an appropriate surfactant may be a solution to enhance the homogeneous dispersion. For the fabrication of flexible devices made of polymer nanocomposites, numerous processes like spin-coating, tape casting, solution casting, hot-pressing, etc. have been adopted. Still, more research should be focused to synthesize PENG devices on a large scale with enhanced performance for practical use.

Robustness of Devices:

Mechanical robustness of PENG is an important requirement for long time utilization. Most research articles contain a short time demonstration of energy harvesting from devices so the long-duration performance efficiency is highly unpredictable. Demonstration of prolonged time is required considering the effect of the structural integrity and adhesion of contacts with the piezoelectric layers during repetitive bending. For mechanical energy harvesting by PENG devices, these devices should have superior stretchability and high mechanical property.

Device Output:

The efficient applications of these PENG devices demand the enhancement of output power. That depends on the piezoelectric co-efficient of materials as well as some other issues which are also required to be furnished. Such as, impedance mismatch is an important thing for the fabrication of a complete device for real-life application. If there is a dissimilarity between the output impedance of the transducer and the input impedance of the external circuit, then significant noise appears which creates an adverse effect on output power. Therefore, during complete device fabrication, this point should be considered. The harvesting power of entire nanogenerator system is fully dependents on presence of all the mechanical dynamics and vibrations, and the energy flow [205]. Therefore, determination of mechanical and electrical impedance through theoretical studies is necessary for output power optimization. Thin piezoelectric layer and parallel connection between multiple harvesters can reduce the impedance matching problem. Decrease in impedance value leads to the increment in capacitance, which further delivers high output power at lower force.

Pulse type a.c. output from PENG is also another negative point. This short-duration power is not effective for continuous power back-up directly. Storing this power after rectification in a capacitor we can fruitfully use. Hence, appropriate steps should be adopted for unipolar constant noiseless power.

PENG Encapsulation:

The performance of PENG is affected by the humidity and other external things like heat, water, etc. Polymer encapsulation can prevent the damage of PENG from these things. PDMS, PVDF, Parylene C are mostly utilized for encapsulation of the devices. However, the robustness, durability, and effective performance for such encapsulation in a harsh environment are not explored so much.

Conclusions

Current civilization demands smart electronic devices which can make their lifestyles easy for them. However, these devices require power all time. Thus, the interest is developing to fabricate self-powered devices. The one key solution is to transfer stray mechanical force to electrical energy through piezoelectric nanogenerators. Through this chapter, detailed information of various materials' behavior as piezoelectric and their exploration as PENG is provided. The next level of development will be focused on the generation of more power towards real-life application accurately. Despite numerous efforts by the researchers, there are still some critical issues those are required to be solved. Considering the high output power generation scientists are now trying to shift from piezoelectric nanogenerator to triboelectric nanogenerator whose pillar is simply made of contact electrification between two layers when they come in contact with each other. This nanogenerator has a greater superior behavior in the case of high output. The long-term stability is also a concerning issue for future studies for all nanogenerators

Acknowledgement

This work was supported by the Basic Science Research Program (2020R1A2C3014237) and National R&D Program (2021M3D1A2049315) of the National Research Foundation of Korea (NRF) by the Ministry of Science and ICT.

References

[1] W.D. Grossmann, I. Grossmann, K. Steininger, Indicators To Determine Winning Renewable Energy Technologies with an Application to Photovoltaics, Environ. Sci. Technol. 44 (2010) 4849-4855. https://doi.org/10.1021/es903434q

[2] T. Wu, Y. Song, Z. Shi, D. Liu, S. Chen, C. Xiong, Q. Yang, High-performance nanogenerators based on flexible cellulose nanofibril/MoS2 nanosheet composite piezoelectric films for energy harvesting, Nano Energy 80 (2021) 105541. https://doi.org/10.1016/j.nanoen.2020.105541

[3] Z.L. Wang, J. Song, Piezoelectric Nanogenerators Based on Zinc Oxide Nanowire Arrays, Science 312 (2006) 242. https://doi.org/10.1126/science.1124005

[4] L. Gu, J. Liu, N. Cui, Q. Xu, T. Du, L. Zhang, Z. Wang, C. Long, Y. Qin, Enhancing the current density of a piezoelectric nanogenerator using a three-dimensional intercalation electrode, Nat. Commun. 11 (2020) 1030. https://doi.org/10.1038/s41467-020-14846-4

[5] D. Hu, M. Yao, Y. Fan, C. Ma, M. Fan, M. Liu, Strategies to achieve high performance piezoelectric nanogenerators, Nano Energy 55 (2019) 288-304. https://doi.org/10.1016/j.nanoen.2018.10.053

[6] F.-R. Fan, Z.-Q. Tian, Z. Lin Wang, Flexible triboelectric generator, Nano Energy 1 (2012) 328-334. https://doi.org/10.1016/j.nanoen.2012.01.004

[7] B. Fatma, S. Gupta, C. Chatterjee, R. Bhunia, V. Verma, A. Garg, Triboelectric generators made of mechanically robust PVDF films as self-powered autonomous sensors for wireless transmission based remote security systems, J. Mater. Chem. A 8 (2020) 15023-15033. https://doi.org/10.1039/D0TA04716C

[8] Y. Yang, W. Guo, K.C. Pradel, G. Zhu, Y. Zhou, Y. Zhang, Y. Hu, L. Lin, Z.L. Wang, Pyroelectric Nanogenerators for Harvesting Thermoelectric Energy, Nano Lett. 12 (2012) 2833-2838. https://doi.org/10.1021/nl3003039

[9] H. Xue, Q. Yang, D. Wang, W. Luo, W. Wang, M. Lin, D. Liang, Q. Luo, A wearable pyroelectric nanogenerator and self-powered breathing sensor, Nano Energy 38 (2017) 147-154. https://doi.org/10.1016/j.nanoen.2017.05.056

[10] J. Ryu, J.-E. Kang, Y. Zhou, S.-Y. Choi, W.-H. Yoon, D.-S. Park, J.-J. Choi, B.-D. Hahn, C.-W. Ahn, J.-W. Kim, Y.-D. Kim, S. Priya, S.Y. Lee, S. Jeong, D.-Y. Jeong, Ubiquitous magneto-mechano-electric generator, Energy Environ. Sci. 8 (2015) 2402-2408. https://doi.org/10.1039/C5EE00414D

[11] M.G. Kang, R. Sriramdas, H. Lee, J. Chun, D. Maurya, G.T. Hwang, J. Ryu, S. Priya, High Power Magnetic Field Energy Harvesting through Amplified Magneto-Mechanical Vibration, Adv. Energy Mater. 8 (2018) 1703313. https://doi.org/10.1002/aenm.201703313

[12] J. He, T. Wen, S. Qian, Z. Zhang, Z. Tian, J. Zhu, J. Mu, X. Hou, W. Geng, J. Cho, J. Han, X. Chou, C. Xue, Triboelectric-piezoelectric-electromagnetic hybrid nanogenerator for high-efficient vibration energy harvesting and self-powered wireless monitoring system, Nano Energy 43 (2018) 326-339. https://doi.org/10.1016/j.nanoen.2017.11.039

[13] T. Quan, X. Wang, Z.L. Wang, Y. Yang, Hybridized Electromagnetic-Triboelectric Nanogenerator for a Self-Powered Electronic Watch, ACS Nano 9 (2015) 12301-12310. https://doi.org/10.1021/acsnano.5b05598

[14] Y. Zhao, P. Deng, Y. Nie, P. Wang, Y. Zhang, L. Xing, X. Xue, Biomolecule-adsorption-dependent piezoelectric output of ZnO nanowire nanogenerator and its application as self-powered active biosensor, Biosens. Bioelectron. 57 (2014) 269-275. https://doi.org/10.1016/j.bios.2014.02.022

[15] Z. Wen, Q. Shen, X. Sun, Nanogenerators for Self-Powered Gas Sensing, Nano-Micro Lett. 9 (2017) 45. https://doi.org/10.1007/s40820-017-0146-4

[16] M. Lee, J. Bae, J. Lee, C.-S. Lee, S. Hong, Z.L. Wang, Self-powered environmental sensor system driven by nanogenerators, Energy Environ. Sci. 4 (2011) 3359-3363. https://doi.org/10.1039/c1ee01558c

[17] Z.-H. Lin, G. Cheng, Y. Yang, Y.S. Zhou, S. Lee, Z.L. Wang, Triboelectric Nanogenerator as an Active UV Photodetector, Adv. Funct. Mater. 24 (2014) 2810-2816. https://doi.org/10.1002/adfm.201302838

[18] Z. Li, Q. Zheng, Z.L. Wang, Z. Li, Nanogenerator-Based Self-Powered Sensors for Wearable and Implantable Electronics, Research 2020 (2020) 8710686. https://doi.org/10.34133/2020/8710686

[19] Z.L. Wang, Entropy theory of distributed energy for internet of things, Nano Energy 58 (2019) 669-672. https://doi.org/10.1016/j.nanoen.2019.02.012

[20] J. Curie, P. Curie, Développement par compression de l'électricité polaire dans les cristaux hémièdres à faces inclinées, Bull. Minéral. 3-4 (1880) 90-93. https://doi.org/10.3406/bulmi.1880.1564

[21] G. Lippmann, Principe de la conservation de l'elecricit'e, Ann de Chemie e de Physique (5 series) 24 (1881) 145.

[22] P. Curie, J. Curie, Contractions et dilatations produites par des tensions 'electriques dans les cristaux h'emi'edres 'a faces inclin'ees., Comptes Rendus (France) 93 (1881) 1137-1140. https://doi.org/10.3406/bulmi.1880.1564

[23] D. Fang, J. Liu, Basic Equations of Piezoelectric Materials, in: D. Fang, J. Liu (Eds.), Fracture mechanics of piezoelectric and ferroelectric solids, Springer Berlin Heidelberg, Berlin, Heidelberg, 2013, pp 77-95. https://doi.org/10.1007/978-3-642-30087-5_4

[24] A.L. Kholkin, N.A. Pertsev, A.V. Goltsev, Piezoelectricity and Crystal Symmetry, in: A. Safari, E.K. Akdoğan (Eds.), Piezoelectric and Acoustic Materials for Transducer Applications, Springer US, Boston, MA, 2008, pp 17-38. https://doi.org/10.1007/978-0-387-76540-2_2

[25] Z.L. Wang, On Maxwell's displacement current for energy and sensors: the origin of nanogenerators, Mater. Today 20 (2017) 74-82. https://doi.org/10.1016/j.mattod.2016.12.001

[26] C. Opoku, A.S. Dahiya, C. Oshman, F. Cayrel, G. Poulin-Vittrant, D. Alquier, N. Camara, Fabrication of ZnO Nanowire Based Piezoelectric Generators and Related Structures, Physics Procedia 70 (2015) 858-862. https://doi.org/10.1016/j.phpro.2015.08.176

[27] M.K. Gupta, J.-H. Lee, K.Y. Lee, S.-W. Kim, Two-Dimensional Vanadium-Doped ZnO Nanosheet-Based Flexible Direct Current Nanogenerator, ACS Nano 7 (2013) 8932-8939. https://doi.org/10.1021/nn403428m

[28] W. Wu, S. Bai, M. Yuan, Y. Qin, Z.L. Wang, T. Jing, Lead Zirconate Titanate Nanowire Textile Nanogenerator for Wearable Energy-Harvesting and Self-Powered Devices, ACS Nano 6 (2012) 6231-6235. https://doi.org/10.1021/nn3016585

[29] X. Chen, S. Xu, N. Yao, Y. Shi, 1.6 V nanogenerator for mechanical energy harvesting using PZT nanofibers, Nano Lett. 10 (2010) 2133-2137. https://doi.org/10.1021/nl100812k

[30] K.-I. Park, S. Xu, Y. Liu, G.-T. Hwang, S.-J.L. Kang, Z.L. Wang, K.J. Lee, Piezoelectric BaTiO3 Thin Film Nanogenerator on Plastic Substrates, Nano Lett. 10 (2010) 4939-4943. https://doi.org/10.1021/nl102959k

[31] S.-H. Shin, Y.-H. Kim, M.H. Lee, J.-Y. Jung, J. Nah, Hemispherically Aggregated BaTiO3 Nanoparticle Composite Thin Film for High-Performance Flexible Piezoelectric Nanogenerator, ACS Nano 8 (2014) 2766-2773. https://doi.org/10.1021/nn406481k

[32] J.H. Jung, C.-Y. Chen, B.K. Yun, N. Lee, Y. Zhou, W. Jo, L.-J. Chou, Z.L. Wang, Lead-free KNbO3 ferroelectric nanorod based flexible nanogenerators and capacitors, Nanotechnol. 23 (2012) 375401. https://doi.org/10.1088/0957-4484/23/37/375401

[33] M.H. Huang, Y. Wu, H. Feick, N. Tran, E. Weber, P. Yang, Catalytic growth of zinc oxide nanowires by vapor transport, Adv. Mater. 13 (2001) 113-116. https://doi.org/10.1002/1521-4095(200101)13:2<113::AID-ADMA113>3.0.CO;2-H

[34] X.Y. Kong, Y. Ding, R. Yang, Z.L. Wang, Single-crystal nanorings formed by epitaxial self-coiling of polar nanobelts, Science 303 (2004) 1348-51. https://doi.org/10.1126/science.1092356

[35] Z.W. Pan, Z.R. Dai, Z.L. Wang, Nanobelts of semiconducting oxides, Science 291 (2001) 1947-9. https://doi.org/10.1126/science.1058120

[36] X.Y. Kong, Z.L. Wang, Spontaneous Polarization-Induced Nanohelixes, Nanosprings, and Nanorings of Piezoelectric Nanobelts, Nano Lett. 3 (2003) 1625-1631. https://doi.org/10.1021/nl034463p

[37] X. Wang, J. Song, J. Liu, Z.L. Wang, Direct-Current Nanogenerator Driven by Ultrasonic Waves, Science 316 (2007) 102. https://doi.org/10.1126/science.1139366

[38] P.X. Gao, J. Song, J. Liu, Z.L. Wang, Nanowire Piezoelectric Nanogenerators on Plastic Substrates as Flexible Power Sources for Nanodevices, Adv. Mater. 19 (2007) 67-72. https://doi.org/10.1002/adma.200601162

[39] R. Yang, Y. Qin, L. Dai, Z.L. Wang, Power generation with laterally packaged piezoelectric fine wires, Nat. Nanotechnol. 4 (2009) 34-39. https://doi.org/10.1038/nnano.2008.314

[40] G. Zhu, A.C. Wang, Y. Liu, Y. Zhou, Z.L. Wang, Functional Electrical Stimulation by Nanogenerator with 58 V Output Voltage, Nano Lett. 12 (2012) 3086-3090. https://doi.org/10.1021/nl300972f

[41] S.N. Cha, J.S. Seo, S.M. Kim, H.J. Kim, Y.J. Park, S.W. Kim, J.M. Kim, Sound-driven piezoelectric nanowire-based nanogenerators, Adv. Mater. 22 (2010) 4726-30. https://doi.org/10.1002/adma.201001169

[42] M.-Y. Choi, D. Choi, M.-J. Jin, I. Kim, S.-H. Kim, J.-Y. Choi, S.Y. Lee, J.M. Kim, S.-W. Kim, Mechanically Powered Transparent Flexible Charge-Generating Nanodevices with Piezoelectric ZnO Nanorods, Adv. Mater. 21 (2009) 2185-2189. https://doi.org/10.1002/adma.200803605

[43] D. Choi, M.-Y. Choi, H.-J. Shin, S.-M. Yoon, J.-S. Seo, J.-Y. Choi, S.Y. Lee, J.M. Kim, S.-W. Kim, Nanoscale Networked Single-Walled Carbon-Nanotube Electrodes

for Transparent Flexible Nanogenerators, J. Phys. Chem. C 114 (2010) 1379-1384. https://doi.org/10.1021/jp909713c

[44] D. Choi, M.-Y. Choi, W.M. Choi, H.-J. Shin, H.-K. Park, J.-S. Seo, J. Park, S.-M. Yoon, S.J. Chae, Y.H. Lee, S.-W. Kim, J.-Y. Choi, S.Y. Lee, J.M. Kim, Fully Rollable Transparent Nanogenerators Based on Graphene Electrodes, Adv. Mater. 22 (2010) 2187-2192. https://doi.org/10.1002/adma.200903815

[45] S. Xu, Y. Qin, C. Xu, Y. Wei, R. Yang, Z.L. Wang, Self-powered nanowire devices, Nat. Nanotechnol. 5 (2010) 366-373. https://doi.org/10.1038/nnano.2010.46

[46] G. Zhu, R. Yang, S. Wang, Z.L. Wang, Flexible High-Output Nanogenerator Based on Lateral ZnO Nanowire Array, Nano Lett. 10 (2010) 3151-3155. https://doi.org/10.1021/nl101973h

[47] M.A. Johar, A. Waseem, M.A. Hassan, J.-H. Kang, J.-S. Ha, J.K. Lee, S.-W. Ryu, Facile growth of high aspect ratio c-axis GaN nanowires and their application as flexible p-n NiO/GaN piezoelectric nanogenerators, Acta Mater. 161 (2018) 237-245. https://doi.org/10.1016/j.actamat.2018.09.030

[48] B. Kumar, D.-H. Lee, S.-H. Kim, B. Yang, S. Maeng, S.-W. Kim, General Route to Single-Crystalline SnO Nanosheets on Arbitrary Substrates, J. Phys. Chem. C 114 (2010) 11050-11055. https://doi.org/10.1021/jp101682v

[49] J.N. Coleman, M. Lotya, A. O'Neill, S.D. Bergin, P.J. King, U. Khan, K. Young, A. Gaucher, S. De, R.J. Smith, I.V. Shvets, S.K. Arora, G. Stanton, H.-Y. Kim, K. Lee, G.T. Kim, G.S. Duesberg, T. Hallam, J.J. Boland, J.J. Wang, J.F. Donegan, J.C. Grunlan, G. Moriarty, A. Shmeliov, R.J. Nicholls, J.M. Perkins, E.M. Grieveson, K. Theuwissen, D.W. McComb, P.D. Nellist, V. Nicolosi, Two-Dimensional Nanosheets Produced by Liquid Exfoliation of Layered Materials, Science 331 (2011) 568. https://doi.org/10.1126/science.1194975

[50] K.-H. Kim, B. Kumar, K.Y. Lee, H.-K. Park, J.-H. Lee, H.H. Lee, H. Jun, D. Lee, S.-W. Kim, Piezoelectric two-dimensional nanosheets/anionic layer heterojunction for efficient direct current power generation, Sci. Rep. 3 (2013) 2017. https://doi.org/10.1038/srep02017

[51] Y. Manjula, R. Rakesh Kumar, P.M. Swarup Raju, G. Anil Kumar, T. Venkatappa Rao, A. Akshaykranth, P. Supraja, Piezoelectric flexible nanogenerator based on ZnO nanosheet networks for mechanical energy harvesting, Chem. Phys. 533 (2020) 110699. https://doi.org/10.1016/j.chemphys.2020.110699

[52] H.-K. Park, K.Y. Lee, J.-S. Seo, J.-A. Jeong, H.-K. Kim, D. Choi, S.-W. Kim, Charge-generating mode control in high-performance transparent flexible piezoelectric

nanogenerators, Adv. Funct. Mater. 21 (2011) 1187-1193.
https://doi.org/10.1002/adfm.201002099

[53] Y. Gao, Z.L. Wang, Equilibrium Potential of Free Charge Carriers in a Bent Piezoelectric Semiconductive Nanowire, Nano Lett. 9 (2009) 1103-1110. https://doi.org/10.1021/nl803547f

[54] C. Soci, A. Zhang, B. Xiang, S.A. Dayeh, D.P.R. Aplin, J. Park, X.Y. Bao, Y.H. Lo, D. Wang, ZnO Nanowire UV Photodetectors with High Internal Gain, Nano Lett. 7 (2007) 1003-1009. https://doi.org/10.1021/nl070111x

[55] Y. Hu, L. Lin, Y. Zhang, Z.L. Wang, Replacing a Battery by a Nanogenerator with 20 V Output, Adv. Mater. 24 (2012) 110-114.
https://doi.org/10.1002/adma.201103727

[56] K.Y. Lee, B. Kumar, J.-S. Seo, K.-H. Kim, J.I. Sohn, S.N. Cha, D. Choi, Z.L. Wang, S.-W. Kim, P-Type Polymer-Hybridized High-Performance Piezoelectric Nanogenerators, Nano Lett. 12 (2012) 1959-1964. https://doi.org/10.1021/nl204440g

[57] T.T. Pham, K.Y. Lee, J.-H. Lee, K.-H. Kim, K.-S. Shin, M.K. Gupta, B. Kumar, S.-W. Kim, Reliable operation of a nanogenerator under ultraviolet light via engineering piezoelectric potential, Energy Environ. Sci. 6 (2013) 841-846.
https://doi.org/10.1039/c2ee23980a

[58] J. Shi, M.B. Starr, H. Xiang, Y. Hara, M.A. Anderson, J.-H. Seo, Z. Ma, X. Wang, Interface Engineering by Piezoelectric Potential in ZnO-Based Photoelectrochemical Anode, Nano Lett. 11 (2011) 5587-5593. https://doi.org/10.1021/nl203729j

[59] Y.Q. Chen, X.J. Zheng, X. Feng, The fabrication of vanadium-doped ZnO piezoelectric nanofiber by electrospinning, Nanotechnol. 21 (2009) 055708.
https://doi.org/10.1088/0957-4484/21/5/055708

[60] P.V. Radovanovic, D.R. Gamelin, High-Temperature Ferromagnetism in Ni2+-Doped ZnO Aggregates Prepared from Colloidal Diluted Magnetic Semiconductor Quantum Dots, Phys. Rev. Lett. 91 (2003) 157202.
https://doi.org/10.1103/PhysRevLett.91.157202

[61] D.A. Schwartz, K.R. Kittilstved, D.R. Gamelin, Above-room-temperature ferromagnetic Ni2+-doped ZnO thin films prepared from colloidal diluted magnetic semiconductor quantum dots, Appl. Phys. Lett. 85 (2004) 1395-1397.
https://doi.org/10.1063/1.1785872

[62] A. Tsukazaki, A. Ohtomo, T. Onuma, M. Ohtani, T. Makino, M. Sumiya, K. Ohtani, S.F. Chichibu, S. Fuke, Y. Segawa, H. Ohno, H. Koinuma, M. Kawasaki, Repeated

temperature modulation epitaxy for p-type doping and light-emitting diode based on ZnO, Nat. Mater. 4 (2005) 42-46. https://doi.org/10.1038/nmat1284

[63] M.-H. Zhao, Z.-L. Wang, S.X. Mao, Piezoelectric Characterization of Individual Zinc Oxide Nanobelt Probed by Piezoresponse Force Microscope, Nano Lett. 4 (2004) 587-590. https://doi.org/10.1021/nl035198a

[64] Y.C. Yang, C. Song, X.H. Wang, F. Zeng, F. Pan, Giant piezoelectric d33 coefficient in ferroelectric vanadium doped ZnO films, Appl. Phys. Lett. 92 (2008) 012907. https://doi.org/10.1063/1.2830663

[65] N. Sinha, S. Goel, A.J. Joseph, H. Yadav, K. Batra, M.K. Gupta, B. Kumar, Y-doped ZnO nanosheets: Gigantic piezoelectric response for an ultra-sensitive flexible piezoelectric nanogenerator, Ceram. Int. 44 (2018) 8582-8590. https://doi.org/10.1016/j.ceramint.2018.02.066

[66] C. Jin, N. Hao, Z. Xu, I. Trase, Y. Nie, L. Dong, A. Closson, Z. Chen, J.X.J. Zhang, Flexible piezoelectric nanogenerators using metal-doped ZnO-PVDF films, Sens. Actuators A 305 (2020) 111912. https://doi.org/10.1016/j.sna.2020.111912

[67] H. Sun, H. Tian, Y. Yang, D. Xie, Y.-C. Zhang, X. Liu, S. Ma, H.-M. Zhao, T.-L. Ren, A novel flexible nanogenerator made of ZnO nanoparticles and multiwall carbon nanotube, Nanoscale 5 (2013) 6117-6123. https://doi.org/10.1039/c3nr00866e

[68] M. Acosta, N. Novak, V. Rojas, S. Patel, R. Vaish, J. Koruza, G.A. Rossetti, J. Rödel, BaTiO3-based piezoelectrics: Fundamentals, current status, and perspectives, Appl. Phys. Rev. 4 (2017) 041305. https://doi.org/10.1063/1.4990046

[69] L.A. Shuvalov, Symmetry aspects of ferroelectricity, J. Phys. Soc. Jpn. 28 (Suppl.) (1970) 38.

[70] F. Jona, G. Shiarane, Ferroelectric Crystals, Dover Publications, 1993, p 402.

[71] R.E. Cohen, H. Krakauer, Lattice dynamics and origin of ferroelectricity in BaTiO3: Linearized-augmented-plane-wave total-energy calculations, Phys. Rev. B: Condens. Matter 42 (1990) 6416-6423. https://doi.org/10.1103/PhysRevB.42.6416

[72] M. Kimura, A. Ando, Y. Sakabe, 2 - Lead zirconate titanate-based piezo-ceramics, in: K. Uchino (Ed.), Advanced Piezoelectric Materials, Woodhead Publishing, 2010, pp. 89-110. https://doi.org/10.1533/9781845699758.1.89

[73] A. Koka, Z. Zhou, H.A. Sodano, Vertically aligned BaTiO3 nanowire arrays for energy harvesting, Energy Environ. Sci. 7 (2014) 288-296. https://doi.org/10.1039/C3EE42540A

[74] R. Kiran, A. Kumar, R. Kumar, R. Vaish, Effect of poling orientation on piezoelectric materials operating in longitudinal mode, Mater. Res. Express 6 (2019) 065711. https://doi.org/10.1088/2053-1591/ab0fd0

[75] J. Yan, Y.G. Jeong, High Performance Flexible Piezoelectric Nanogenerators based on BaTiO3 Nanofibers in Different Alignment Modes, ACS Appl. Mater. Interfaces 8 (2016) 15700-15709. https://doi.org/10.1021/acsami.6b02177

[76] E.L. Tsege, G.H. Kim, V. Annapureddy, B. Kim, H.-K. Kim, Y.-H. Hwang, A flexible lead-free piezoelectric nanogenerator based on vertically aligned BaTiO3 nanotube arrays on a Ti-mesh substrate, RSC Adv. 6 (2016) 81426-81435. https://doi.org/10.1039/C6RA13482C

[77] Z. Zhou, X. Du, Z. Zhang, J. Luo, S. Niu, D. Shen, Y. Wang, H. Yang, Q. Zhang, S. Dong, Interface modulated 0-D piezoceramic nanoparticles/PDMS based piezoelectric composites for highly efficient energy harvesting application, Nano Energy 82 (2021) 105709. https://doi.org/10.1016/j.nanoen.2020.105709

[78] B. Chen, H. Li, W. Tian, C. Zhou, PZT Based Piezoelectric Sensor for Structural Monitoring, J. Electron. Mater. 48 (2019) 2916-2923. https://doi.org/10.1007/s11664-019-07034-8

[79] G.L. Smith, J.S. Pulskamp, L.M. Sanchez, D.M. Potrepka, R.M. Proie, T.G. Ivanov, R.Q. Rudy, W.D. Nothwang, S.S. Bedair, C.D. Meyer, R.G. Polcawich, PZT-Based Piezoelectric MEMS Technology, J. Am. Ceram. Soc. 95 (2012) 1777-1792. https://doi.org/10.1111/j.1551-2916.2012.05155.x

[80] X. Niu, W. Jia, S. Qian, J. Zhu, J. Zhang, X. Hou, J. Mu, W. Geng, J. Cho, J. He, X. Chou, High-Performance PZT-Based Stretchable Piezoelectric Nanogenerator, ACS Sustainable Chem. Eng. 7 (2019) 979-985. https://doi.org/10.1021/acssuschemeng.8b04627

[81] G. Shirane, A. Takeda, Phase Transitions in Solid Solutions of PbZrO3 and PbTiO3 (I) Small Concentrations of PbTiO3, J. Phys. Soc. Jpn. 7 (1952) 5-11. https://doi.org/10.1143/JPSJ.7.5

[82] G. Shirane, K. Suzuki, A. Takeda, Phase Transitions in Solid Solutions of PbZrO3 and PbTiO3 (II) X-ray Study, J. Phys. Soc. Jpn. 7 (1952) 12-18. https://doi.org/10.1143/JPSJ.7.12

[83] G. Shirane, K. Suzuki, Crystal structure of Pb(Zr-Ti)O3, J. Phys. Soc. Jpn. 7 (1952) 333-333. https://doi.org/10.1143/JPSJ.7.333

[84] E. Sawaguchi, Ferroelectricity versus Antiferroelectricity in the Solid Solutions of PbZrO3 and PbTiO3, J. Phys. Soc. Jpn. 8 (1953) 615-629. https://doi.org/10.1143/JPSJ.8.615

[85] B. Jaffe, R.S. Roth, S. Marzullo, Piezoelectric Properties of Lead Zirconate-Lead Titanate Solid-Solution Ceramics, J. Appl. Phys. 25 (1954) 809-810. https://doi.org/10.1063/1.1721741

[86] R.E. Cohen, Origin of ferroelectricity in perovskite oxides, Nature 358 (1992) 136-138. https://doi.org/10.1038/358136a0

[87] Y. Kuroiwa, S. Aoyagi, A. Sawada, J. Harada, E. Nishibori, M. Takata, M. Sakata, Evidence for Pb-O covalency in tetragonal PbTiO3, Phys. Rev. Lett. 87 (2001) 217601. https://doi.org/10.1103/PhysRevLett.87.217601

[88] H. Lee, H. Kim, D.Y. Kim, Y. Seo, Pure Piezoelectricity Generation by a Flexible Nanogenerator Based on Lead Zirconate Titanate Nanofibers, ACS Omega 4 (2019) 2610-2617. https://doi.org/10.1021/acsomega.8b03325

[89] N. Cui, W. Wu, Y. Zhao, S. Bai, L. Meng, Y. Qin, Z.L. Wang, Magnetic Force Driven Nanogenerators as a Noncontact Energy Harvester and Sensor, Nano Lett. 12 (2012) 3701-3705. https://doi.org/10.1021/nl301490q

[90] L. Gu, N. Cui, L. Cheng, Q. Xu, S. Bai, M. Yuan, W. Wu, J. Liu, Y. Zhao, F. Ma, Y. Qin, Z.L. Wang, Flexible Fiber Nanogenerator with 209 V Output Voltage Directly Powers a Light-Emitting Diode, Nano Lett. 13 (2013) 91-94. https://doi.org/10.1021/nl303539c

[91] R. Bhunia, S. Gupta, B. Fatma, Prateek, R.K. Gupta, A. Garg, Milli-Watt Power Harvesting from Dual Triboelectric and Piezoelectric Effects of Multifunctional Green and Robust Reduced Graphene Oxide/P(VDF-TrFE) Composite Flexible Films, ACS Appl. Mater. Interfaces 11 (2019) 38177-38189. https://doi.org/10.1021/acsami.9b13360

[92] X. Zhou, K. Parida, O. Halevi, Y. Liu, J. Xiong, S. Magdassi, P.S. Lee, All 3D-printed stretchable piezoelectric nanogenerator with non-protruding kirigami structure, Nano Energy 72 (2020) 104676. https://doi.org/10.1016/j.nanoen.2020.104676

[93] R. Bhunia, S. Das, S. Dalui, S. Hussain, R. Paul, R. Bhar, A.K. Pal, Flexible nano-ZnO/polyvinylidene difluoride piezoelectric composite films as energy harvester, Appl. Phys. A 122 (2016) 637. https://doi.org/10.1007/s00339-016-0161-1

[94] S. Gong, B. Zhang, J. Zhang, Z.L. Wang, K. Ren, Biocompatible poly(lactic acid)-based hybrid piezoelectric and electret nanogenerator for electronic skin applications, Adv. Funct. Mater. 30 (2020) 1908724. https://doi.org/10.1002/adfm.201908724

[95] L. Lu, W. Ding, J. Liu, B. Yang, Flexible PVDF based piezoelectric nanogenerators, Nano Energy 78 (2020) 105251. https://doi.org/10.1016/j.nanoen.2020.105251

[96] H. Kawai, The Piezoelectricity of Poly (vinylidene Fluoride), Jpn. J. Appl. Phys. 8 (1969) 975-976. https://doi.org/10.1143/JJAP.8.975

[97] W.P. Mason, Piezoelectricity, its history and applications, J. Acoust. Soc. Am. 70 (1981) 1561-1566. https://doi.org/10.1121/1.387221

[98] K. Al Abdullah, M.A. Batal, R. Hamdan, T. Khalil, J. Zaraket, M. Aillerie, C. Salame, The Enhancement of PVDF Pyroelectricity (Pyroelectric Coefficient and Dipole Moment) by Inclusions, Energy Procedia 119 (2017) 545-555. https://doi.org/10.1016/j.egypro.2017.07.074

[99] S.N. Fedosov, H. von Seggern, Pyroelectricity in polyvinylidene fluoride: Influence of polarization and charge, J. Appl. Phys. 103 (2008) 014105. https://doi.org/10.1063/1.2824940

[100] R. Gregorio, R.C. CapitãO, Morphology and phase transition of high melt temperature crystallized poly(vinylidene fluoride), J. Mater. Sci. 35 (2000) 299-306. https://doi.org/10.1023/A:1004737000016

[101] V. Sencadas, R. Gregorio, S. Lanceros-Méndez, α to β Phase Transformation and Microestructural Changes of PVDF Films Induced by Uniaxial Stretch, J. Macromol. Sci. Part B Phys. 48 (2009) 514-525. https://doi.org/10.1080/00222340902837527

[102] A. Itoh, Y. Takahashi, T. Furukawa, H. Yajima, Solid-state calculations of poly(vinylidene fluoride) using the hybrid DFT method: spontaneous polarization of polymorphs, Polym. J. 46 (2014) 207-211. https://doi.org/10.1038/pj.2013.96

[103] G. Zhu, Z. Zeng, L. Zhang, X. Yan, Piezoelectricity in β-phase PVDF crystals: A molecular simulation study, Comput. Mater. Sci. 44 (2008) 224-229. https://doi.org/10.1016/j.commatsci.2008.03.016

[104] S. Qin, X. Zhang, Z. Yu, F. Zhao, Polarization study of poly(vinylidene fluoride) films under cyclic electric fields, Polym. Eng. Sci. 60 (2020) 645-656. https://doi.org/10.1002/pen.25323

[105] S.K. Mahadeva, J. Berring, K. Walus, B. Stoeber, Effect of poling time and grid voltage on phase transition and piezoelectricity of poly(vinyledene fluoride) thin films using corona poling, J. Phys. D: Appl. Phys. 46 (2013) 285305. https://doi.org/10.1088/0022-3727/46/28/285305

[106] A. Salimi, A.A. Yousefi, Analysis method: FTIR studies of β-phase crystal formation in stretched PVDF films, Polym. Test. 22 (2003) 699-704. https://doi.org/10.1016/S0142-9418(03)00003-5

[107] L. Li, M. Zhang, M. Rong, W. Ruan, Studies on the transformation process of PVDF from α to β phase by stretching, RSC Adv. 4 (2014) 3938-3943. https://doi.org/10.1039/C3RA45134H

[108] H. Zhang, H. Lu, Z. Liu, L. Li, Preparation of High-Performance Polyvinylidene Fluoride Films by the Combination of Simultaneous Biaxial Stretching and Solid-State Shear Milling Technologies, Ind. Eng. Chem. Res. 59 (2020) 18539-18548. https://doi.org/10.1021/acs.iecr.0c03383

[109] S. Gupta, R. Bhunia, B. Fatma, D. Maurya, D. Singh, Prateek, R. Gupta, S. Priya, R.K. Gupta, A. Garg, Multifunctional and Flexible Polymeric Nanocomposite Films with Improved Ferroelectric and Piezoelectric Properties for Energy Generation Devices, ACS Appl. Energy Mater. 2 (2019) 6364-6374. https://doi.org/10.1021/acsaem.9b01000

[110] D. Singh, Deepak, A. Garg, An efficient route to fabricate fatigue-free P(VDF-TrFE) capacitors with enhanced piezoelectric and ferroelectric properties and excellent thermal stability for sensing and memory applications, Phys. Chem. Chem. Phys. 19 (2017) 7743-7750. https://doi.org/10.1039/C7CP00275K

[111] L. Jiang, H. Xie, Y. Hou, S. Wang, Y. Xia, Y. Li, G.-H. Hu, Q. Yang, C. Xiong, Z. Gao, Enhanced piezoelectricity of a PVDF-based nanocomposite utilizing high-yield dispersions of exfoliated few-layer MoS2, Ceram. Int. 45 (2019) 11347-11352. https://doi.org/10.1016/j.ceramint.2019.02.213

[112] C.-T. Pan, S.-Y. Wang, C.-K. Yen, A. Kumar, S.-W. Kuo, J.-L. Zheng, Z.-H. Wen, R. Singh, S.P. Singh, M.T. Khan, R.K. Chaudhary, X. Dai, A. Chandra Kaushik, D.-Q. Wei, Y.-L. Shiue, W.-H. Chang, Polyvinylidene Fluoride-Added Ceramic Powder Composite Near-Field Electrospinned Piezoelectric Fiber-Based Low-Frequency Dynamic Sensors, ACS Omega 5 (2020) 17090-17101. https://doi.org/10.1021/acsomega.0c00805

[113] H. Parangusan, D. Ponnamma, M.A.A. AlMaadeed, Toward High Power Generating Piezoelectric Nanofibers: Influence of Particle Size and Surface Electrostatic Interaction of Ce-Fe2O3 and Ce-Co3O4 on PVDF, ACS Omega 4 (2019) 6312-6323. https://doi.org/10.1021/acsomega.9b00243

[114] J.R. Gregorio, M. Cestari, Effect of crystallization temperature on the crystalline phase content and morphology of poly(vinylidene fluoride), J. Polym. Sci., Part B: Polym. Phys. 32 (1994) 859-870. https://doi.org/10.1002/polb.1994.090320509

[115] V. Sencadas, R. Gregorio Filho, S. Lanceros-Mendez, Processing and characterization of a novel nonporous poly(vinilidene fluoride) films in the β phase, J. Non-Cryst. Solids 352 (2006) 2226-2229. https://doi.org/10.1016/j.jnoncrysol.2006.02.052

[116] P. Ueberschlag, PVDF piezoelectric polymer, Sens. Rev. 21 (2001) 118-125. https://doi.org/10.1108/02602280110388315

[117] Z. Pi, J. Zhang, C. Wen, Z.-b. Zhang, D. Wu, Flexible piezoelectric nanogenerator made of poly(vinylidenefluoride-co-trifluoroethylene) (PVDF-TrFE) thin film, Nano Energy 7 (2014) 33-41. https://doi.org/10.1016/j.nanoen.2014.04.016

[118] A.J. Lovinger, Ferroelectric polymers, Science 220 (1983) 1115-1121. https://doi.org/10.1126/science.220.4602.1115

[119] S. Khadtare, E.J. Ko, Y.H. Kim, H.S. Lee, D.K. Moon, A flexible piezoelectric nanogenerator using conducting polymer and silver nanowire hybrid electrodes for its application in real-time muscular monitoring system, Sens. Actuators A 299 (2019) 111575. https://doi.org/10.1016/j.sna.2019.111575

[120] L. Persano, C. Dagdeviren, Y.W. Su, Y.H. Zhang, S. Girardo, D. Pisignano, Y.G. Huang, J.A. Rogers, High performance piezoelectric devices based on aligned arrays of nanofibers of poly(vinylidenefluoride-co-trifluoroethylene), Nat. Commun. 4 (2013) 1633. https://doi.org/10.1038/ncomms2639

[121] K. Maity, S. Garain, K. Henkel, D. Schmeißer, D. Mandal, Self-Powered Human-Health Monitoring through Aligned PVDF Nanofibers Interfaced Skin-Interactive Piezoelectric Sensor, ACS Appl. Polym. Mater. 2 (2020) 862-878. https://doi.org/10.1021/acsapm.9b00846

[122] R.K. Singh, S.W. Lye, J. Miao, Measurement of impact characteristics in a string using electrospun PVDF nanofibers strain sensors, Sens. Actuators A 303 (2020) 111841. https://doi.org/10.1016/j.sna.2020.111841

[123] M. García-Iglesias, B.F.M. de Waal, A.V. Gorbunov, A.R.A. Palmans, M. Kemerink, E.W. Meijer, A Versatile Method for the Preparation of Ferroelectric Supramolecular Materials via Radical End-Functionalization of Vinylidene Fluoride Oligomers, J. Am. Chem. Soc. 138 (2016) 6217-6223. https://doi.org/10.1021/jacs.6b01908

[124] S. Dey, M. Purahmad, S.S. Ray, A.L. Yarin, M. Dutta, Investigation of PVDF-TrFE nanofibers for energy harvesting, 2012 IEEE Nanotech. Mater. Dev. Conf. (NMDC2012), 2012 21-24.

[125] S. You, L. Zhang, J. Gui, H. Cui, S. Guo, A Flexible Piezoelectric Nanogenerator Based on Aligned P(VDF-TrFE) Nanofibers, Micromachines 10 (2019) 302. https://doi.org/10.3390/mi10050302

[126] M.S. Singhvi, S.S. Zinjarde, D.V. Gokhale, Polylactic acid: synthesis and biomedical applications, J. Appl. Microbiol. 127 (2019) 1612-1626. https://doi.org/10.1111/jam.14290

[127] E.J. Curry, K. Ke, M.T. Chorsi, K.S. Wrobel, A.N. Miller, A. Patel, I. Kim, J. Feng, L. Yue, Q. Wu, C.-L. Kuo, K.W.H. Lo, C.T. Laurencin, H. Ilies, P.K. Purohit, T.D. Nguyen, Biodegradable piezoelectric force sensor, Proc. Natl. Acad. Sci. U.S.A. 115 (2018) 909. https://doi.org/10.1073/pnas.1710874115

[128] Y. Tajitsu, Development of environmentally friendly piezoelectric polymer film actuator having multilayer structure, Jpn. J. Appl. Phys. 55 (2016) 04EA07. https://doi.org/10.7567/JJAP.55.04EA07

[129] M. Ando, H. Kawamura, H. Kitada, Y. Sekimoto, T. Inoue, Y. Tajitsu, Pressure-Sensitive Touch Panel Based on Piezoelectric Poly(L-lactic acid) Film, Jpn. J. Appl. Phys. 52 (2013) 09KD17. https://doi.org/10.7567/JJAP.52.09KD17

[130] M. Varga, J. Morvan, N. Diorio, E. Buyuktanir, J. Harden, J.L. West, A. Jákli, Direct piezoelectric responses of soft composite fiber mats, Appl. Phys. Lett. 102 (2013) 153903. https://doi.org/10.1063/1.4802593

[131] C. Zhao, J. Zhang, Z.L. Wang, K. Ren, A Poly(l-Lactic Acid) Polymer-Based Thermally Stable Cantilever for Vibration Energy Harvesting Applications, Adv. Sustainable Syst. 1 (2017) 1700068. https://doi.org/10.1002/adsu.201700068

[132] J. Zhang, S. Gong, X. Li, J. Liang, Z.L. Wang, K. Ren, A Wind-Driven Poly(tetrafluoroethylene) Electret and Polylactide Polymer-Based Hybrid Nanogenerator for Self-Powered Temperature Detection System, Adv. Sustainable Syst. 5 (2021) 2000192. https://doi.org/10.1002/adsu.202000192

[133] B. Medronho, A. Romano, M.G. Miguel, L. Stigsson, B. Lindman, Rationalizing cellulose (in)solubility: reviewing basic physicochemical aspects and role of hydrophobic interactions, Cellulose 19 (2012) 581-587. https://doi.org/10.1007/s10570-011-9644-6

[134] S. Park, J.O. Baker, M.E. Himmel, P.A. Parilla, D.K. Johnson, Cellulose crystallinity index: measurement techniques and their impact on interpreting cellulase

performance, Biotechnol. Biofuels 3 (2010) 10. https://doi.org/10.1186/1754-6834-3-10

[135] A.C. O'Sullivan, Cellulose: the structure slowly unravels, Cellulose 4 (1997) 173-207. https://doi.org/10.1023/A:1018431705579

[136] E. Fukada, Piezoelectricity of Wood, J. Phys. Soc. Jpn. 10 (1955) 149-154. https://doi.org/10.1143/JPSJ.10.149

[137] I. Chae, C.K. Jeong, Z. Ounaies, S.H. Kim, Review on Electromechanical Coupling Properties of Biomaterials, ACS Appl. Bio Mater. 1 (2018) 936-953. https://doi.org/10.1021/acsabm.8b00309

[138] N. Sriplai, R. Mangayil, A. Pammo, V. Santala, S. Tuukkanen, S. Pinitsoontorn, Enhancing piezoelectric properties of bacterial cellulose films by incorporation of MnFe2O4 nanoparticles, Carbohydr. Polym. 231 (2020) 115730. https://doi.org/10.1016/j.carbpol.2019.115730

[139] S. Yoon, J.W. Kim, H.C. Kim, J. Kim, Effect of Process Orientation on the Mechanical Behavior and Piezoelectricity of Electroactive Paper, Mater. 13 (2020) 204. https://doi.org/10.3390/ma13010204

[140] M.M. Alam, D. Mandal, Native Cellulose Microfiber-Based Hybrid Piezoelectric Generator for Mechanical Energy Harvesting Utility, ACS Appl. Mater. Interfaces 8 (2016) 1555-1558. https://doi.org/10.1021/acsami.5b08168

[141] Q. Zheng, H. Zhang, H. Mi, Z. Cai, Z. Ma, S. Gong, High-performance flexible piezoelectric nanogenerators consisting of porous cellulose nanofibril (CNF)/poly(dimethylsiloxane) (PDMS) aerogel films, Nano Energy 26 (2016) 504-512. https://doi.org/10.1016/j.nanoen.2016.06.009

[142] B. Baytekin, H.T. Baytekin, B.A. Grzybowski, Retrieving and converting energy from polymers: deployable technologies and emerging concepts, Energy Environ. Sci. 6 (2013) 3467-3482. https://doi.org/10.1039/c3ee41360h

[143] H.Y. Choi, Y.G. Jeong, Microstructures and piezoelectric performance of eco-friendly composite films based on nanocellulose and barium titanate nanoparticle, Compos. B. Eng. 168 (2019) 58-65. https://doi.org/10.1016/j.compositesb.2018.12.072

[144] M. Pusty, P.M. Shirage, Gold nanoparticle-cellulose/PDMS nanocomposite: a flexible dielectric material for harvesting mechanical energy, RSC Adv. 10 (2020) 10097-10112. https://doi.org/10.1039/C9RA10811D

[145] S. Bauer, Piezo-, pyro- and ferroelectrets: Soft transducer materials for electromechanical energy conversion, IEEE Trans. Dielectr. Electr. Insul. 13 (2006) 953-962. https://doi.org/10.1109/TDEI.2006.247819

[146] T. Greeshma, R. Balaji, S. Jayakumar, PVDF Phase Formation and Its Influence on Electrical and Structural Properties of PZT-PVDF Composites, Ferroelectr. Lett. 40 (2013) 41-55. https://doi.org/10.1080/07315171.2013.814460

[147] S.F. Mendes, C.M. Costa, C. Caparros, V. Sencadas, S. Lanceros-Méndez, Effect of filler size and concentration on the structure and properties of poly(vinylidene fluoride)/BaTiO3 nanocomposites, J. Mater. Sci. 47 (2012) 1378-1388. https://doi.org/10.1007/s10853-011-5916-7

[148] M. Pusty, L. Sinha, P.M. Shirage, A flexible self-poled piezoelectric nanogenerator based on a rGO-Ag/PVDF nanocomposite, New J. Chem. 43 (2019) 284-294. https://doi.org/10.1039/C8NJ04751K

[149] R. Bhunia, R. Dey, S. Das, S. Hussain, R. Bhar, A. Kumar Pal, Enhanced piezo-electric property induced in graphene oxide/polyvinylidene fluoride composite flexible thin films, Polym. Compos. 39 (2018) 4205-4216. https://doi.org/10.1002/pc.24493

[150] Y. Ahn, J.Y. Lim, S.M. Hong, J. Lee, J. Ha, H.J. Choi, Y. Seo, Enhanced Piezoelectric Properties of Electrospun Poly(vinylidene fluoride)/Multiwalled Carbon Nanotube Composites Due to High β-Phase Formation in Poly(vinylidene fluoride), J. Phys. Chem. C 117 (2013) 11791-11799. https://doi.org/10.1021/jp4011026

[151] V. Bhavanasi, V. Kumar, K. Parida, J. Wang, P.S. Lee, Enhanced Piezoelectric Energy Harvesting Performance of Flexible PVDF-TrFE Bilayer Films with Graphene Oxide, ACS Appl. Mater. Interfaces 8 (2016) 521-529. https://doi.org/10.1021/acsami.5b09502

[152] V. Jella, S. Ippili, J.-H. Eom, J. Choi, S.-G. Yoon, Enhanced output performance of a flexible piezoelectric energy harvester based on stable MAPbI3-PVDF composite films, Nano Energy 53 (2018) 46-56. https://doi.org/10.1016/j.nanoen.2018.08.033

[153] A. Sultana, P. Sadhukhan, M.M. Alam, S. Das, T.R. Middya, D. Mandal, Organo-Lead Halide Perovskite Induced Electroactive β-Phase in Porous PVDF Films: An Excellent Material for Photoactive Piezoelectric Energy Harvester and Photodetector, ACS Appl. Mater. Interfaces 10 (2018) 4121-4130. https://doi.org/10.1021/acsami.7b17408

[154] N. Chamankar, R. Khajavi, A.A. Yousefi, A. Rashidi, F. Golestanifard, A flexible piezoelectric pressure sensor based on PVDF nanocomposite fibers doped with PZT

particles for energy harvesting applications, Ceram. Int. 46 (2020) 19669-19681. https://doi.org/10.1016/j.ceramint.2020.03.210

[155] U. Yaqoob, R.M. Habibur, M. Sheeraz, H.C. Kim, Realization of self-poled, high performance, flexible piezoelectric energy harvester by employing PDMS-rGO as sandwich layer between P(VDF-TrFE)-PMN-PT composite sheets, Compos. B. Eng. 159 (2019) 259-268. https://doi.org/10.1016/j.compositesb.2018.09.102

[156] S. Ippili, V. Jella, J.-H. Eom, J. Kim, S. Hong, J.-S. Choi, V.-D. Tran, N. Van Hieu, Y.-J. Kim, H.-J. Kim, S.-G. Yoon, An eco-friendly flexible piezoelectric energy harvester that delivers high output performance is based on lead-free MASnI3 films and MASnI3-PVDF composite films, Nano Energy 57 (2019) 911-923. https://doi.org/10.1016/j.nanoen.2019.01.005

[157] A.D. Hussein, R.S. Sabry, O. Abdul Azeez Dakhil, R. Bagherzadeh, Effect of Adding BaTiO3 to PVDF as Nano Generator, J. Phys. Conf. Ser. 1294 (2019) 022012. https://doi.org/10.1088/1742-6596/1294/2/022012

[158] H.H. Singh, M. Singh, K. Gangwal, M. Faraz, N. Khare, BaTiO3-PVDF composite film for piezoelectric nanogenerator, AIP Conf. Proc. 2265 (2020) 030642. https://doi.org/10.1063/5.0017115

[159] R. Naik, S.R. T, Self-powered flexible piezoelectric nanogenerator made of poly (vinylidene fluoride)/Zirconium oxide nanocomposite, Mater. Res. Express 6 (2019) 115330. https://doi.org/10.1088/2053-1591/ab49b3

[160] K. Maity, S. Garain, K. Henkel, D. Schmeißer, D. Mandal, Natural Sugar-Assisted, Chemically Reinforced, Highly Durable Piezoorganic Nanogenerator with Superior Power Density for Self-Powered Wearable Electronics, ACS Appl. Mater. Interfaces 10 (2018) 44018-44032. https://doi.org/10.1021/acsami.8b15320

[161] C. Kumar, A. Gaur, S. Tiwari, A. Biswas, S.K. Rai, P. Maiti, Bio-waste polymer hybrid as induced piezoelectric material with high energy harvesting efficiency, Compos. Commun. 11 (2019) 56-61. https://doi.org/10.1016/j.coco.2018.11.004

[162] A. Gaur, S. Tiwari, C. Kumar, P. Maiti, Retracted Article: A bio-based piezoelectric nanogenerator for mechanical energy harvesting using nanohybrid of poly(vinylidene fluoride), Nanoscale Adv. 1 (2019) 3200-3211. https://doi.org/10.1039/C9NA00214F

[163] A. Sultana, S.K. Ghosh, M.M. Alam, P. Sadhukhan, K. Roy, M. Xie, C.R. Bowen, S. Sarkar, S. Das, T.R. Middya, D. Mandal, Methylammonium Lead Iodide Incorporated Poly(vinylidene fluoride) Nanofibers for Flexible Piezoelectric-

Pyroelectric Nanogenerator, ACS Appl. Mater. Interfaces 11 (2019) 27279-27287. https://doi.org/10.1021/acsami.9b04812

[164] R. Bhunia, B. Ghosh, D. Ghosh, S. Hussain, R. Bhar, A.K. Pal, Free-standing nanocrystalline-Cadmium sulfide/Polyvinylidene fluoride composite thin film: synthesis and characterization, J. Polym. Res. 22 (2015) 71. https://doi.org/10.1007/s10965-015-0712-8

[165] G. Tian, W. Deng, Y. Gao, D. Xiong, C. Yan, X. He, T. Yang, L. Jin, X. Chu, H. Zhang, W. Yan, W. Yang, Rich lamellar crystal baklava-structured PZT/PVDF piezoelectric sensor toward individual table tennis training, Nano Energy 59 (2019) 574-581. https://doi.org/10.1016/j.nanoen.2019.03.013

[166] S. Xu, Y.-w. Yeh, G. Poirier, M.C. McAlpine, R.A. Register, N. Yao, Flexible Piezoelectric PMN-PT Nanowire-Based Nanocomposite and Device, Nano Lett. 13 (2013) 2393-2398. https://doi.org/10.1021/nl400169t

[167] K. Liu, H.J. Choi, B.K. Kim, D.B. Kim, C.S. Han, S.W. Kim, H.B. Kang, J.-W. Park, Y.S. Cho, Piezoelectric energy harvesting and charging performance of Pb(Zn1/3Nb2/3)O3-Pb(Zr0.5Ti0.5)O3 nanoparticle-embedded P(VDF-TrFE) nanofiber composite sheets, Compos. Sci. Technol. 168 (2018) 296-302. https://doi.org/10.1016/j.compscitech.2018.10.012

[168] J. Liu, B. Yang, L. Lu, X. Wang, X. Li, X. Chen, J. Liu, Flexible and lead-free piezoelectric nanogenerator as self-powered sensor based on electrospinning BZT-BCT/P(VDF-TrFE) nanofibers, Sens. Actuators A 303 (2020) 111796. https://doi.org/10.1016/j.sna.2019.111796

[169] J. Li, C. Zhao, K. Xia, X. Liu, D. Li, J. Han, Enhanced piezoelectric output of the PVDF-TrFE/ZnO flexible piezoelectric nanogenerator by surface modification, Appl. Surf. Sci. 463 (2019) 626-634. https://doi.org/10.1016/j.apsusc.2018.08.266

[170] S. Bairagi, S.W. Ali, Investigating the role of carbon nanotubes (CNTs) in the piezoelectric performance of a PVDF/KNN-based electrospun nanogenerator, Soft Matter 16 (2020) 4876-4886. https://doi.org/10.1039/D0SM00438C

[171] N.A. Shepelin, P.C. Sherrell, E. Goudeli, E.N. Skountzos, V.C. Lussini, G.W. Dicinoski, J.G. Shapter, A.V. Ellis, Printed recyclable and self-poled polymer piezoelectric generators through single-walled carbon nanotube templating, Energy Environ. Sci. 13 (2020) 868-883. https://doi.org/10.1039/C9EE03059J

[172] Z.W. Ouyang, E.C. Chen, T.M. Wu, Enhanced piezoelectric and mechanical properties of electroactive polyvinylidene fluoride/iron oxide composites, Mater. Chem. Phys. 149 (2015) 172-178. https://doi.org/10.1016/j.matchemphys.2014.10.003

[173] W.C. Gan, W.H.A. Majid, Effect of TiO2on enhanced pyroelectric activity of PVDF composite, Smart Mater. Struct. 23 (2014) 045026. https://doi.org/10.1088/0964-1726/23/4/045026

[174] L. Yang, T. Qiu, M. Shen, H. He, H. Huang, Metal-organic frameworks Co3[Co(CN)6]2: A promising candidate for dramatically reinforcing the piezoelectric activity of PVDF, Compos. Sci. Technol. 196 (2020) 108232. https://doi.org/10.1016/j.compscitech.2020.108232

[175] B. Jaleh, A. Jabbari, Evaluation of reduced graphene oxide/ZnO effect on properties of PVDF nanocomposite films, Appl. Surf. Sci. 320 (2014) 339-347. https://doi.org/10.1016/j.apsusc.2014.09.030

[176] S.K. Karan, R. Bera, S. Paria, A.K. Das, S. Maiti, A. Maitra, B.B. Khatua, An approach to design highly durable piezoelectric nanogenerator based on self-poled PVDF/AlO-rGO flexible nanocomposite with high power density and energy conversion efficiency, Adv. Energy Mater. 6 (2016) 1601016. https://doi.org/10.1002/aenm.201601016

[177] R.S. Sabry, A.D. Hussein, PVDF:ZnO/BaTiO3 as high out-put piezoelectric nanogenerator, Polym. Test. 79 (2019) 106001. https://doi.org/10.1016/j.polymertesting.2019.106001

[178] B. Saravanakumar, R. Mohan, K. Thiyagarajan, S.-J. Kim, Fabrication of a ZnO nanogenerator for eco-friendly biomechanical energy harvesting, RSC Adv. 3 (2013) 16646-16656. https://doi.org/10.1039/c3ra40447a

[179] H. Kim, S.M. Kim, H. Son, H. Kim, B. Park, J. Ku, J.I. Sohn, K. Im, J.E. Jang, J.-J. Park, O. Kim, S. Cha, Y.J. Park, Enhancement of piezoelectricity via electrostatic effects on a textile platform, Energy Environ. Sci. 5 (2012) 8932-8936. https://doi.org/10.1039/c2ee22744d

[180] C.K. Jeong, K.-I. Park, J. Ryu, G.-T. Hwang, K.J. Lee, Large-Area and Flexible Lead-Free Nanocomposite Generator Using Alkaline Niobate Particles and Metal Nanorod Filler, Adv. Funct. Mater. 24 (2014) 2620-2629. https://doi.org/10.1002/adfm.201303484

[181] S. Xu, B.J. Hansen, Z.L. Wang, Piezoelectric-nanowire-enabled power source for driving wireless microelectronics, Nat. Commun. 1 (2010) 93. https://doi.org/10.1038/ncomms1098

[182] C.K. Jeong, I. Kim, K.-I. Park, M.H. Oh, H. Paik, G.-T. Hwang, K. No, Y.S. Nam, K.J. Lee, Virus-Directed Design of a Flexible BaTiO3 Nanogenerator, ACS Nano 7 (2013) 11016-11025. https://doi.org/10.1021/nn404659d

[183] K.-I. Park, S.B. Bae, S.H. Yang, H.I. Lee, K. Lee, S.J. Lee, Lead-free BaTiO3 nanowires-based flexible nanocomposite generator, Nanoscale 6 (2014) 8962-8968. https://doi.org/10.1039/C4NR02246G

[184] S. Lee, J. Lee, W. Ko, S. Cha, J. Sohn, J. Kim, J. Park, Y. Park, J. Hong, Solution-processed Ag-doped ZnO nanowires grown on flexible polyester for nanogenerator applications, Nanoscale 5 (2013) 9609-9614. https://doi.org/10.1039/c3nr03402j

[185] Y. Hu, Y. Zhang, C. Xu, L. Lin, R.L. Snyder, Z.L. Wang, Self-Powered System with Wireless Data Transmission, Nano Lett. 11 (2011) 2572-2577. https://doi.org/10.1021/nl201505c

[186] X. Xue, S. Wang, W. Guo, Y. Zhang, Z.L. Wang, Hybridizing Energy Conversion and Storage in a Mechanical-to-Electrochemical Process for Self-Charging Power Cell, Nano Lett. 12 (2012) 5048-5054. https://doi.org/10.1021/nl302879t

[187] H. He, Y. Fu, T. Zhao, X. Gao, L. Xing, Y. Zhang, X. Xue, All-solid-state flexible self-charging power cell basing on piezo-electrolyte for harvesting/storing body-motion energy and powering wearable electronics, Nano Energy 39 (2017) 590-600. https://doi.org/10.1016/j.nanoen.2017.07.033

[188] X. Xue, P. Deng, B. He, Y. Nie, L. Xing, Y. Zhang, Z.L. Wang, Flexible Self-Charging Power Cell for One-Step Energy Conversion and Storage, Adv. Energy Mater. 4 (2014) 1301329. https://doi.org/10.1002/aenm.201301329

[189] P. Thakur, A. Kool, N.A. Hoque, B. Bagchi, F. Khatun, P. Biswas, D. Brahma, S. Roy, S. Banerjee, S. Das, Superior performances of in situ synthesized ZnO/PVDF thin film based self-poled piezoelectric nanogenerator and self-charged photo-power bank with high durability, Nano Energy 44 (2018) 456-467. https://doi.org/10.1016/j.nanoen.2017.11.065

[190] A. Rasheed, W. He, Y. Qian, H. Park, D.J. Kang, Flexible Supercapacitor-Type Rectifier-free Self-Charging Power Unit Based on a Multifunctional Polyvinylidene Fluoride-ZnO-rGO Piezoelectric Matrix, ACS Appl. Mater. Interfaces 12 (2020) 20891-20900. https://doi.org/10.1021/acsami.9b22362

[191] Y. Zhang, Y. Zhang, X. Xue, C. Cui, B. He, Y. Nie, P. Deng, Z. Lin Wang, PVDF-PZT nanocomposite film based self-charging power cell, Nanotechnol. 25 (2014) 105401. https://doi.org/10.1088/0957-4484/25/10/105401

[192] P. Pazhamalai, K. Krishnamoorthy, V.K. Mariappan, S. Sahoo, S. Manoharan, S.-J. Kim, A High Efficacy Self-Charging MoSe2 Solid-State Supercapacitor Using Electrospun Nanofibrous Piezoelectric Separator with Ionogel Electrolyte, Adv. Mater. Interfaces 5 (2018) 1800055. https://doi.org/10.1002/admi.201800055

[193] G. Wei, Z. Wang, R. Zhu, H. Kimura, PVDF/BCT-BZT Nanocomposite Film for a Piezo-Driven Self-Charging Power Cell, J. Electrochem. Soc. 165 (2018) A1238-A1246. https://doi.org/10.1149/2.0401807jes

[194] A. Maitra, S.K. Karan, S. Paria, A.K. Das, R. Bera, L. Halder, S.K. Si, A. Bera, B.B. Khatua, Fast charging self-powered wearable and flexible asymmetric supercapacitor power cell with fish swim bladder as an efficient natural bio-piezoelectric separator, Nano Energy 40 (2017) 633-645. https://doi.org/10.1016/j.nanoen.2017.08.057

[195] D. Zhou, F. Wang, X. Zhao, J. Yang, H. Lu, L.-Y. Lin, L.-Z. Fan, Self-Chargeable Flexible Solid-State Supercapacitors for Wearable Electronics, ACS Appl. Mater. Interfaces 12 (2020) 44883-44891. https://doi.org/10.1021/acsami.0c14426

[196] K. Krishnamoorthy, P. Pazhamalai, V.K. Mariappan, S.S. Nardekar, S. Sahoo, S.-J. Kim, Probing the energy conversion process in piezoelectric-driven electrochemical self-charging supercapacitor power cell using piezoelectrochemical spectroscopy, Nat. Commun. 11 (2020) 2351. https://doi.org/10.1038/s41467-020-15808-6

[197] N. Wang, W. Dou, S. Hao, Y. Cheng, D. Zhou, X. Huang, C. Jiang, X. Cao, Tactile sensor from self-chargeable piezoelectric supercapacitor, Nano Energy 56 (2019) 868-874. https://doi.org/10.1016/j.nanoen.2018.11.065

[198] D. Zhou, L. Xue, L. Wang, N. Wang, W.-M. Lau, X. Cao, Self-chargeable sodium-ion battery for soft electronics, Nano Energy 61 (2019) 435-441. https://doi.org/10.1016/j.nanoen.2019.04.068

[199] R. Song, H. Jin, X. Li, L. Fei, Y. Zhao, H. Huang, H. Lai-Wa Chan, Y. Wang, Y. Chai, A rectification-free piezo-supercapacitor with a polyvinylidene fluoride separator and functionalized carbon cloth electrodes, J. Mater. Chem. A 3 (2015) 14963-14970. https://doi.org/10.1039/C5TA03349G

[200] K. Parida, V. Bhavanasi, V. Kumar, J. Wang, P.S. Lee, Fast charging self-powered electric double layer capacitor, J. Power Sources 342 (2017) 70-78. https://doi.org/10.1016/j.jpowsour.2016.11.083

[201] S. Sahoo, K. Krishnamoorthy, P. Pazhamalai, V.K. Mariappan, S. Manoharan, S.-J. Kim, High performance self-charging supercapacitors using a porous PVDF-ionic liquid electrolyte sandwiched between two-dimensional graphene electrodes, J. Mater. Chem. A 7 (2019) 21693-21703. https://doi.org/10.1039/C9TA06245A

[202] D. Zhou, N. Wang, T. Yang, L. Wang, X. Cao, Z.L. Wang, A piezoelectric nanogenerator promotes highly stretchable and self-chargeable supercapacitors, Mater. Horiz. 7 (2020) 2158-2167. https://doi.org/10.1039/D0MH00610F

[203] Z.L. Wang, W. Wu, Nanotechnology-enabled energy harvesting for self-powered micro-/nanosystems, Angew. Chem. Int. Ed. 51 (2012) 11700-21. https://doi.org/10.1002/anie.201201656

[204] A. Pfenniger, D. Obrist, A. Stahel, V.M. Koch, R. Vogel, Energy harvesting through arterial wall deformation: design considerations for a magneto-hydrodynamic generator, Med. Biol. Eng. Comput. 51 (2013) 741-55. https://doi.org/10.1007/s11517-012-0989-2

[205] L. Junrui, L. Wei-Hsin, Impedance matching for improving piezoelectric energy harvesting systems, SPIE Conf. Proc. Proc. 2010.

[206] N. Wang, R. Daniels, L. Connelly, M. Sotzing, C. Wu, R. Gerhard, G.A. Sotzing, Y. Cao, All-organic flexible ferroelectret nanogenerator with fabric-based electrodes for self-powered body area networks, Small 17 (2021) 2103161. https://doi.org/10.1002/smll.202103161

[207] X. Ma, X. Zhang, Low cost electrostatic vibration energy harvesters based on negatively-charged polypropylene cellular films with a folded structure, Smart Mater. Struct. 26 (2017) 085001. https://doi.org/10.1088/1361-665X/aa75f5

[208] Z. Luo, J. Shi, S.P. Beeby, Novel thick-foam ferroelectret with engineered voids for energy harvesting applications, J. Phys. Conf. Ser. 773 (2016) 012030. https://doi.org/10.1088/1742-6596/773/1/012030

[209] P. Adhikary, S. Garain, D. Mandal, The co-operative performance of a hydrated salt assisted sponge like P(VDF-HFP) piezoelectric generator: an effective piezoelectric based energy harvester, Phys. Chem. Chem. Phys. 17 (2015) 7275-7281. https://doi.org/10.1039/C4CP05513F

[210] B. Mahanty, S.K. Ghosh, S. Garain, D. Mandal, An effective flexible wireless energy harvester/sensor based on porous electret piezoelectric polymer, Mater. Chem. Phys. 186 (2017) 327-332. https://doi.org/10.1016/j.matchemphys.2016.11.003

[211] X. Zhang, L. Wu, G.M. Sessler, Energy scavenging from vibration with two-layer laminated fluoroethylenepropylene piezoelectret films, Joint IEEE Intern. Symp. on the Applications of Ferroelectric (ISAF) 2015, pp 24-27. https://doi.org/10.1109/ISAF.2015.7172659

[212] Y. Wang, L. Wu, X. Zhang, Energy harvesting from vibration using flexible floroethylenepropylene piezoelectret films with cross-tunnel structure, IEEE Trans. Dielectr. Electr. Insul. 22 (2015) 1349-1354. https://doi.org/10.1109/TDEI.2015.7116321

[213] G.M. Sessler, P. Pondrom, X. Zhang, Stacked and folded piezoelectrets for vibration-based energy harvesting, Phase Transitions 89 (2016) 667-677. https://doi.org/10.1080/01411594.2016.1202408

[214] A. Kachroudi, S. Basrour, L. Rufer, F. Jomni, Air-spaced PDMS piezo-electret cantilevers for vibration energy harvesting, J. Phys. Conf. Ser. 773 (2016) 012072. https://doi.org/10.1088/1742-6596/773/1/012072

[215] J. Zhong, Q. Zhong, G. Chen, B. Hu, S. Zhao, X. Li, N. Wu, W. Li, H. Yu, J. Zhou, Surface charge self-recovering electret film for wearable energy conversion in a harsh environment, Energy Environ. Sci. 9 (2016) 3085-3091. https://doi.org/10.1039/C6EE02135B

[216] Y. Mao, P. Zhao, G. McConohy, H. Yang, Y. Tong, X. Wang, Sponge-Like Piezoelectric Polymer Films for Scalable and Integratable Nanogenerators and Self-Powered Electronic Systems, Adv. Energy Mater. 4 (2014) 1301624. https://doi.org/10.1002/aenm.201301624

[217] M. Choi, G. Murillo, S. Hwang, J.W. Kim, J.H. Jung, C.-Y. Chen, M. Lee, Mechanical and electrical characterization of PVDF-ZnO hybrid structure for application to nanogenerator, Nano Energy 33 (2017) 462-468. https://doi.org/10.1016/j.nanoen.2017.01.062

[218] B. Dudem, D.H. Kim, L.K. Bharat, J.S. Yu, Highly-flexible piezoelectric nanogenerators with silver nanowires and barium titanate embedded composite films for mechanical energy harvesting, Appl. Energy 230 (2018) 865-874. https://doi.org/10.1016/j.apenergy.2018.09.009

[219] L. Zhang, J. Gui, Z. Wu, R. Li, Y. Wang, Z. Gong, X. Zhao, C. Sun, S. Guo, Enhanced performance of piezoelectric nanogenerator based on aligned nanofibers and three-dimensional interdigital electrodes, Nano Energy 65 (2019) 103924. https://doi.org/10.1016/j.nanoen.2019.103924

[220] X. Guan, B. Xu, J. Gong, Hierarchically architected polydopamine modified BaTiO3@P(VDF-TrFE) nanocomposite fiber mats for flexible piezoelectric nanogenerators and self-powered sensors, Nano Energy 70 (2020) 104516. https://doi.org/10.1016/j.nanoen.2020.104516

[221] J. Liu, B. Yang, J. Liu, Development of environmental-friendly BZT-BCT/P(VDF-TrFE) composite film for piezoelectric generator, J. Mater. Sci.: Mater. Electron. 29 (2018) 17764-17770. https://doi.org/10.1007/s10854-018-9883-5

Advanced Functional Piezoelectric Materials and Applications   Materials Research Forum LLC
Materials Research Foundations 131 (2022) 117-137   https://doi.org/10.21741/978164490209-4

Chapter 4

Piezoelectric Materials based Phototronics

Mamata Singh[1], Ahumuza Benjamin[2], N.P. Singh[3], Vivek Mishra[4*]

[1]Department of Chemistry, GITAM School of Sciences, Gandhi Institute of Technology and Management (Deemed to be University), Bangalore, India

[2]Department of Mechanical Engineering, GITAM Institute of Technology, Gandhi Institute of Technology and Management (Deemed to be University), Bangalore, India

[3] Centre for Nano Science and Engineering (CeNSE), Indian Institute of Science, Bangalore, karnataka, India".

[4]Amity Institute of Click Chemistry Research and Studies (AICCRS), Amity University Uttar Pradesh, India

Abstract

In 2010, the fundamentals of piezo-phototronics were introduced. By applying stress, a piezo potential is formed in a non-central symmetrical crystal due to ion polarisation. Due to the co-existence of semiconductor and piezoelectricity characteristics, piezo potential induced in the crystal is of significant impact on charge transfer at the junction/interface. The phototronic product uses the piezo capability to manage carrier formation, separation, transportation, and recombination to improve the performance of optoelectronic devices such as photon detectors, solar cells and LED. Today, most of the unique applications in this field can be found in sensing, human-computer interfacing, actuating nanorobotics by effectively integrating piezo-phototronic devices and piezotronic with silicon-based CMOS technology. This chapter gives an insight into the fundamentals of piezo-phototronics.

Keywords

Piezoelectricity, Piezo-Phototronics, Semiconductor, Photoexcitation, Piezotronics

Contents

1.   Introduction

1.1   Piezoelectric effect

This effect is a coupling of electrical polarization and mechanics and was first observed in quartz crystals by Jacques and Pierre Curie in 1880 [1–3]. Cations and anions are created on both opposed surfaces following the deformation of a dielectric material along its asymmetry direction by an applied force [4–7]. When the deforming agent is removed, the dielectric material reverts to its initial state of being uncharged. Furthermore, as the direction of the applied force varies, the polarity of the charge changes as well. Under the same study comes an inverse piezoelectric effect where the crystal undergoes deformation in response to the application of the electric field [8–10]. The piezoelectric effect in wurtzite materials resorts to the non-centrosymmetric structure. Wurtzite structures like GaN, ZnO, ZnS, and InN exhibit piezoelectric capabilities, but aren't as widely employed as PZT due to their low piezoelectric coefficient [11–13]. The most classic piezoelectric material is $Pb(Zr, Ti)O3$ (PZT), used in strain sensors, actuators, and energy-harvesting devices with great success[14–17]. PZT, on the other hand, has an excessively high electrical resistance, making it unsuitable for use in electronic equipment. Perovskites, such as PZT, have good mechanical and piezoelectric capabilities which is why the study of piezoelectric effects and materials which exhibit this effect has traditionally been more targeted to ceramic materials [18–21].

1.2   Piezotronic effect

Static polarisation ions are induced at the semi-conductor surface of non-centrosymmetric crystals that are piezoelectric when a uniform strain is applied or inside if the strain is not homogeneous [22–26]. The imposed strain affects photonic processes as well as electronic transport and can therefore be utilized to statically or dynamically alter material properties [27–30]. Taking an example of a metal-semiconductor interface, the static polarisation interface charges induced by piezoelectric forces aren't entirely screened if the semiconductor is non-centrosymmetric and with considerable doping. The applied strain (tensile or compressive) can then regulate the height of the Schottky barrier, allowing the piezoelectric effect to tune the barrier electronic-transport properties [31–35]. A lower barrier height can improve electron transmission, and vice versa, similar to a diode. Thus, the piezotronics effect, where piezoelectric polarisation charges act as a "gating" voltage to tune transmission of electrons across the surface interface or connection junction. Further experimentation with logic gates and strain-gated piezotronic transistors led to the development of more observations and conclusions, giving more confirmation to this effect [36–40].

2.   Piezo-phototronic effect

This effect works on a principle that, at interfaces or junctions, the piezoelectric potential is used to regulate the carrier production, transportation, separation, recombination processes, etc. [26,34,35]. The discipline and fundamental principle of piezo-phototronics were first proposed in 2010 by Wang for materials that have semiconductor, piezoelectric, and photoexcitation characteristics all at the same time [41]. The investigation of the relationship between photoexcitation The Piezo-phototronics field arose from photonic qualities and piezoelectric capabilities [42–44]. Furthermore, optoelectronics is a well-known topic that explores the interaction of semiconductor properties with photoexcitation properties.

The piezophotonic effect, when first predicted, indicated the emission of photons when a strong local piezoelectric field is applied. Later, Hao's group employed a two-step procedure to investigate this effect. Photoemission was achieved due to period straining of a luminous object by a mechanical straining stage. From this, the piezophotonic effect for the ZnS: Mn system was established due to a series of processes, including liberation of surface state trapped electrons and the transition of an electron to lower energy levels [46,47]. The resulting energy from these processes is responsible for the excitation of trapped quasi-free electrons, which later decay to emit photons. As a result of piezophotonic coupling, which induces and controls the mechano-luminescence process during the conversion from mechanical stress to light emission, strain-induced polarisation charges can influence luminous processes from phosphors [48–50].

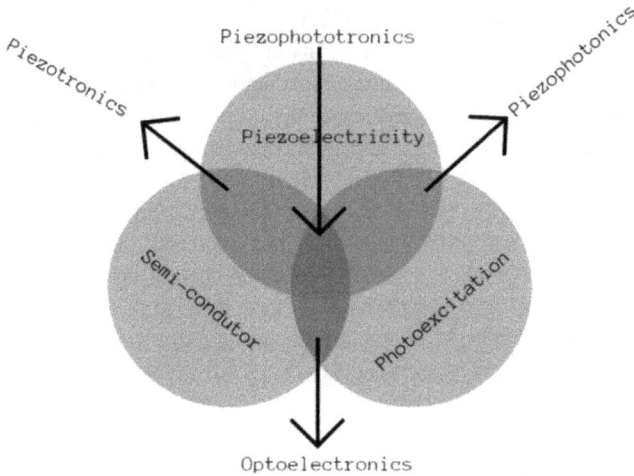

Fig. 1 3-way coupling among piezoelectricity, semiconductor and photoexcitation [45].

The building blocks of electronics and optoelectronics, i.e., the p–n heterojunction and the M–S contact, must thus be contrasted and analyzed in-depth to better comprehend piezotronic piezo-phototronic effects. To modify the features of charge carrier transport, piezoelectric polarisation charges positioned at the M–S contact's interfacial region can efficiently modulate the Schottky barrier height (SBH). Suppose PEMs at the M-S contact are n-type. Anions repelling electrons are formed at the Metal–Semiconductor contact interface during compressive strain, thus forming a larger SBH. Same as the variations of the SBH at Metal–Semiconductor contacts, also modified by the associated piezoelectric ions, the band of energy at the heterojunction varies with strain. The piezoelectric potential, also known as the piezo-phototronic effect, tunes and controls optoelectronic carriers' generation, transport, separation, and interface/junction recombination. When tensile strain is applied to typical piezoelectric semiconductors, the +c direction undergoes positive polarization. Near the p-n junction contact, a dip appears in the local band diagram. Negative piezoelectric polarisation charges caused by compression-strain oppose electrons away from the M-S interface, thus lowering local band structure. Further study in this upcoming and lucrative field has resulted in the development of innovative piezoelectric-photon-electronic nanodevices based on piezoelectric potential and semiconductor characteristics.

The polarisation charges from the piezoelectric material produce light emission from the phosphor in a typical piezophotonic device, resulting in a novel sort of interaction between piezoelectric and photonic features. The luminous intensity of composite phosphor based luminescence piezophototronic devices has also been illustrated, with the luminescence intensity highly influenced by the induced strain rate. The mechano-luminescence process, in which mechanical action on a material generates light emission, can be piezophotonic engineered to design and deploy adaptive sensor array devices for non-static pressure mapping. The piezophotonic coupling has the ability to change the luminescence wavelength, opening up new possibilities in various sensor and imaging applications. Devices like solar cells, photodetectors, and LEDs falling under the optoelectronics category can all benefit from piezo-phototronics.

3. Piezoelectric semiconductor NWs

Both NWs and thin films are affected by the underlying mechanics of the piezotronic and piezophototronic phenomena. NWS, on the other hand, have more significant benefits as explained below [40,51–53]:

NWS could be made cheaply and in huge quantities using wet-chemical synthesis at <100°C with an arbitrary shape; on the other hand, growing high-quality thin films under the same conditions is problematic.

- This is due to the small diameter and high slenderness ratios; Nanowires are extremely flexible and soft. According to numerical simulations, they can withstand significant mechanical deformation (approximately 6% deformation under tension) without fracture or breaking while thin films will quickly crack or break down.

- The toughness and robustness of the Nano Wire structure can be improved by their small size making it almost fatigue-free.

- Because even a modest external force can cause mechanical agitation, using NWs to design ultrasensitive devices is very useful and vital.

- Nano wires have a high piezoelectric coefficient as compared to thin films

Because they can withstand a considerable external mechanical strain/stress, 1-Dimensional nanobelts and NWs and nanobelts are ideal architectures for piezotronic and piezophototronic experiments and research. For piezotronic applications, Wurtzite GaN, ZnO, and doped ferroelectrics are promising possibilities. Because of the following reasons,

- ZnO NWs have been the most thoroughly studied piezotronics and piezo-phototronics materials. ZnO NWs may be easily manufactured in huge quantities via solid vapor processing or low-temperature wet chemical synthesis.

- More significantly, Nanowires of ZnO are biocompatible and eco-friendly.

- NWs as a result of HTCVD processes typically have fewer flaws, making them the best candidates for piezotronics and piezophototronics research. NWS formed using a low-temperature wet chemical technique, on the other hand, exhibit high concentrations of various sorts of defects, which are often beneficial in piezoelectric nanogenerators.

4.   Effect on 2D materials

The main focus for clarifying and using piezo-phototronic and piezotronic effects has been one-dimensional nanomaterials and thin-film structures [54,55]. Due to their non-centrosymmetric designs, 2Dmaterials (e.g., monolayer transition metal dichalcogenides [TMDCs]) are shown to have substantial piezoelectricity in recent investigations. This, combined with their excellent semiconducting, superior mechanical properties, and high crystallinity, suggests that 2D piezotronic materials can exhibit high-performance piezotronic characteristics, as evidenced by conclusions of the report about the piezotronic coupling in atomically thin MoS_2. Furthermore, significant progress has been made in the study of the piezotronic effect and piezoelectricity in a variety of 2-Dimension materials. Piezoelectricity and piezotronic products in 2D materials are still being researched [53,54,56]. Liu et al. describe the accomplishments made in relevant topics in their essay in this edition. These preliminary findings show that exploring the piezotronic effect and its connection to critical scientific phenomena such as topological properties and quantum transport in systems of materials with significantly controlled and reduced dimensionality is feasible and has a lot of potentials. These fundamental studies could pave the way for nanodevice powering, optoelectronics, and flexible electronics with components as thin as a few atomic layers [57,58]. Meanwhile, realizing the full promise of 2D materials in piezotronics and piezophototronics necessitates advancements in the scalable and reliable production of high-quality 2D materials.

5.   Effect on 3rd generation semiconductors

Because of their outstanding material properties, such as high-switching frequency, high-voltage resistance, high-radiation resistance, and high-temperature resistance, third-generation semiconductor materials, such as wideband gap SiC and GaN, have inspired a strong interest in developing consumer electronics technology, electric vehicles, defense applications, and 5G communication systems [45,52,59]. The broad bandgap nature of these materials and strong piezoelectric characteristics suggests that piezotronic and piezophototronic couplings can be of great significance, making them the most favorable platforms for investigating the essential coupling between a number of interesting processes and piezoelectricity, for example high-frequency transport and high-field operation. The knowledge gained from these fundamental investigations is projected to have a positive impact on the future technology of related devices with improved efficiency and performance (e.g., insulated-gate bipolar transistors and high-electron-mobility transistors), which are essential for pervasive technologies in society [60–62]. Furthermore, the viability and maturity of relevant technological processes in manufacturing and integrating of these materials on a commercial scale are expected to advance and boost R&D efforts on piezophotonic, piezophototronic, and piezotronic effects on 3rd-generation semiconductors, as well as associated fundamental interests and technological potential.

Piezotronics and piezo-phototronics are third-generation semiconductor disciplines that dynamically couple polarisation potential induced by strain with the moving charge-carrier transport behavior. Photonic and electronic devices used in human-machine interaction, mechanosensation, robotics, AI, etc. are at risk [63–66]. For further improvements in these sectors, a fundamental understanding of quantum mechanical computations, technological implementations of piezo-phototronic and piezotronic devices and physics are all necessary. Advanced grouping and identification methods, such as determination of the distribution of piezoelectric polarisation charges and, are needed to provide a dependable understanding of essential structual properties and materials linked to piezo-phototronic and piezotronic phenomena. Study and optimization of materials, together with design, production, and characterization of piezotronic devices' arrays, are necessary to facilitate applications at a basic single machine system level to a complete system component. Combining piezo-phototronic and piezotronic effects with instruments having existing optoelectronics, quantum and electronics devices, we believe, shall have transformative implications on sensors, artificial intelligence, energy, and human-integrated technologies.

6.   Piezo-phototronic effect on LED

A p–n junction, for example, is the basic structure of a standard LED. Under forward bias, there are too many carriers above their equilibrium values. Radiative recombination will occur as a result of the injection of minority carriers [67–70]. The LED's working idea is to obtain radiative coupling between holes and electrons at the junction by using forward bias. The main idea is that piezoelectric-charges produced at the interface substantially

Advanced Functional Piezoelectric Materials and Applications   Materials Research Forum LLC
Materials Research Foundations 131 (2022) 117-137   https://doi.org/10.21741/978164490209-4

affect the band structure, the system enables over carrier production, transportation, and recombination at the p–n or M–S interface [63,71].

7.   Piezo-phototronic effect on solar cell

The solar cell's efficiency depends on the effective separation of the electrons and holes at the junction [65,72]. This effect can be utilized to change the band structure at the junction. By designing the arrangements of NWs and the strain to be purposefully induced in the solar cells' packing, the piezo-phototronics effect offers a novel notion for enhancing the conversation efficiency of solar energy [73–75]. One of the most essential disciplines in optoelectronics is solar cell physics. Photon-generated e[-1]-hole pairs in semiconductors are detached using a barrier at an M-S interface. An inner field formed at a depletion zone of the charge at a p–n junction, resulting in current output. Piezoelectric charges at the junction can sharply influence the ability of the junction to separate charges. Open-circuit-voltage, photocurrent, max- power output, ideal conversion efficiency and fill factor are all important metrics to consider when evaluating the performance of a solar cell. The fill factor is calculated by multiplying the product of photocurrent and open-circuit voltage by max. output power.

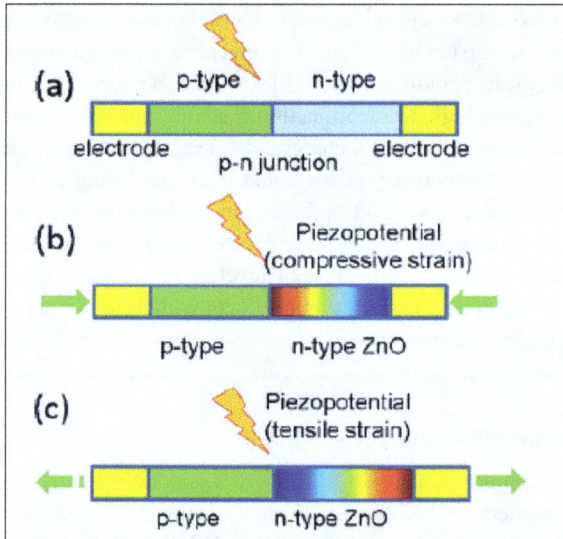

Fig. 2 (a) An universal NW piezoelectric solar cell (PSC) with a p-n junction. (b) PSCs under compression, (c) PSCs under tensile strain, The color code represents the distribution of the piezo potential at the n-type semiconductor NWs [35,65,72].

Optimizing solar cell performance from the device structure perspective of a material can be achieved in 2 ways, i.e., building novel structures and improving the material's energy-efficiency. The MIS structure has been employed in a silicon solar cell to reduce the saturation current density [76–78]. The thermodynamic theory was used to explore the thickness dependency of solar cells with a p–n junction to optimize open-circuit voltages. Solar cells made of polymer PBDTTT4 having a low-bandgap structure were developed to achieve 0.76 volts of open-circuit voltage and 6.77% power conversion efficiency. In order to understand the cause of the polymer based solar cells' open circuit voltage, a number of tests and theories have been advanced [72,79–81].

A fresh new method for improving the organic solar cells performance has recently been created combining piezoelectric and ferroelectric materials: ferroelectric polymer solar cell and PSC (Piezoelectric solar cell). The main idea is to increase charge separation by utilizing the material's intrinsic-potential. It is essential to develop a basic theory to comprehend the experimental occurrences. A p–n junction, also known as a metal-semiconductor (M–S) contact, is the basic structure of a conventional nano/microwire solar cell. The working principle of the solar cell is to practice a high electric field in the depletion region to aid in the separation of e^{-1}-hole pairs generated by incident-photons. Solar cells' performance can be adjusted and controlled using the piezoelectric charges generated at the junction area under strain. The primary piezo-phototronic effect is modifying the band structure at the interface by the piezo potential thus controlling over carrier production, dissociation, and transport junction or interface [56,82,83].

8.   Piezo-phototronics in luminescence applications

A number of optic technologies can transfer a great capacity of digital data at a very fast rate and very efficiently. A pressure-sensor array of InGaN–GaNMQW nanopillars was recently described based on piezophotonically controlled photoluminescence-imaging [59,71,84,85]. When strain is applied on these nanopillars, the photoluminescence intensity may be regulated linearly by mechanical strain, and piezoelectric polarisation created at the quantum well Metal-Semiconductor interface can control the recombination and excitation processes [86–89]. The distribution of pressure may be traced in at a given instant in real-time by monitoring the signals of photoluminescence from the nanopillar array. The piezoelectric force from the PMN–PT substrate resulted into emission of light from the Mn doped ZnS phosphor, indicating a connection between photonic and piezoelectric processes. After that, the structure was refined by combining a transparent polymer with metal ion-doped ZnS to create adaptive composite phosphors. White light was observed when tension was applied to the flexible luminescent device. The conduction and valance-bands of ZnS are tilted by piezoelectric polarisation charges, permitting bound electrons in impurity at the conduction-band edge to pass to holes in the valance-band, resulting in energy release. The released energy excites the Mn2+dopant ions, and photon emission occurs once the excited Mn2+ ions revert to the ground state. The magnetic field was linked to the piezo-phototronic technique, which emitted green and white light, to create a magnetic-induced-luminescence device made of flexible composite laminates. Light emission might be managed in an unilateral and dynamic manner via strain-mediated coupling and modulation of the weak magnetic field. Energy harvesting, non-destructive sensing, and stimuli-responsive multimodal bioimaging are all possibilities with piezo-phototronic luminescence systems like those described above.

9.   Piezo-phototronics in other applications

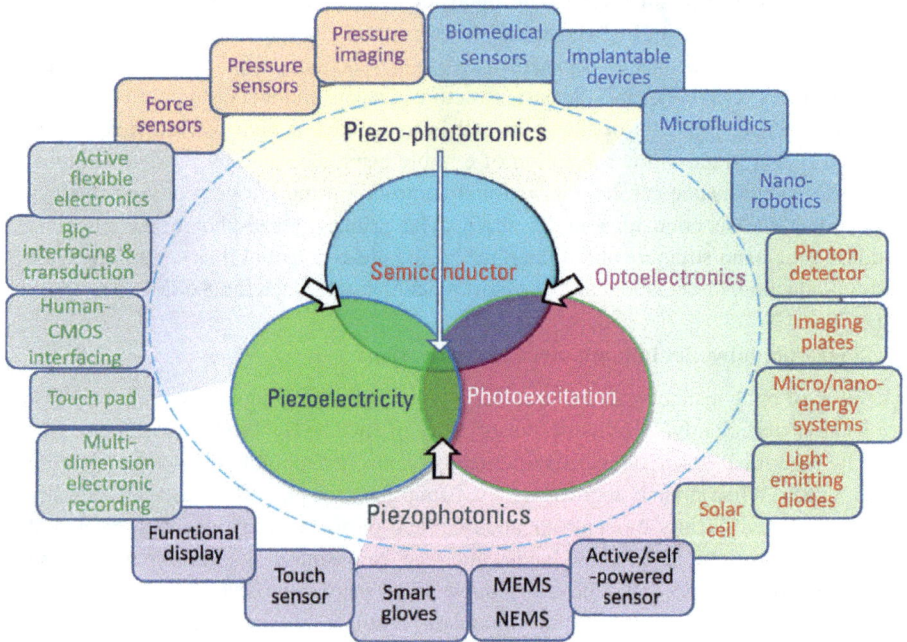

Fig. 3 A number of fields for technological implementation and research in piezo phototronics and piezotronics [26].

Advanced Functional Piezoelectric Materials and Applications   Materials Research Forum LLC
Materials Research Foundations 131 (2022) 117-137   https://doi.org/10.21741/978164490209-4

Other uses for the piezo-phototronic approach include the fabrication of exceptionally efficient photo electro chemical photo anodes. A new finding demonstrated and produced oxygen evolution reaction by physically diverting the photoanode and altering the interfacial band-structure of a ZnO–Ni(OH)$_2$ hetero p-n junction through the phototronic effect, resulting in dramatically higher photocurrent density as a result of the oxidation processes. In this circumstance, polarisation charges at the M-S junction interface accelerate extra photoexcited charges from ZnO towards the Oxygen evolution reaction interface [45,52]. This could open up a novel pathway to enhance the performance of low-cost catalysts for solar fuel. TiO$_2$ nanoparticles were assembled on piezoelectric ZnO nanoplatelets in another study to create a photocatalyst which emerges as a catalyst. The photocatalysis process was enhanced by around 20% by managing the cooling of the mixed photo-catalyst and altering the thermal stress in ZnO. The induced piezoelectric polarisation charges successfully separated the photogenerated MO charge carriers at the heterojunction interface between ZnO and TiO$_2$. The piezo-phototronic phenomenon has also been employed in information storage and calculations involving mechanical-electro-optical logic [90–92]. Polarization charges induced by strain at the electrode/CdS interface can control charge carrier optoelectronic processes and electrical transport, allowing optical - mechanical stimuli to be transformed into electronic signals via the piezo-phototronic effect. These characteristics aren't currently present in current technologies, which gives adaptive electronics and optoelectronics a lot of flexibility and potential for applications in optical nano electro mechanical systems and human-machine interfaces. Solar cells based on p-type polymers and sputtered ZnO films and are also being studied for piezo-phototronic coupling [23,24,93–95].

Si, II-VI, and III–V semiconducting materials are employed in LEDs, CMOS technology, solar cells and photon detectors, in today's electronics [96]. In contrast, piezoelectricity is reliant on PZT perovskite materials, that are hardly employed in optoelectronics. The overlapping across piezoelectricity and optoelectronics is fairly limited due to major differences in materials systems. We merged piezoelectricity with optoelectronic excitation processes and built a few new domains using wurtzite materials including GaN, ZnO, and InN, which have piezoelectric and piezoelectric semiconductor capabilities. The core is based on the piezoelectric potential, which is created when a load is given to a piezoelectric material by the polarisation of ions. The piezo potential is used as a "gate" voltage to tube/control charge carrier transport at contact or junction, known as piezotronics. The piezo-phototronic effect is the use of the piezo potential to influence carrier formation, separation, transport, and/or recombination in optoelectronic devices such as photon detectors, solar cells, and LEDs to improve their performance. Z.L. Piezotronics have a similar role in human–CMOS technology interface as mechano-sensation in physiology.

Mechano-sensation is a mechanical stimulus-response mechanism [97–101]. The physiological foundation for touch, balance, pain, and hearing experiences is the conversion of mechanical inputs into brain impulses. Human-machine interaction, sensor networks, bioscience, and integration, and energy sciences are all expected to benefit from piezotronics and piezo-phototronics.

References

[1] J. Curie, P. Curie, No Title, (1880). https://doi.org/10.3406/bulmi.1880.1564. https://doi.org/10.3406/bulmi.1880.1564

[2] Introduction to Piezoelectricity and Piezoelectric Ceramics - Kadco Ceramics, (n.d.). https://www.kadcoceramics.com/introduction-to-piezoelectricity-and-piezoelectric-ceramics/ (accessed June 26, 2021).

[3] Piezoelectricity - Wikipedia, (n.d.). https://en.wikipedia.org/wiki/Piezoelectricity#Application (accessed June 11, 2021).

[4] A. Ramadoss, B. Saravanakumar, S.W. Lee, Y.S. Kim, S.J. Kim, Z.L. Wang, Piezoelectric-Driven Self-Charging Supercapacitor Power Cell, ACS Nano. 9 (2015) 4337-4345. https://doi.org/10.1021/acsnano.5b00759

[5] What is Piezoelectricity? | OnScale, (n.d.). https://onscale.com/piezoelectricity/what-is-piezoelectricity/ (accessed June 26, 2021).

[6] T. Bailey, J.E. Ubbard, Distributed piezoelectric-polymer active vibration control of a cantilever beam, Journal of Guidance, Control, and Dynamics. 8 (1985) 605-611. https://doi.org/10.2514/3.20029

[7] K.S. Hong, H. Xu, H. Konishi, X. Li, Piezoelectrochemical effect: A new mechanism for azo dye decolorization in aqueous solution through vibrating piezoelectric microfibers, Journal of Physical Chemistry C. 116 (2012) 13045-13051. https://doi.org/10.1021/jp211455z

[8] Piezoelectricity - Engineering LibreTexts, (n.d.). https://eng.libretexts.org/Bookshelves/Materials_Science/Supplemental_Modules_(Materials_Science)/Electronic_Properties/Piezoelectricity (accessed June 11, 2021).

[9] The Piezoelectric Effect - Piezoelectric Motors & Motion Systems, (n.d.). https://www.nanomotion.com/nanomotion-technology/piezoelectric-effect/ (accessed June 11, 2021).

[10] Piezoelectric Material - an overview | ScienceDirect Topics, (n.d.). https://www.sciencedirect.com/topics/materials-science/piezoelectric-material (accessed June 11, 2021).

[11] M.A. Migliorato, D. Powell, A.G. Cullis, T. Hammerschmidt, G.P. Srivastava, Composition and strain dependence of the piezoelectric coefficients in Inx Ga1-x As alloys, Physical Review B - Condensed Matter and Materials Physics. 74 (2006).

[12] O. Thomas, J.F. Deü, J. Ducarne, Vibrations of an elastic structure with shunted piezoelectric patches: Efficient finite element formulation and electromechanical coupling coefficients, International Journal for Numerical Methods in Engineering. 80 (2009) 235-268. https://doi.org/10.1002/nme.2632

[13] M.A. Dubois, P. Muralt, Measurement of the effective transverse piezoelectric coefficient e31,f of AlN and Pb(Zrx,Ti1-x)O3 thin films, Sensors and Actuators, A: Physical. 77 (1999) 106-112. https://doi.org/10.1016/S0924-4247(99)00070-9

[14] H.Y.S. Al-Zahrani, J. Pal, M.A. Migliorato, Non-linear piezoelectricity in wurtzite ZnO semiconductors, Nano Energy. 2 (2013) 1214-1217. https://doi.org/10.1016/j.nanoen.2013.05.005

[15] N. Izyumskaya, Y.I. Alivov, S.J. Cho, H. Morkoç, H. Lee, Y.S. Kang, Processing, structure, properties, and applications of PZT thin films, Critical Reviews in Solid State and Materials Sciences. 32 (2007) 111-202. https://doi.org/10.1080/10408430701707347

[16] M.G. Kang, W.S. Jung, C.Y. Kang, S.J. Yoon, Recent progress on PZT based piezoelectric energy harvesting technologies, Actuators. 5 (2016). https://doi.org/10.3390/act5010005

[17] K. il Park, J.H. Son, G.T. Hwang, C.K. Jeong, J. Ryu, M. Koo, I. Choi, S.H. Lee, M. Byun, Z.L. Wang, K.J. Lee, Highly-efficient, flexible piezoelectric PZT thin film nanogenerator on plastic substrates, Advanced Materials. 26 (2014) 2514-2520. https://doi.org/10.1002/adma.201305659

[18] V. Jella, S. Ippili, J.H. Eom, S.V.N. Pammi, J.S. Jung, V.D. Tran, V.H. Nguyen, A. Kirakosyan, S. Yun, D. Kim, M.R. Sihn, J. Choi, Y.J. Kim, H.J. Kim, S.G. Yoon, A comprehensive review of flexible piezoelectric generators based on organic-inorganic metal halide perovskites, Nano Energy. 57 (2019) 74-93. https://doi.org/10.1016/j.nanoen.2018.12.038

[19] J. Hao, W. Li, J. Zhai, H. Chen, Progress in high-strain perovskite piezoelectric ceramics, Materials Science and Engineering R: Reports. 135 (2019) 1-57. https://doi.org/10.1016/j.mser.2018.08.001

[20] S. Zhang, R. Xia, T.R. Shrout, G. Zang, J. Wang, Piezoelectric properties in perovskite 0.948(K0.5Na 0.5)NbO3-0.052LiSbO3 lead-free ceramics, Journal of Applied Physics. 100 (2006). https://doi.org/10.1063/1.2382348

[21] A.S. Bhalla, R. Guo, R. Roy, The perovskite structure - A review of its role in ceramic science and technology, Materials Research Innovations. 4 (2000) 3-26. https://doi.org/10.1007/s100190000062

[22] Z.L. Wang, From nanogenerators to piezotronicsa-A decade-long study of ZnO nanostructures, MRS Bulletin. 37 (2012) 814-827. https://doi.org/10.1557/mrs.2012.186

[23] L. Wang, Z.L. Wang, Advances in piezotronic transistors and piezotronics, Nano Today. 37 (2021) 101108. https://doi.org/10.1016/j.nantod.2021.101108

[24] L. Wang, Z.L. Wang, Advances in piezotronic transistors and piezotronics, Nano Today. 37 (2021). https://doi.org/10.1016/j.nantod.2021.101108

[25] Q. Zhang, S. Zuo, P. Chen, C. Pan, Piezotronics in two-dimensional materials, InfoMat. 3 (2021) 987-1007. https://doi.org/10.1002/inf2.12220

[26] Z.L. Wang, W. Wu, Piezotronics and piezo-phototronics: Fundamentals and applications, National Science Review. 1 (2014) 62-90. https://doi.org/10.1093/nsr/nwt002

[27] S. Xu, W. Guo, S. Du, M.M.T. Loy, N. Wang, Piezotronic effects on the optical properties of ZnO nanowires, Nano Letters. 12 (2012) 5802-5807. https://doi.org/10.1021/nl303132c

[28] S. Xu, W. Guo, S. Du, M.M.T. Loy, N. Wang, Piezotronic effects on the optical properties of ZnO nanowires, Nano Letters. 12 (2012) 5802-5807. https://doi.org/10.1021/nl303132c

[29] R. Yu, L. Dong, C. Pan, S. Niu, H. Liu, W. Liu, S. Chua, D. Chi, Z.L. Wang, Piezotronic effect on the transport properties of GaN nanobelts for active flexible electronics, Advanced Materials. 24 (2012) 3532-3537. https://doi.org/10.1002/adma.201201020

[30] X. Li, W. Yu, X. Gao, H. Liu, N. Han, X. Zhang, PVDF microspheres@PLLA nanofibers-based hybrid tribo/piezoelectric nanogenerator with excellent electrical

output properties, Materials Advances. 2 (2021) 6011-6019.
https://doi.org/10.1039/D1MA00464F

[31] R.T. Tung, Recent advances in Schottky barrier concepts, Materials Science and Engineering: R: Reports. 35 (2001) 1-138. https://doi.org/10.1016/S0927-796X(01)00037-7

[32] L.J. Brillson, Y. Lu, ZnO Schottky barriers and Ohmic contacts, Journal of Applied Physics. 109 (2011). https://doi.org/10.1063/1.3581173

[33] J. Meng, Z. Li, Schottky-Contacted Nanowire Sensors, Advanced Materials. 32 (2020). https://doi.org/10.1002/adma.202000130

[34] Z.L. Wang, W. Wu, Piezotronics and piezo-phototronics: fundamentals and applications, National Science Review. 1 (2014) 62-90. https://doi.org/10.1093/nsr/nwt002

[35] Y. Zhang, Y. Yang, Z.L. Wang, Piezo-phototronics effect on nano/microwire solar cells, Energy and Environmental Science. 5 (2012) 6850-6856. https://doi.org/10.1039/c2ee00057a

[36] W. Han, Y. Zhou, Y. Zhang, C.Y. Chen, L. Lin, X. Wang, S. Wang, Z.L. Wang, Strain-gated piezotronic transistors based on vertical zinc oxide nanowires, ACS Nano. 6 (2012) 3760-3766. https://doi.org/10.1021/nn301277m

[37] P. Puneetha, S.P.R. Mallem, Y.W. Lee, J.H. Lee, J. Shim, Strain-induced piezotronic effects in nano-sized GaN thin films, Nano Energy. 88 (2021). https://doi.org/10.1016/j.nanoen.2021.106305

[38] Z. Gao, J. Zhou, Y. Gu, P. Fei, Y. Hao, G. Bao, Z.L. Wang, Effects of piezoelectric potential on the transport characteristics of metal-ZnO nanowire-metal field effect transistor, Journal of Applied Physics. 105 (2009). https://doi.org/10.1063/1.3125449

[39] W. Liu, M. Lee, L. Ding, J. Liu, Z.L. Wang, Piezopotential gated nanowire-nanotube hybrid field-effect transistor, Nano Letters. 10 (2010) 3084-3089. https://doi.org/10.1021/nl1017145

[40] C.R. Crowell, S.M. Sze, Current transport in metal-semiconductor barriers, Solid State Electronics. 9 (1966) 1035-1048. https://doi.org/10.1016/0038-1101(66)90127-4

[41] Z.L. Wang, Piezopotential gated nanowire devices: Piezotronics and piezo-phototronics, Nano Today. 5 (2010) 540-552. https://doi.org/10.1016/j.nantod.2010.10.008

[42] C. Pan, M. Chen, R. Yu, Q. Yang, Y. Hu, Y. Zhang, Z.L. Wang, Progress in Piezo-Phototronic-Effect-Enhanced Light-Emitting Diodes and Pressure Imaging, Advanced Materials. 28 (2016) 1535-1552. https://doi.org/10.1002/adma.201503500

[43] S. Tu, Y. Guo, Y. Zhang, C. Hu, T. Zhang, T. Ma, H. Huang, Piezocatalysis and Piezo-Photocatalysis: Catalysts Classification and Modification Strategy, Reaction Mechanism, and Practical Application, Advanced Functional Materials. 30 (2020). https://doi.org/10.1002/adfm.202005158

[44] Y. Liu, Y. Zhang, Q. Yang, S. Niu, Z.L. Wang, Fundamental theories of piezotronics and piezo-phototronics, Nano Energy. 14 (2014) 257-275. https://doi.org/10.1016/j.nanoen.2014.11.051

[45] C. Pan, J. Zhai, Z.L. Wang, Piezotronics and Piezo-phototronics of Third Generation Semiconductor Nanowires, Chemical Reviews. 119 (2019) 9303-9359. https://doi.org/10.1021/ACS.CHEMREV.8B00599/ASSET/IMAGES/ACS.CHEMRE V.8B00599.SOCIAL.JPEG_V03. https://doi.org/10.1021/acs.chemrev.8b00599

[46] Q. Yang, W. Wang, S. Xu, Z.L. Wang, Enhancing light emission of ZnO microwire-based diodes by piezo-phototronic effect, Nano Letters. 11 (2011) 4012-4017. https://doi.org/10.1021/nl202619d

[47] Q. Yang, W. Wang, S. Xu, Z.L. Wang, Enhancing Light Emission of ZnO Microwire-Based Diodes by Piezo-Phototronic Effect, Nano Letters. 11 (2011) 4012-4017. https://doi.org/10.1021/nl202619d

[48] J. Zhou, P. Fei, Y. Gu, W. Mai, Y. Gao, R. Yang, G. Bao, Z.L. Wang, Piezoelectric-potential-controlled polarity-reversible Schottky diodes and switches of ZnO wires, Nano Letters. 8 (2008) 3973-3977. https://doi.org/10.1021/nl802497e

[49] L. Zhu, Q. Lai, W. Zhai, B. Chen, Z.L. Wang, Piezo-phototronic effect enhanced polarization-sensitive photodetectors based on cation-mixed organic-inorganic perovskite nanowires, Materials Today. 37 (2020) 56-63. https://doi.org/10.1016/j.mattod.2020.02.018

[50] V. Consonni, A.M. Lord, Polarity in ZnO nanowires: A critical issue for piezotronic and piezoelectric devices, Nano Energy. 83 (2021). https://doi.org/10.1016/j.nanoen.2021.105789

[51] G.S. Weldegrum, P. Singh, B.R. Huang, T.Y. Chiang, K.W. Tseng, C.J. Yu, C. Ji, J.P. Chu, ZnO-NWs/metallic glass nanotube hybrid arrays: Fabrication and material characterization, Surface and Coatings Technology. 408 (2021). https://doi.org/10.1016/j.surfcoat.2020.126785

[52] C. Pan, J. Zhai, Z.L. Wang, Piezotronics and Piezo-phototronics of Third Generation Semiconductor Nanowires, Chemical Reviews. 119 (2019) 9303-9359. https://doi.org/10.1021/acs.chemrev.8b00599

[53] M. Nehra, N. Dilbaghi, G. Marrazza, A. Kaushik, R. Abolhassani, Y.K. Mishra, K.H. Kim, S. Kumar, 1D semiconductor nanowires for energy conversion, harvesting and storage applications, Nano Energy. 76 (2020). https://doi.org/10.1016/j.nanoen.2020.104991

[54] Y. Peng, M. Que, J. Tao, X. Wang, J. Lu, G. Hu, B. Wan, Q. Xu, C. Pan, Progress in piezotronic and piezo-phototronic effect of 2D materials, 2D Materials. 5 (2018). https://doi.org/10.1088/2053-1583/aadabb

[55] Q. Cheng, J. Pang, D. Sun, J. Wang, S. Zhang, F. Liu, Y. Chen, R. Yang, N. Liang, X. Lu, Y. Ji, J. Wang, C. Zhang, Y. Sang, H. Liu, W. Zhou, WSe2 2D p-type semiconductor-based electronic devices for information technology: Design, preparation, and applications, InfoMat. 2 (2020) 656-697. https://doi.org/10.1002/inf2.12093

[56] X. Zhao, Q. Li, L. Xu, Z. Zhang, Z. Kang, Q. Liao, Y. Zhang, Interface Engineering in 1D ZnO-Based Heterostructures for Photoelectrical Devices, Advanced Functional Materials. 32 (2022). https://doi.org/10.1002/adfm.202106887

[57] Y. He, A. Sobhani, S. Lei, Z. Zhang, Y. Gong, Z. Jin, W. Zhou, Y. Yang, Y. Zhang, X. Wang, B. Yakobson, R. Vajtai, N.J. Halas, B. Li, E. Xie, P. Ajayan, Layer Engineering of 2D Semiconductor Junctions, Advanced Materials. 28 (2016) 5126-5132. https://doi.org/10.1002/adma.201600278

[58] M. Collet, M. Ouisse, M.N. Ichchou, R. Ohayon, Semi-active optimization of 2D wave dispersion into shunted piezo-composite systems for controlling acoustic interaction, Smart Materials and Structures. 21 (2012). https://doi.org/10.1088/0964-1726/21/9/094002

[59] Z.L. Wang, W. Wu, C. Falconi, Piezotronics and piezo-phototronics with third-generation semiconductors, MRS Bulletin. 43 (2018) 922-927. https://doi.org/10.1557/mrs.2018.263

[60] X. Chen, S. Shen, L. Guo, S.S. Mao, Semiconductor-based photocatalytic hydrogen generation, Chemical Reviews. 110 (2010) 6503-6570. https://doi.org/10.1021/cr1001645

[61] X. Chen, S. Shen, L. Guo, S.S. Mao, Semiconductor-based photocatalytic hydrogen generation, Chemical Reviews. 110 (2010) 6503-6570. https://doi.org/10.1021/cr1001645

[62] T. Pham, S. Kommandur, H. Lee, D. Zakharov, M.A. Filler, F.M. Ross, One-dimensional twisted and tubular structures of zinc oxide by semiconductor-catalyzed vapor-liquid-solid synthesis, Nanotechnology. 32 (2021). https://doi.org/10.1088/1361-6528/abc452

[63] B.G. Crutchley, I.P. Marko, J. Pal, M.A. Migliorato, S.J. Sweeney, Optical properties of InGaN-based LEDs investigated using high hydrostatic pressure dependent techniques, Physica Status Solidi (B) Basic Research. 250 (2013) 698-702. https://doi.org/10.1002/pssb.201200514

[64] M. El-Salamony, M.A. Aziz, Solar Panel Effect on Low-Speed Airfoil Aerodynamic Performance, Unmanned Systems. 9 (2021) 333-347. https://doi.org/10.1142/S2301385021500175

[65] B. Dai, G.M. Biesold, M. Zhang, H. Zou, Y. Ding, Z.L. Wang, Z. Lin, Piezo-phototronic effect on photocatalysis, solar cells, photodetectors and light-emitting diodes, Chemical Society Reviews. 50 (2021) 13646-13691. https://doi.org/10.1039/D1CS00506E

[66] D. Hong, W. Zang, X. Guo, Y. Fu, H. He, J. Sun, L. Xing, B. Liu, X. Xue, High Piezo-photocatalytic Efficiency of CuS/ZnO Nanowires Using Both Solar and Mechanical Energy for Degrading Organic Dye, ACS Applied Materials and Interfaces. 8 (2016) 21302-21314. https://doi.org/10.1021/acsami.6b05252

[67] L. Pan, S. Sun, Y. Chen, P. Wang, J. Wang, X. Zhang, J.J. Zou, Z.L. Wang, Advances in Piezo-Phototronic Effect Enhanced Photocatalysis and Photoelectrocatalysis, Advanced Energy Materials. 10 (2020). https://doi.org/10.1002/aenm.202000214

[68] Y. Hu, Y. Zhang, L. Lin, Y. Ding, G. Zhu, Z.L. Wang, Piezo-phototronic effect on electroluminescence properties of p-type GaN thin films, Nano Letters. 12 (2012) 3851-3856. https://doi.org/10.1021/nl301879f

[69] Y. Hu, Y. Zhang, L. Lin, Y. Ding, G. Zhu, Z.L. Wang, Piezo-phototronic effect on electroluminescence properties of p-type GaN thin films, Nano Letters. 12 (2012) 3851-3856. https://doi.org/10.1021/nl301879f

[70] Z. Huo, Y. Zhang, X. Han, W. Wu, W. Yang, X. Wang, M. Zhou, C. Pan, Piezo-phototronic effect enhanced performance of a p-ZnO NW based UV-Vis-NIR photodetector, Nano Energy. 86 (2021). https://doi.org/10.1016/j.nanoen.2021.106090

[71] W. Wu, Z.L. Wang, Piezotronics and piezo-phototronics for adaptive electronics and optoelectronics, Nature Reviews Materials. 1 (2016). https://doi.org/10.1038/natrevmats.2016.31

[72] C. Pan, S. Niu, Y. Ding, L. Dong, R. Yu, Y. Liu, G. Zhu, Z.L. Wang, Enhanced Cu 2S/CdS coaxial nanowire solar cells by piezo-phototronic effect, Nano Letters. 12 (2012) 3302-3307. https://doi.org/10.1021/nl3014082

[73] J. Wallentin, N. Anttu, D. Asoli, M. Huffman, I. Åberg, M.H. Magnusson, G. Siefer, P. Fuss-Kailuweit, F. Dimroth, B. Witzigmann, H.Q. Xu, L. Samuelson, K. Deppert, M.T. Borgström, InP nanowire array solar cells achieving 13.8% efficiency by exceeding the ray optics limit, Science. 339 (2013) 1057-1060. https://doi.org/10.1126/science.1230969

[74] H. Hoopes, Good vibrations lead to efficient excitations in hybrid solar cells, Gizmag.com, 2013. http://www.gizmag.com/vibration-sound-efficient-hybrid-solar-cell-arrays/29679/ (accessed June 11, 2021).

[75] H. Li, Y. Sang, S. Chang, X. Huang, Y. Zhang, R. Yang, H. Jiang, H. Liu, Z.L. Wang, Enhanced Ferroelectric-Nanocrystal-Based Hybrid Photocatalysis by Ultrasonic-Wave-Generated Piezophototronic Effect, Nano Letters. 15 (2015) 2372-2379. https://doi.org/10.1021/nl504630j

[76] F.B. Madsen, A.E. Daugaard, S. Hvilsted, A.L. Skov, The Current State of Silicone-Based Dielectric Elastomer Transducers, Macromolecular Rapid Communications. 37 (2016) 378-413. https://doi.org/10.1002/marc.201500576

[77] L. Kleinman, Deformation potentials in silicon. I. Uniaxial strain, Physical Review. 128 (1962) 2614-2621. https://doi.org/10.1103/PhysRev.128.2614

[78] B.S. Lee, J. Yoon, C. Jung, D.Y. Kim, S.Y. Jeon, K.H. Kim, J.H. Park, H. Park, K.H. Lee, Y.S. Kang, J.H. Park, H. Jung, W.R. Yu, S.G. Doo, Silicon/Carbon Nanotube/BaTiO3 Nanocomposite Anode: Evidence for Enhanced Lithium-Ion Mobility Induced by the Local Piezoelectric Potential, ACS Nano. 10 (2016) 2617-2627. https://doi.org/10.1021/acsnano.5b07674

[79] Y. Yang, W. Guo, Y. Zhang, Y. Ding, X. Wang, Z.L. Wang, Piezotronic effect on the output voltage of P3HT/ZnO micro/nanowire heterojunction solar cells, Nano Letters. 11 (2011) 4812-4817. https://doi.org/10.1021/nl202648p

[80] Y. Yang, W. Guo, Y. Zhang, Y. Ding, X. Wang, Z.L. Wang, Piezotronic Effect on the Output Voltage of P3HT/ZnO Micro/Nanowire Heterojunction Solar Cells, Nano Letters. 11 (2011) 4812-4817. https://doi.org/10.1021/nl202648p

[81] G. Michael, Y. Zhang, J. Nie, D. Zheng, G. Hu, R. Liu, M. Dan, L. Li, Y. Zhang, High-performance piezo-phototronic multijunction solar cells based on single-type two-dimensional materials, Nano Energy. 76 (2020). https://doi.org/10.1016/j.nanoen.2020.105091

[82] J. Shi, M.B. Starr, H. Xiang, Y. Hara, M.A. Anderson, J.H. Seo, Z. Ma, X. Wang, Interface engineering by piezoelectric potential in ZnO-based photoelectrochemical anode, Nano Letters. 11 (2011) 5587-5593. https://doi.org/10.1021/nl203729j

[83] S. Xia, Y. Diao, C. Kan, Electronic and optical properties of two-dimensional GaN/ZnO heterojunction tuned by different stacking configurations, Journal of Colloid and Interface Science. 607 (2022) 913-921. https://doi.org/10.1016/j.jcis.2021.09.050

[84] J. Zhang, Tunable local and global piezopotential properties of graded InGaN nanowires, Nano Energy. 86 (2021). https://doi.org/10.1016/j.nanoen.2021.106125

[85] Z.L. Wang, R. Yang, J. Zhou, Y. Qin, C. Xu, Y. Hu, S. Xu, Lateral nanowire/nanobelt based nanogenerators, piezotronics and piezo-phototronics, Materials Science and Engineering R: Reports. 70 (2010) 320-329. https://doi.org/10.1016/j.mser.2010.06.015

[86] Y. Liu, Q. Yang, Y. Zhang, Z. Yang, Z.L. Wang, Nanowire piezo-phototronic photodetector: Theory and experimental design, Advanced Materials. 24 (2012) 1410-1417. https://doi.org/10.1002/adma.201104333

[87] C. Gong, M.H.W. Lam, H. Yu, The fabrication of a photoresponsive molecularly imprinted polymer for the photoregulated uptake and release of caffeine, Advanced Functional Materials. 16 (2006) 1759-1767. https://doi.org/10.1002/adfm.200500907

[88] C.F. Tan, W.L. Ong, G.W. Ho, Self-Biased Hybrid Piezoelectric-Photoelectrochemical Cell with Photocatalytic Functionalities, ACS Nano. 9 (2015) 7661-7670. https://doi.org/10.1021/acsnano.5b03075

[89] H. Chen, G. Noirbent, Y. Zhang, K. Sun, S. Liu, D. Brunel, D. Gigmes, B. Graff, F. Morlet-Savary, P. Xiao, F. Dumur, J. Lalevée, Photopolymerization and 3D/4D applications using newly developed dyes: Search around the natural chalcone scaffold in photoinitiating systems, Dyes and Pigments. 188 (2021). https://doi.org/10.1016/j.dyepig.2021.109213

[90] A. Babaeinesami, H. Tohidi, S.M. Seyedaliakbar, Designing a data-driven leagile sustainable closed-loop supply chain network, International Journal of Management Science and Engineering Management. 16 (2021) 14-26. https://doi.org/10.1080/17509653.2020.1811794

[91] Y. Ji, Y. Liu, Y. Yang, Multieffect Coupled Nanogenerators, Research. 2020 (2020) 1-24. https://doi.org/10.34133/2020/6503157

[92] C. Ren, K.F. Wang, B.L. Wang, Adjusting the electromechanical coupling behaviors of piezoelectric semiconductor nanowires via strain gradient and flexoelectric effects, Journal of Applied Physics. 128 (2020). https://doi.org/10.1063/5.0028923

[93] C. An, H. Qi, L. Wang, X. Fu, A. Wang, Z.L. Wang, J. Liu, Piezotronic and piezo-phototronic effects of atomically-thin ZnO nanosheets, Nano Energy. 82 (2021). https://doi.org/10.1016/j.nanoen.2020.105653

[94] S. Liu, M. Han, X. Feng, Q. Yu, L. Gu, L. Wang, Y. Qin, Z.L. Wang, Statistical Piezotronic Effect in Nanocrystal Bulk by Anisotropic Geometry Control, Advanced Functional Materials. 31 (2021). https://doi.org/10.1002/adfm.202010339

[95] E.K. Lee, H. Yoo, Self-powered sensors: New opportunities and challenges from two-dimensional nanomaterials, Molecules. 26 (2021). https://doi.org/10.3390/molecules26165056

[96] R.S. Dahiya, A. Adami, C. Collini, L. Lorenzelli, POSFET tactile sensing arrays using CMOS technology, Sensors and Actuators, A: Physical. 202 (2013) 226-232. https://doi.org/10.1016/j.sna.2013.02.007

[97] M. Ma, Y. Huang, J. Liu, K. Liu, Z. Wang, C. Zhao, S. Qu, Z. Wang, Engineering the photoelectrochemical behaviors of ZnO for efficient solar water splitting, Journal of Semiconductors. 41 (2020). https://doi.org/10.1088/1674-4926/41/9/091702

[98] M. Chen-Glasser, P. Li, J. Ryu, S. Hong, No Title, InTech, 2018. https://doi.org/10.5772/intechopen.76963. https://doi.org/10.5772/intechopen.76963

[99] B. Mika, Design and testing of piezoelectric sensors, PhD Thesis, Texas A&M University. (2009) 109. http://repositories.tdl.org/tdl-ir/handle/1969.1/ETD-TAMU-1565 (accessed June 11, 2021).

[100] K.K. Sappati, S. Bhadra, Piezoelectric polymer and paper substrates: A review, Sensors (Switzerland). 18 (2018) 3605. https://doi.org/10.3390/s18113605

[101] S.K. Ghosh, T.K. Sinha, B. Mahanty, D. Mandal, Self-poled Efficient Flexible "Ferroelectretic" Nanogenerator: A New Class of Piezoelectric Energy Harvester, Energy Technology. 3 (2015) 1190-1197. https://doi.org/10.1002/ente.201500167

Advanced Functional Piezoelectric Materials and Applications   Materials Research Forum LLC
Materials Research Foundations 131 (2022) 138-164   https://doi.org/10.21741/978164490209-5

Chapter 5

Piezoelectric Composites and their Applications

M. Ramesh*, M. Muthukrishnan

Department of Mechanical Engineering, KIT-Kalaignarkarunanidhi Institute of Technology, Coimbatore, Tamil Nadu, India

* mramesh97@gmail.com

Abstract

The advancement of electronics is inextricably linked to breakthroughs in materials science. Due to their diverse mechanical, physical, and chemical properties, piezoelectric composites have sparked a lot of scientific interest in recent years. Because of their great tailorability, piezoelectric composites made of a piezoelectric ceramic and polymers are intriguing materials. Furthermore, mechanical strength, cheap cost, bio-compatibility, and ease of fabrication make organic materials superior to inorganic materials and make them as better alternative materials in modern manufacturing technologies. This chapter provides a detailed review of current research and advancements on piezoelectric composites and their creative applications.

Keyword

Piezoelectric Composites, Ceramics, Polymers, Biocompatibility, Applications

Contents

1.   Introduction

Piezoelectric materials are known for its transformation of electrical energy to mechanical energy and are broadly used as sensors, actuators, and transducers. Since the advent of piezoelectric effect discovered by 1880, the direct piezoelectric effect refers to the materials which generate electrical charges in proportional to the applied induced stresses [1]. The word piezo is a Greek word peizin meaning 'squeeze' or 'pressure'. Hence, piezoelectric is also referring to pressure electricity [2]. Piezoelectric effect is a reversible process which is presented in Fig. 1. The piezoelectric materials are also found to exhibit converse effect, which refers to the production of mechanical strain for the applied electric field. This phenomenon was discovered by Jonas Ferdinand Gabriel Lippmann in 1881. The first practical application of piezoelectric effect comes from Paul Langevin in 1917 who invented submarine detector which brings improvement in submarine ultrasonic echo location detection. In the following years, more applications of piezoelectric effect in the form of accelerometers, ultrasonic transducers, microphones were developed [3].

Advanced Functional Piezoelectric Materials and Applications | Materials Research Forum LLC
Materials Research Foundations 131 (2022) 138-164 | https://doi.org/10.21741/978164490209-5

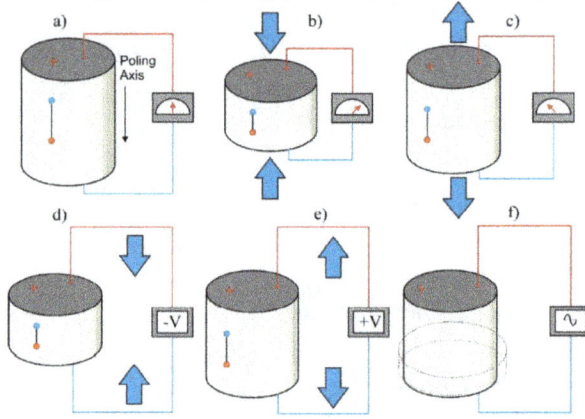

Fig. 1 Scheme of direct piezoelectric effect: (a-c) converse piezoelectric effect and (d-f) extended scale for clarity [3].

2. The mechanism of piezoelectricity and principle of PZT–polymer composites

Piezoelectricity involves production of polarization (P) by piezoelectric crystal and is proportional to the stress (S) developed due to squeezing, bending or twisting. The general equation [4] of the piezoelectricity is

$$P_{pe} = p \times S \tag{1}$$

Where p is the piezoelectric coefficient which varies with different piezoelectric materials. For example, the piezoelectric coefficient for quartz is 3×10^{-12}. In converse, the piezoelectric effect is given by:

$$S_{pe} = d \times E \tag{2}$$

S_{pe} is the mechanical strain; E refers to the magnitude of the electric field and 'd' refers to piezoelectric coefficient or piezoelectric modulus.

$$P_{pe} = d \times T = d \times c \times S = e \times S \tag{3}$$

$$T_{pe} = c \times S_{pe} = c \times d \times E = e \times E \tag{4}$$

Where c represents the elastic constant, s is the coefficient of compliance which relates the stress and the deformation produced by it ($S = s \times T$) and piezoelectric stress constant is referred by 'e'.

$$\text{Piezoelectric voltage constant } (g_{33}) = d_{33}/\varepsilon_\circ K_3 \tag{5}$$

In composite materials g_{33} is the ratio of the maximized piezoelectric coefficient (d_{33}) over the minimized dielectric permittivity (ε_{33}). Electro-mechanical factor, k_3 is an indicator of the energy conversion effectiveness.

The piezoelectric governing constituent equations are given by:

$$\begin{vmatrix} \delta \\ D \end{vmatrix} = \begin{vmatrix} S^E & d^t \\ d & \epsilon^T \end{vmatrix} \begin{vmatrix} \sigma \\ E \end{vmatrix} \tag{6}$$

Where stress (σ) and strain (δ), electric field magnitude (E) the elastic compliance (s), dielectric constant (ϵ), electric displacement (D) and piezoelectric coefficient (d) are the components of the governing equations. The superscripts E represents constants estimate at constant electric field and T denote constants at constant stress. The polar axis of most piezoelectric materials and the direction of the stress in relation to the coordinates are well-defined. For piezoelectric materials like PZT, PMT-PT and PVDF the polar coordinate is defined by moving direction while other non-ferro materials like ZnO or AlN, the polar axis depends in the orientation of the crystals.

PZT–polymer composite structures are generally diphasic composites. Each phase's connection is represented by two numbers, m (active phase) and n (inactive phase). For example, PZT is an active phase in an inactive phase of a polymer. PZT–polymer composite structures are typically diphasic, with the connectivity of each phase denoted by the values m-n, where m denotes the connectivity of an active phase and n denotes the connectivity of an inactive phase. Where the connectivity of each phase is identified by two numbers m-n, m stands for the connectivity of an active phase and n for an inactive phase. Generally, 10 types of diphasic combinations are possible. If the phase is connected in 1 direction, it is represented by 1 and if it's connected in three directions of x, y and z axis, it is called as 3. The applied stress can be directed either parallel (3-direction) or perpendicular (1-direction), to the polar axis. For example, a 1-3 composite has an assembly of connectivity's where PZT rods (1D) and arranged and connected to connected polymer matrix (3D) and can be in vice versa. In 33-mode, voltage is created along the same axis as of the compressive stress or strain 3D axis, while the whereas in 31 mode, generated voltage points in the direction of the applied force while the applied stress is perpendicular to the polar axis with the generated. Generally, epoxy resin is injected into an array of piezoelectric rods organised in a specific rack. In further cooling of the samples; the specimens are cut to the required specification dimensions. They are electrically poled and further polished and electrode at the top and bottom [5].

In this case, with the external electric field E_{33}, the piezoelectric coefficients d_{33}, of the material remains the same d_{33} of the normal piezoelectric materials when the electric field is applied. The elongation or the value of d_{33} of the piezoelectric rods is not affected on the application of the electric field. Under the external loading the embedded PZT rods will withstand the load and as the electric field is increases, stress is increased and is inversely related to the volume fraction. This larger 'g' values are associated with the high voltages.

3.   Piezoelectric materials

Generally, materials that exhibit piezoelectric characteristics are categorized into four categories: (i) natural piezoelectric crystals, (ii) synthetic piezoelectric crystals, (iii) piezoelectric ceramics and (iv) polymer composite piezoelectric films [6]. The natural piezoelectric materials are crystals like quartz (stable crystal and frequency reference crystals) which is widely known and used for several decades in the industry. Other naturally available piezoelectric materials are topaz, sucrose, Rochelle salt, Tourmaline tendon, etc. Also, organic piezoelectric materials like bone, wood, silk and certain ceramics dentine enamel are widely used. Other than natural crystals all other piezoelectric crystals come under synthetic crystals. Lead zirconate titanate (PZT), Barium titanate ($BaTiO_3$), Zinc oxide (ZnO) [7], Lead titanate ($PbTiO_3$), Potassium niobate ($KNbO_3$), Gallium orthophosphate ($GaPO_4$), [8], Langasite ($La_3Ga_5SiO_{14}$), Sodium potassium niobate (NaKNb) [9] are some of the piezoelectric crystals categorized under synthetic crystals. Bismuth ferrite ($BiFeO_3$), Triglycine sulphate [10] and Sodium niobate ($NaNbO_3$) are categorized under lead free piezoelectric consists of minimum quantity of titanium or zirconium tetravalent metal ion within the lattice of lead or barium, and O_2 ions. The types of piezoelectric materials are presented in Fig. 2 [11].

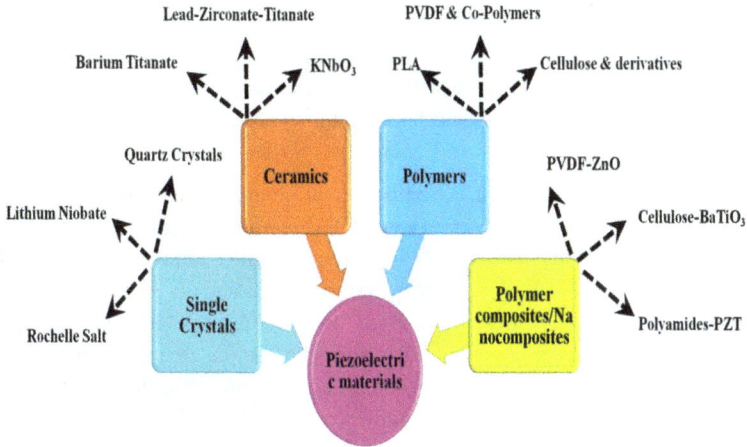

Fig. 2 Different classes of piezoelectric materials [11].

Piezoelectric ceramics comprises of perovskite ceramic crystals. These crystals consist of tetravalent zirconium or titanium metal ions in a lattice of lead or barium ions and O_2 ions. The piezoceramic elements are of different shapes like discs, rods, plates, etc. They convert the applied mechanical force into electrical energy where the value of compressive stress or is directly proportional to field strength. Piezoelectric polymers can be fabricated into variety of shapes and can be mass produced. They exhibit high voltage sensitivity, low acoustic and mechanical impedance which is desirable characteristic for various piezoelectric applications. Polymer composites like PVDF (polyvinylidene-difluoride), Polyvinylide-netrifluoroethylene (PVDF-TrFe), polyvinylidene fluoride tetrafluoroethylene and their copolymers have advantages over other materials because of its tailorable properties for various applications. Thus, PVDF and its copolymers can be extruded or moulded into desired shapes and they are used as coating on the substrate to induce the desired piezoelectric effect.

Piezoelectric composites are the upgraded version of piezo-polymers. They are developed by immersing a piezoelectric material in a passive electricity matrix of piezo-polymer. Piezo-composites are made by using two different ceramics. For example, piezo-composite of $BaTiO_3$ fibers reinforcement in PZT matrix. Piezoelectric thermoset composite (PTC) is the type of piezoelectric ceramics and piezo-polymers materials that are widely known for its direct and inverse piezoelectric effects. Piezoelectric ceramic/polymer composites by virtue of their superior properties like high strength, high coupling factor, high stiffness, modulus and strain coefficients, mechanical stability and flexibility, low acoustic impedance, easy to process at lower temperature and pressure, good to human tissue, broad bandwidth, nominal dielectric constant shows better choice than single phase materials [12, 13].

The composite's thickness-mode electromechanical coupling can exceed the constituent ceramic's kt (0.40–0.50), approaching practically the value of that ceramic's rod-based electro-mechanical coupling, k_{33} (0.70–0.80). When ceramics are replaced with polymers, the acoustic match of traditional piezo-ceramics to tissue or water is substantially improved. Another fascinating product effect called magneto electric effect which is developed when magneto-strictive and piezoelectric ceramics are mixed. In the magneto-electric effect, an electric field in response to an applied magnetic field is produced. PZT is an inorganic compound with fine grain structure. It is formed under very high temperature by combining titanate with lead and zirconium [14]. PZT samples like $PbTiO_3$ and $PbZrO_3$ mixture exhibits higher superior physical strength, chemically inertness, high dielectric constant, high coupling coefficient, high operating temperatures, high density, easy to manufacture cost and less operational cost. PZT composites shows high piezoelectric voltage constant (g_{33}), high values of kp and d which makes it highly acceptable for embedding over polymer composites than the other piezoelectric ceramics.

When compressed, PZT develops a potential difference across two of its faces and gives noise-free response that makes its effective sensor tool inflow or level sensors, ultrasonic ultrasound transducers, sonar devices, etc. Also, when an external electric field is applied PZT changes its shape physically which makes it as a suitable candidate for development of actuators [15, 16]. PZT are widely used as heat sensors for its pyroelectric characteristics where it can develop voltage difference across two of its faces when heated or cooled. PZT is used to make high-value ceramic capacitors and ceramic resonators for reference timing in electronic circuitry. Sakamoto et al. [17] developed a piezoelectric composite (PZT/C/PU), made from ferroelectric ceramic PZT (49 vol. %) and polyurethane polymer (50 vol. %). These composites are doped with graphite filler (1 vol. %) to overcome the voltage drops across the phases due to resistance and to have an efficient polarisation of dipoles. This increase the conductivity because most and the resulting composite exhibited higher pyroelectric coefficient in comparison with other composition. The PZT/C/PU composites sensors are used for detection of structural damages by detecting the dynamic stress waves that are developed due to energy release due to failure. Under simulated acoustic emission (AE) they are able to detect both bending and extensional modes of AE for up to 8 m from the source [18]. Dielectric and piezoelectric constants of PZT-based polymer composites is presented in Table 1 [19].

Table 1 Dielectric and piezoelectric constants of PZT-based polymer composites [19].

Connectivity	Material	Density ρ $(10^3$ kg $m^{-3})$	Elastic constant c_{33} (GPa)	Dielectric constant ϵ_3 (GPa)	Piezoelectric constants		
					d_{33} $(10^{-12}CN^{-1})$	g_{33} $(10^{-3}\,m\,VN^{-1})$	g_h $(10^{-3}\,m\,VN^{-1})$
	PZT(501 A) Single phase	7.9	81	2000	400	20	3
3-1	PZT–Epoxy	3.0	19	400	300	75	40
3-3	PZT–Silicone rubber (Replica type)	3.3	3	40	110	280	80
	PZT–Silicone rubber (Ladder type)	4.5	19	400	250	60	
3-0	PZT–PVDF	5.5	2.6	120	90	85	
	PZT–Rubber	6.2	0.08	73	52	140	30
	PZT–Chloroprene rubber		40			90	
	Extended PVDF	1.8	3	13	20	160	80

On the other hand, viscoelastic properties of the polymer matrix in the PZT composites under applied DC field are proven to display high hysteresis and creep. Also, the dissipation of heat during the applied field is the major issues of piezo-ceramic composites which have detrimental effects on the piezoelectricity property of the materials. Lead magnesium niobate-lead titanate (PMN-PT) materials possess excellent piezoelectric properties because of which it is widely used as actuators. They also exhibit superior characteristics like least hysteresis, low creep and lack of high voltage over piezoelectric ceramics and piezo-films and it is preferred material for the actuators in the medical area and in hearing aid [19, 20].

4   Applications of piezoelectric composite materials

The mutual energy conversion features of piezoelectricity facilitate piezoelectric materials to work as sensors, capacitors, actuators or transducers [21]. If they are utilised for sensor applications, the mechanical energy they receive is converted into an electrical energy that may be analysed using electrical equipment. As actuators, the behaviour of the substrate in response to an applied voltage is controlled by the strain development. For the transducer applications, the electrical input signal is converted into a mechanical energy. Piezoelectric composites material has the potential market size of $267.40 million by 2025.

4.1   Energy harvesting applications

Energy harvesting/storing technologies are also called as scavenging technologies that utilises the ambient wasted energy like wind, water, radio frequency (RF), acoustic waves, temperature gradients, vibrations, heat emissions, luminosity, radiation and kinetic energy into electrical energy. Generally, energy harvesting system comprises three parts like (a) source of energy where the electrical power is scavenged, (b) harvesting mechanism that converts available ambient energy to electrical signal and vice versa, and (c) the sink that is used to store the produced energy [22]. Piezoelectric energy harvesting (PEH) have advantages over other harvesting technologies due to their, high power density, simple architecture, flexible, stretchable ability to be meshed to form hybrid composite materials to suit for a broad range of voltages and higher scalability into micro and nanoscale devices which makes it as suitable candidate to give continuous supply for low power appliances like actuators, sensors, medical implants, etc. [23, 24].

Piezoelectric polymer composites, cellulose $BaTiO_3$, polyimides-PZT are mechanically flexible and are able to withstand high strain [25]. Similarly piezoelectric ceramics like $BaTiO_3$, PZT, potassium niobate ($KNbO_3$) are used for energy harvesting applications. With the advancement of technologies in integrated circuits, wireless networks, and mobile communication the demand for high power consumption is greatly reduced that provides further opportunities to explore in energy harvesting technologies [26]. PEH is one such area which makes use of the kinetic energy which is available in the form of vibrations, displacements that are random due to structural vibrations from vehicles, industrial machines, buildings and forces developed due to human activities like respiration, heartbeat, walking or jogging, etc. [27-31]. Thus, every mechanical movement can be utilised as a potential energy source of kinetic energy to provide autonomy in power for devices [32]. For example, an amount of energy can be harvested by movements of upper limbs while breathing and walking generates around 100 mW and 1W respectively [33]. For mechanical energy, the harvested power density (P_{res}) based on the motion frequency and magnitude.

$$P_{res} = 4\pi^3 m f_{res}^3 y Zmax \qquad (7)$$

Where m represents inertial mass, Z_{max} indicates the maximum displacement, f_{res} = resonance frequency; and y = amplitude of vibration. The frequency level of various energy resources are given in Fig. 3

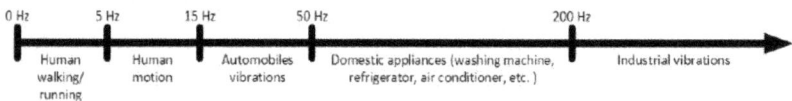

Fig. 3 Frequency of different mechanical energy sources [34].

Advanced Functional Piezoelectric Materials and Applications   Materials Research Forum LLC
Materials Research Foundations 131 (2022) 138-164   https://doi.org/10.21741/978164490209-5

For conversion of mechanical energy into electrical signal, of all the choices available like Electromagnetic, electrostatic, and piezoelectric, piezoelectricity is more attractive for applications that require higher values of energy density, voltage and capacitance with little mechanical damping [35, 36]. PEH systems comprises of three phases (a) mechanical–mechanical energy conversion phase that involves high stresses and impedance, (b) mechanical to electrical energy conversion phase which includes piezoelectric coupling factor and piezoelectric coefficients, and (c) electrical-electrical energy conversion that is associated with electrical impedance matching (Fig. 4) [37].

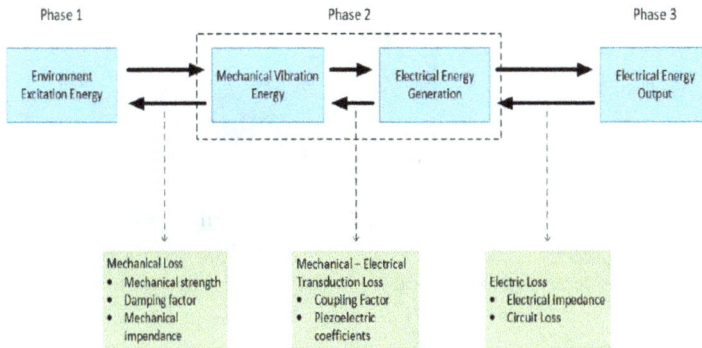

Fig. 4 Phases of piezoelectric energy harvesting technologies [37].

The design of piezoelectric materials based on derivation of kinetic energy by kinetic energy harvesting through strain, vibration and fluid. The piezoelectric strain energy harvester utilises mechanical stimulus and movement of the mechanical sources like compression of the tyres on the roads, muscle extension, pulse, heartbeat, etc. Strain energy harvester does not utilise the effects of vibration or inertia forces. In case of vibration energy harvesting (VEH), the vibration is converted into electrical signal. Recent methodologies like electromagnetic (EM) VEH uses vibration sources of harmonic and random excitations for producing higher voltage and power.The piezoelectric energy harvester models are based on linear spring mass damper system which involves the generation of simple harmonic motion when the system is excited by the application of forces which results in transfer of mechanical power to electrical power. Maximum power output occurs at resonance condition which depends on mass, damping ratio, mechanical damping factor, resonance frequency, the input acceleration, etc.

4.2   Medical applications of piezoelectric materials

The studies on the effect of organic-based piezoelectric materials in the human body is mostly attributed to Eiichi Fukada who identified the presence of biological piezoelectric effects on bone, muscles, tendons, intestines, aorta, etc. [38-41]. The organic piezoelectric effect on human body is owing to the non-symmetric organic molecules and in particular a protein seems to be the major driving force of the piezoelectric effect. For proteins, amino acids form the basic building blocks which makes up of further molecules like keratin, elastin and collagen. Collegen is found to be the major source of piezoelectricity in bones. The presence of dipoles in amino acids that are derived from polar side group attribute for their distinct piezoelectric properties [42].

DNA, M13 bacteriophage and polymeric L-lactic acid are also found to exhibit piezoelectric properties. Piezoelectric properties in DNA and M13 bacteriophage are attributed to the presence of phosphate groups and extruding protein groups respectively. In case of lactic acid, carbon–oxygen double bond forms the cause for the piezoelectric effect [43]. In case of skin, each layers like epidermis, horny layer, and dermis have their own piezoelectric property. The dermis is the middle layer comprises of blood vessels, nerves, hair follicles, various glands and a protein layer collagen. The stratum corneum is the outmost layer of epidermis is also called as horny layer because of its hardened or cornified cells comprises of different cells like keratinocytes and corneocytes, etc. Keratinocytes produces protein element keratin filaments [44].

The presence of piezoelectric molecules aids in understanding the mechanics of the human body and is used to find the biomedical solutions in the microscopic and macroscopic levels which creates interest in the development of biosensors. These biosensors are widely used in the biomedical fields like health monitoring and injury analysis by detecting the diseases that affects the distribution and amount of proteins in the particular tissues. For example, electromechanical coupling factor of collagen is used to detect the presence of cancer cells in the breast [45]. Similarly, induced piezoelectric field around the injured bones accelerate the growth of neurons and bone repair [46]. Taking advantage of induced piezoelectric field, development of synthetic and prosthetic bones is widely used for its potential of increasing the speed of osteo conduction and subsequently bone repair.

4.2.1   Piezoelectric medical devices

The purpose of developing piezoelectric medical devices is to mimic the biology based piezoelectric phenomena in the human body. The selection of material as medical device is determined by the cost and strength of piezoelectric effect. PZT, ceramics, like quartz, barium titanate that are natively biocompatible and potassium sodium niobate, lithium niobate, aluminium nitride, lithium tantalite are more biocompatible by processing and they do not contain lead. But the ceramic materials lack flexibility which the medical field demands for implantation. Biocompatible polymers like PVDF is used for monitoring heartbeat and respiratory issues and they operate with frequency of 0.1–2 Hz. PVDF is also used as monitoring system the swallowing pattern of food [47]. PLLA and its copolymers are used in bio-mechanical energy harvesting systems, sensors, capacitors, and scaffolds [48]. They are used for monitoring eye pressure, brain pressure and lung pressure.

4.2.2 Piezoelectric sensors

Piezoelectric materials with its ability of converting mechanical input to electrical signal is widely used to monitor the human body indicators such as dynamic pressure changers, heartbeat, breathing, etc. These piezoelectric sensors can be made for use in both lower pressure system (1-10 kPa) and high pressure system (10 KPa – 100 Pa) that corresponds to blood pressure monitoring other bodily movements [49]. With the advent of internet of things (IoT) many wearable device and implanted devices with sensors are developed and is even remotely monitored by the health officials for collecting data regarding health and early warning signs [50, 51].

Smart medical sensors developed to detect tissue tension are prepared by using polyaniline and polyvinylidene fluoride (PANI/PVDF). They detect strain up to 11% with 185% linear responses [52]. Similarly, PVDF-trifluoroethylene biosensors are capable of high sensitivity and are used for monitoring breath and pulse [53]. Biosensors made from PVDF and reduced graphene films are able to monitor temperature variations [54]. Using integrated circuit technology developed polymer based piezoelectric biosensor diaphragm that detects changes in nucleic acid for disease diagnosis [55]. Similarly, biosensors are designed to detect nitric oxide release, able to capture pathogens and aid in faster wound recovery.

4.2.3 Piezoelectric prosthetic skin

Mechanoreceptors for human beings provide details about touch, temperature range or humidity, pressure, and can sense vibration with frequency range from 3 to 400 Hz. In short, these mechanoreceptors sense mechanical forces. Polymers being flexible in nature can be moulded to form artificial skin that could emulate fingerprints with enhanced sensitivity [56]. However, the conventional tactile skin sensors based on metals and semiconductors lacks stretchability and portability. With the advancement of technologies, new materials like polydimethylsiloxane, carbon nanotubes, and grapheme are used for preparing skin sensors which mimics the origin skin in terms of stretchability, compressibility and spatial resolution of touch [57]. This gives robot more improved sense of touch and improve its performance. The prosthetic skin comprises of nano-wire piezotronic transistors functional arrays. Each transistor contains about 1,500 nanowires which are around 500 nm in diameter. The nano-wires are composed of zinc oxide that provides the transistor with both piezoelectric and semiconducting properties. The transistors are capable of sensing the polarity changes due to the pressure applied.

4.2.4 Cochlear implants

The basilar membrane system is an important in the inner ear. It is positioned inside the chochlea and forms the base of the organ of corti where sensory receptors of hearing are placed. This delicate structure is essential structure to our ability to hear. Basilar membrane supports the organ of corti while the sound vibration passes the ear hair (Fig. 5) [58]. The breakdown of ear cells which results in hearing loss, which is usually treated with cochlear implants (CIs). These implants are generally considered as the treatment of severe-to-profound sensor in hearing loss. Although new technologies for cochlar implants are available to cure deafness, even the new developed fully implantable CIs (FICIs) are not compatible with water, high ambient noise and consume more power which demands frequent recharging [59]. Also FICIs demands ultra-low-power audio processing and energy-efficient neural stimulation.

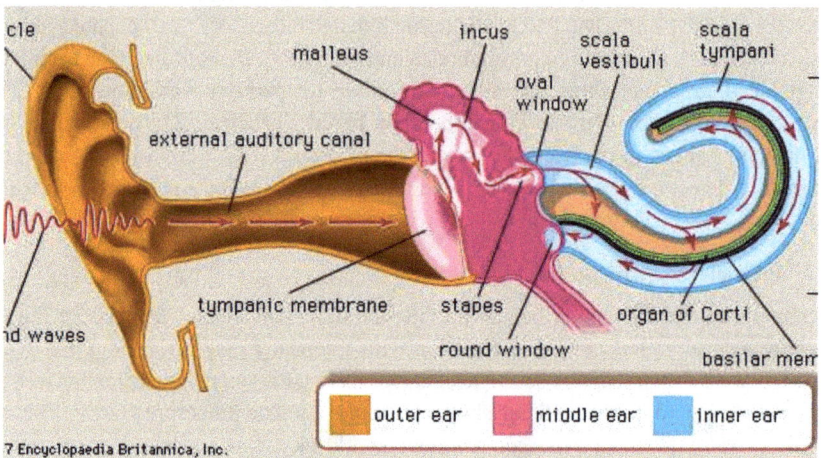

Fig. 5 Anatomy of basilar membrane [58].

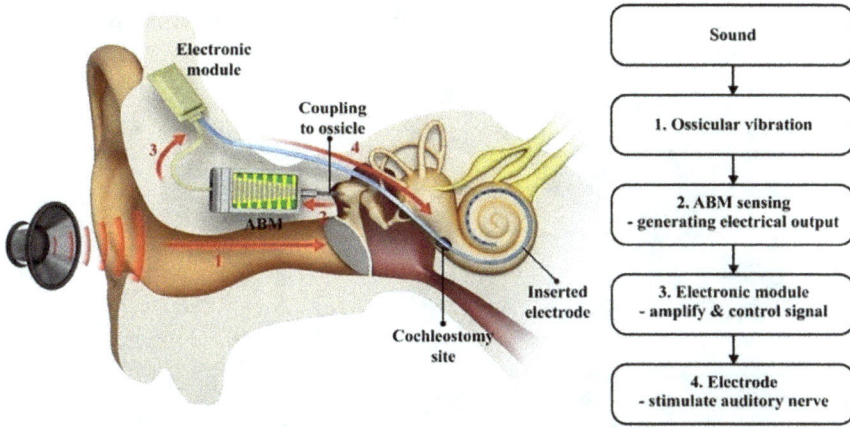

Fig. 6 Schematic representation of artificial basilar membrane [60].

To overcome these disadvantages, a bioelectronics middle ear microphone is developed by Maniglia et al. [61] that eliminates the requirement for a subcutaneous microphone. Furthermore, a piezoelectric device that is auto powered can be utilised to reduce the power consumption. Thus, piezoelectric materials are used in creating artificial basilar membrane (ABM) (Fig. 6). Similarly, many researchers like Jun et al. [62] developed a Si_3N_4-based ABM membrane with ZnO piezo-nanopillar array. By varying the parameters of rigidity and thickness of the ABMs, these AMBs perform its function of frequency selector for the cochlea. Frequency range of hearing for the human beings is from 20 Hz–20 kHz. Piezoelectric materials like PZT, AlN films and polymers like PVDF are fabricated to cantilever arrays with varying length depend on various frequencies. The ABMs are designed with the frequency between 100 Hz to 10 kHz to accommodate the human vocal sound [63].

4.2.5 Piezoelectric surgery

Piezoelectric bone surgery is a process of cutting bone tissue by piezoelectric vibrations. The process was invented in 1988 by modifying the traditional instruments of ultrasound technology and currently it is widely used for periodontal bone surgery, maxillofacial, cranial, oral and ENT surgery. This method utilizes metallic tip vibration of high frequency (25–50 kHz) and is widely used for use in procedures. These piezoelectric metallic tips (Fig. 7 A,B) at high frequency are able to cut with precise incisions over hard tissue and mineralised structures without affecting the soft tissue surrounding them as separation between interfaces can be easily accomplished.

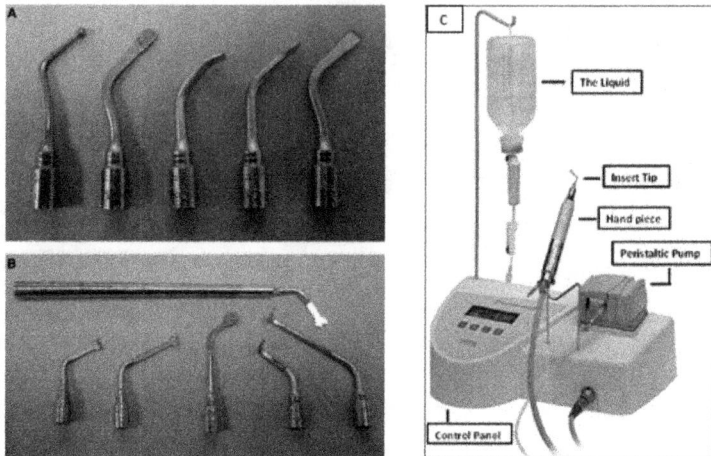

Fig. 7 (A) Diamond-coated and scalpel piezoelectric tips for piezotome, (B) serrated tips of blade or shank [64], and (C) piezosurgery unit [65].

The advantages of piezoelectric surgery over traditional methods are selective precision thin line cutting at any angles and direction, fewer traumas to soft tissue, reduced neurological trauma, reduced blood loss and thus better visibility in surgical area. The disadvantages involve cost of the tools, slow process and frequent tool tip breakages. The piezoelectric devices unlike implanted devices don't have contact with the human cells. They are made of PZT materials coated with titanium nitrate coating or diamond coating to improve the surface hardness for better life. The inserted tips are classified as sharp, blunt and smooth inserted tips. A typical piezosurgery unit (Fig. 7C) consists of control panel, dynamometric wrench that is used to tighten the tool tips. A peristotic pump serves to irrigate the tool tip at a predetermined rate of flow at a range of 0-60 ml at -4 °C to remove the detritus present during the surgery.

4.2.6 Ultrasonic dental scaling

A piezoelectric ultrasonic dental scaling is a method of removing biofilms from tooth surface and hardened calculus (tartar) from the gum pockets. It's a dental procedure to clean the tooth enamel surfaces and smoothen the roots to prevent further build-ups. Generally, ultrasonic dental scalar cleaning operates at a frequency of 25-50 Hz canals and the piezoelectric scalar work sat a range of 28,000 to 36,000 cycles per second. The piezoelectric tip vibrates at high speed back and forth against the tooth while the irrigation water is sprayed through built-in-sprayer which forms micro and nanosized bubbles that provide antibacterial rinse and clean the debris of work [66].

4.2.7 Microdosing

Microdosing is the method of dispensing medicines to study the behaviour of drugs in human body. The drug that is administered is on the order of microdose level (1 to 100 micrograms) that will have only cellular response and not the whole body. This method is also called as 'Phase O' that is followed by pharmaceutical companies to test their drugs during their initial stage. Also, it reduces the potential wastage of resources and reduces the test on living animals. In this method, vibrations break the drug into powder and can be aerosolized and can be inhaled [67]. In case of solid, micro dispensing is achieved by single oscillation of the stacked actuator. A tube is fitted to the actuator and with an input electrical signal; the stacked actuators forces the micro-dose of the solid with a single push to the target receiver [68]. Piezoelectric stacked actuators made from PZT or PVDF polymers can be used for continuous fluid administration of eye drops or insulin dispensation using diaphragm pumps. These pumps comprise of four chambers and they operate in tandem with the connection of electrodes. The opening and closing of the particular chamber if determined is certain pre-set order sequence. The parameters to be considered for the precise dispensation are the degree of diaphragm membrane fluctuation which is controlled by the voltage and the properties and phase of the material which governs the direction of flow [69].

4.2.8 Energy harvesting

Piezoelectric energy harvesting involves both external and internal power sensors and in case of internal sensors like that are implanted in the body requires continuous source of energy for working. It is not possible to have a battery powered biocompatible sensor as the power cannot be replenished or recharged. It limits the number of the sensors that can be implanted within the body for a longer period of time. It is possible to have continuous energy source for the sensor using piezoelectric devices. The implant using piezoelectric source should be as small as possible without affecting the nearby organs. Also, these energy harvesters would be biocompatible, non-lethal and sealable. Further, existing mode of energy harvester like piezoelectric vibration cantilever beams cannot be used for biological energy harvesting due to high resonant frequency per peak power generation but human motions generally involves low frequency like walking (1 Hz) and heartbeat rate

(39 Hz). It is the main limitation because only at the resonance frequency of the film or diaphragm, maximum energy is generated and also these energy harvesters are of low efficiency in power output [70]. Piezoelectric materials like ceramic nano-ribbons made of PMN-PT, PZT, or $BaTiO_3$ attached with flexible films like polyethylene terephthalate, polyethylene naphthalate or polymide are used as energy harvesters in implants. The consistent movement of body parts like piezoelectric films wrapped around expanding artery, lungs or heart are used as the source of energy for these energy harvesters. Pacemakers for heart can make use of these devices as power source. PVDF polymers can also be used for making these films [71].

4.2.9 Catheter applications

Catheters are very important medical instruments made of thin tubes inserted inside the body to treat diseases, for drug delivery, to collect information during surgery, prosthesis stents, drainage of air from lungs, collection of body fluids and to drain urine from urinary bladder. These tubes are biocompatible, made of medical grade materials and are available for various levels of stiffness based on the applications. It can be inserted into a body cavity, skin. Urinary duct, blood vessel, brain, skin ducts, etc. PVDF materials are the most suitable materials as catheter materials because of its excellent piezoelectric and adhesive properties. PVDF nanofibers (shell) added with polyethylene dioxythiophene (PEDOT) (core) is used as an electrode in catheter devices [72].

4.2.10 Neural stimulators

Li et al. [73] attempted to activate spinal cord neural circuits using an ultrasound-driven $BaTiO_3$ piezoelectric stimulator with silicone coating for locomotion restoration of a rat affected with spinal cord injury (SCI). These simulator systems can be either wired or wireless and both the systems require and piezoelectric sensor implanted to the nervous systems.

4.2.11 Healthcare monitoring

Piezoelectric sensors implants are used as health monitoring devices. They gave real time data of the disease by monitoring interstitial fluids, tears, saliva, sweat, etc. [74]. Similarly wearable sensors like electronic-skin made from composite material made from PVDF and reduced graphene are used for detecting various stimuli like pressure and temperature [75]. Similarly piezoelectric composites of PVDF and gold (Au) biosensors are used for monitoring variation in pressure. PVDF/carbon nanotubes (CNT) are highly efficient as actuators and sensor devices and some are used for bone tissue engineering applications [76].

Advanced Functional Piezoelectric Materials and Applications   Materials Research Forum LLC
Materials Research Foundations 131 (2022) 138-164   https://doi.org/10.21741/978164490209-5

5.   Structural health monitoring and repair

The evaluation of overall efficiency of building structures in terms of life, durability and structural failure is a major concern as it increases vibration level and reduces load carrying ability. Rehabilitation and reconstruction can be facilitated by early detection of irregularities in a structure, such as cracks or damages. Thus, structural health monitoring and structural repair are the interrelated research topics which involves scientific techniques like non-destructive evaluation (NDE) techniques like impulse-echo technique, ultrasonic inspection technique, etc. [77-79]. These NDE techniques generally use wave propagation signals for structural health monitoring with respect to parameters like natural frequency, modal shapes and damping. Also recent imaging techniques like thermo-graphic inspection, radiographic inspection techniques use infra-red and laser respectively for structural analysis [80].

Piezoelectric materials due to excellent sensing properties are used in actuators and sensors for structural assessment of beam, plates and pipes. Furthermore, they are more reliable with consistent output in determining the damage location. The projected waves can be modelled as pulse when it is impacted on the solid. Piezoelectric patches made from carbon fiber reinforced polymers that is embedded with lead zirocondate titanate composites are used for structural monitoring electrical impedance spectroscopy. Moon et al. [81] developed a piezoelectric device of piezo-polymer/metal thin plate composite structure with various combinations of ply angles and electrode patterns. The piezo-polymer composite device under electric field is used to detect deformation of bending and torsion. The study of wave propagation inside the piezoelectric materials involves the study of time delay effect which occurs when applied electric signal is converted to elastic wave. It was found that different metallic layers have distinct wave projection characteristics. Many researches are based on studying the lector-mechanical effect of piezoelectric composite structures of regular shape, beams, cylinders, pipes and shells under different boundary conditions like fixed-ended, simply-supported and cantilever [82, 83].

The dynamics response data from the piezoelectric sensors results in these wave signals which possess information about the timings of the boundary reflected waves, direct, damage and reflected waves that can be further analysed by signal processing technique to further assess the damage locations (Fig. 8). Piezoelectric sensors and actuators are used to study the effect of delamination in beams by the studying the variation in velocities of the projected waves in intact and delaminated beams. In case of plates and pipes, the position, orientation and the extent of the crack are studied by attenuation in the strength of the direct wave incidence at the piezoelectric sensor [85].

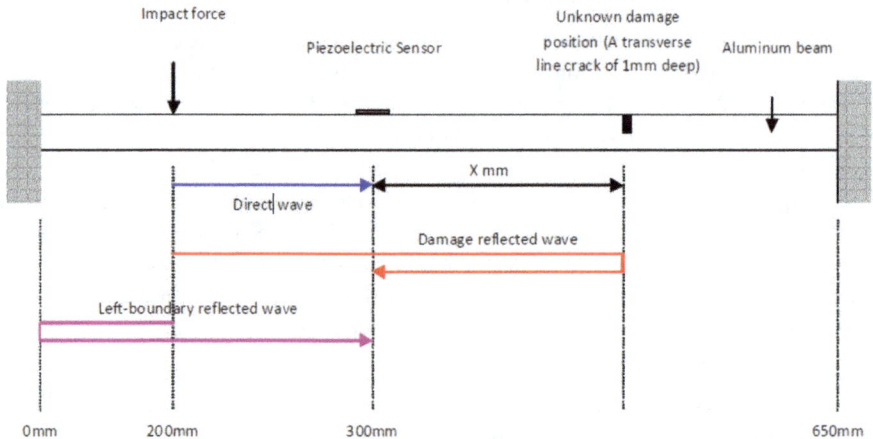

Fig. 8 Study of crack by wave propagation in beams [84].

Conclusion

This chapter focuses on composites made of piezoelectric materials. Underwater monitoring and medical diagnostic transducer applications benefit greatly from piezoelectric composite materials. High connecting factors, low acoustic impedance, good fixation with human tissue, mechanical modulus, and so on are some of the benefits of these composites. Long-fiber based composites with the greatest piezoelectric capacity and interpenetrating composites with the highest charge coefficients are discovered as the most suitable for numerous applications by building piezoelectric composite materials. Furthermore, in the case of reinforced composites, grain size adjustments of the component phases can greatly improve the piezoelectric properties of a material, while retaining appropriate acoustic impedance. All of these themes together give comprehensive coverage of the most significant aspects of piezo-electric composite materials, guiding readers to a better understanding of the phenomena and resulting in the concept and development of superior components and devices.

Advanced Functional Piezoelectric Materials and Applications
Materials Research Foundations 131 (2022) 138-164

Materials Research Forum LLC
https://doi.org/10.21741/978164490209-5

References

[1] A. Manbachi, R.S.C. Cobbold, Development and application of piezoelectric materials for ultrasound generation and detection, Ultrasound. 19 (2011) 187-96. https://doi.org/10.1258/ult.2011.011027

[2] Harper, Douglas. "Piezoelectric". Online Etymology Dictionary.

[3] P. Dineva, D. Gross, R. Müller, T. Rangelov, Dynamic fracture of piezoelectric materials. Springer, 2014, 212, XIV, 249. https://doi.org/10.1007/978-3-319-03961-9

[4] ANSI/IEEE, IEEE standard on piezoelectricity. IEEE Standard (1987) 176-187.

[5] K.A. Klicker, J.V. Biggers, R.E. Newnham, J. Am. Ceram. Soc. 64 (1981) 5-9. https://doi.org/10.1111/j.1151-2916.1981.tb09549.x

[6] H. Liu, J. Zhong, C. Lee, S.W. Lee, L. Lin, A comprehensive review on piezoelectric energy harvesting technology: Materials, mechanisms, and applications. Appl. Phy. Rev. 5 (2018) 041306. https://doi.org/10.1063/1.5074184

[7] M. Zgonik, P. Bernasconi, M. Duelli, R. Schlesser, P. Günter, M.H. Garrett, D. Rytz, Y. Zhu, X. Wu, Dielectric, elastic, piezoelectric, electro-optic, and elasto-optic tensors of BaTiO3 crystals. Phys. Rev. B 50 (1994) 5941-5949. https://doi.org/10.1103/PhysRevB.50.5941

[8] S. Zhang, L. Laurent, S. Rhee, C.A. Randall, T.R. Shrout, Shear-mode piezoelectric properties of Pb(Yb1/2Nb1/2)O3-PbTiO3 single crystals. Appl. Phy. Lett. 81 (2002) 892-894. https://doi.org/10.1063/1.1497435

[9] T.Y. Ke, H.A. Chen, H.S. Sheu, J.W. Yeh, H.N. Lin, C.Y. Lee, H.T. Chiu, Sodium niobate nanowire and its piezoelectricity. J. Phys. Chem. C 112 (2008) 8827-8831. https://doi.org/10.1021/jp711598j

[10] T. Ikeda, Y. Tanaka, H. Toyoda, Piezoelectric properties of triglycine sulphate. J. Phys. Soc. Japan 16 (1961) 2593-2594. https://doi.org/10.1143/JPSJ.16.2593

[11] S. Mishra, L. Unnikrishnan, S.K. Nayak, S. Mohanty, Advances in piezoelectric polymer composites for energy harvesting applications: A systematic review. Macromol. Mater. Eng. 304 (2018) 1800463. https://doi.org/10.1002/mame.201800463

[12] M. Ramesh, A. Ravanan. One-dimensional nanomaterials for supercapacitors. Morphology Design Paradigms for Supercapacitors, 2020, pp. 33-58. https://doi.org/10.1201/9780429263347-2

[13] M. Ramesh, M. Muthukrishnan. Bio-Based Magnetic metal organic framework nanocomposites. Metal-Organic Framework Nanocomposites: From Design to Applications, 2020, pp. 167-190. https://doi.org/10.1201/9780429346262-6

[14] A. Yousefi-Koma, Piezoelectric ceramics as intelligent materials, Fundamentals of Smart Materials, The Royal Society of Chemistry, (2018) p. 233.

[15] S. Das, A.K. Biswal, A. Roy, Fabrication of flexible piezoelectric PMN-PT based composite films for energy harvesting. IOP Conf Ser: Mater Sci Eng 178 (2017) 012020. https://doi.org/10.1088/1757-899X/178/1/012020

[16] M. Ramesh, M. Muthukrishnan, A. Khan. Metal-organic frameworks and permeable natural polymers for reasonable carbon dioxide fixation. Metal-Organic Frameworks for Chemical Reactions, 2021, pp. 417-440. https://doi.org/10.1016/B978-0-12-822099-3.00017-4

[17] W.K. Sakamoto, P. Marin-Franch, D. Tunnicliffe, D.K. Das-Gupta, Lead zirconatetitanate/polyurethane (PZT/PU) composite for acoustic emission sensors. Annual Report Conference on Electrical Insulation and Dielectric Phenomena (Cat. No.01CH37225), 2001, doi:10.1109/ceidp.2001.963479.

[18] M.R. Gorman, W.H. Prosser, AE source orientation by plate wave analysis. J. Acoustic Emission, 9 (1990) 283-288.

[19] Uchino K., Piezoelectro Composites. In: Saleem Hashmi (ed), Reference Module in Materials Science andMaterials Engineering. Oxford: Elsevier; 2016, pp. 1-12.

[20] M. Ramesh, N. Kuppuswamy, S. Praveen, Metal-organic framework for batteries and supercapacitors. Metal-Organic Frameworks for Chemical Reactions, 2021, pp.19-36. https://doi.org/10.1016/B978-0-12-822099-3.00002-2

[21] J. Jin, Q. Wang, S.T. Quek, Lamb wave propagation in a metallic semi-infinite medium covered with piezoelectric layer. Int. J. Solids Struct. 39 (2002) 2547-2556. https://doi.org/10.1016/S0020-7683(02)00091-4

[22] S. Sudevalayam, P. Kulkarni, Energy harvesting sensor nodes: survey and implications. IEEE Commun. Surv. Tutor. 13 (2011) 443-461. https://doi.org/10.1109/SURV.2011.060710.00094

[23] H. Li, C. Tian, Z. D. Deng, Energy harvesting from low frequency applications using piezoelectric materials. Appl. Phys. Rev. 1 (2014) 041301. https://doi.org/10.1063/1.4900845

[24] A. Toprak, O. Tigli, Piezoelectric energy harvesting: State-of-the-art and challenges. Appl. Phys. Rev. 1 (2014) 031104. https://doi.org/10.1063/1.4896166

[25] S. Mishra, L. Unnikrishnan, S.K. Nayak, S. Mohanty, Advances in piezoelectric polymer composites for energy harvesting applications: a systematic review. Macromol. Mater. Eng. 304 (2018) 1800463. https://doi.org/10.1002/mame.201800463

[26] T.W. Brown, T. Bischof-Niemz, K. Blok, C. Breyer, H. Lund, B.V. Mathiesen, Response to 'Burden of proof: A comprehensive review of the feasibility of 100% renewable-electricity systems. Renew. Sustain. Energy Rev. 92 (2018) 834-847. https://doi.org/10.1016/j.rser.2018.04.113

[27] J. He, T. Wen, S. Qian, Z. Zhang, Z. Tian, J. Zhu, J. Mu, X. Hou, W. Geng, J. Cho, Triboelectric-piezoelectric-electromagnetic hybrid nanogenerator for high-efficient vibration energy harvesting and self-powered wireless monitoring system. Nano Energy 43 (2018) 326-339. https://doi.org/10.1016/j.nanoen.2017.11.039

[28] D. P. Arnold, Review of microscale magnetic power generation. IEEE Trans. Magn. 43 (2007) 3940-3951. https://doi.org/10.1109/TMAG.2007.906150

[29] Y. Suzuki, Recent progress in MEMS electret generator for energy harvesting. IEE J Trans. Electr. Electron. Eng. 6 (2011) 101-111. https://doi.org/10.1002/tee.20631

[30] H. Liu, K. H. Koh, C. Lee, Ultra-wide frequency broadening mechanism for micro-scale electromagnetic energy harvester. Appl. Phys. Lett. 104 (2014) 053901. https://doi.org/10.1063/1.4863565

[31] H. Liu, Z. Ji, T. Chen, L. Sun, S. C. Menon, C. Lee, An intermittent self-powered energy harvesting system from low-frequency hand shaking. IEEE Sens. J. 15 (2015) 4782-4790. https://doi.org/10.1109/JSEN.2015.2411313

[32] A. Rjafallah, A. Hajjaji, D. Guyomar, K. Kandoussi, F. Belhora, Y. Boughaleb, Modeling of polyurethane/lead zirconatetitanate composites for vibration energy harvesting. J. Compos. Mater. 53 (2018) 613-623. https://doi.org/10.1177/0021998318788604

[33] P. Mitcheson, T. Green, E. Yeatman, A. Holmes, Architectures for vibration-driven micropower generators. J. Microelectromech. Syst. 13 (2004) 429-440. https://doi.org/10.1109/JMEMS.2004.830151

[34] B. Maamer, A. Boughamoura, A.M.F. El-Bab, L.A. Francis, F. Tounsi, A review on design improvements and techniques for mechanical energy harvesting using piezoelectric and electromagnetic schemes. Energy Convers. Manag. 199 (2019) 111973. https://doi.org/10.1016/j.enconman.2019.111973

[35] S. Khalid, I. Raouf, A. Khan, N. Kim, H.S. Kim, A Review of human-powered energy harvesting for smart electronics: recent progress and challenges. Int. J. Precis. Eng. Manuf. Green Technol. 6 (2019) 821-851. https://doi.org/10.1007/s40684-019-00144-y

[36] S. Roundy, P.K. Wright, J.M. Rabaey, Energy scavenging for wireless sensor networks; Springer: Boston, MA, USA, 2004. https://doi.org/10.1007/978-1-4615-0485-6

[37] K. Uchino, T. Ishii, Energy flow analysis in piezoelectric energy harvesting systems. Ferroelectrics 400 (2010) 305-320. https://doi.org/10.1080/00150193.2010.505852

[38] E. Fukada, Piezoelectric properties of organic polymers. Annals of the New York Acad. Sci. 238 (1974) 7-25. https://doi.org/10.1111/j.1749-6632.1974.tb26776.x

[39] E. Fukada, Piezoelectricity of natural biomaterials. Ferroelect. 60 (1984) 285-296. https://doi.org/10.1080/00150198408017529

[40] E. Fukada, K. Hara, Piezoelectric effect in blood vessel walls. J. Phys. Soc. Japan. 26 (1969) 777-780. https://doi.org/10.1143/JPSJ.26.777

[41] E. Fukada, H. Ueda, Piezoelectric effect in muscle. Japan. J. Appl. Phy. 9 (1970) 844. https://doi.org/10.1143/JJAP.9.844

[42] V.V. Lemanov, S.N. Popov, G.A. Pankova, Piezoelectric properties of crystals of some protein amino acids and their related compounds. Phy. Sol. Sta. 44 (2002) 1929-1935. https://doi.org/10.1134/1.1514783

[43] Y. Zhu, S. Zhang, J. Wen, Influence of orientation on the piezoelectric properties of deoxyribonucleic acid. Ferroelect. 101 (1990) 129-139. https://doi.org/10.1080/00150199008016509

[44] D. De Rossi, C. Domenici, P. Pastacaldi. Piezoelectric properties of dry human skin. IEEE Trans. Elect. Insul. El-21 (1986) 511-517. https://doi.org/10.1109/TEI.1986.349102

[45] K. Park, W. Chen, M.A. Chekmareva, D.J. Foran, J.P. Desai. Electromechanical coupling factor of breast tissue as a biomarker for breast cancer. IEEE Trans. Biomed. Eng. 65 (2018) 96-103. https://doi.org/10.1109/TBME.2017.2695103

[46] N. More, G. Kapusetti, Piezoelectric material - A promising approach for bone and cartilagere generation. Medi. Hypoth. 108 (2017) 10-16. https://doi.org/10.1016/j.mehy.2017.07.021

[47] E.S. Sazonov, J.M. Fontana. A sensor system for automatic detection of food intake through non-invasive monitoring of chewing. IEEE Sens. J. 12 (2012) 1340-1348. https://doi.org/10.1109/JSEN.2011.2172411

[48] A.H. Rajabi, M. Jaffe, T.L. Arinzeh, Piezoelectric materials for tissue regeneration: A review. Acta Biomater. 24 (2015) 12-23. https://doi.org/10.1016/j.actbio.2015.07.010

[49] Y. Zhang, F. Zhang, D. Zhu, Advances of flexible pressure sensors toward artificial intelligence and health care applications. Mater. Horizon. 2 (2015) 133-254. https://doi.org/10.1039/C4MH00147H

[50] S. Seneviratne, Y. Hu, T. Nguyen, G. Lan, S. Khalifa, K. Thilakarathna, et al. A survey of wearable devices and challenges. IEEE Communi. Sur. Tutor. 19 (2017) 2573-2620. https://doi.org/10.1109/COMST.2017.2731979

[51] H.A. Sonar, J. Paik, Soft pneumatic actuator skin with piezoelectric sensors for vibrotactile feedback. Front. Robot. AI 2 (2016) 1-11. https://doi.org/10.3389/frobt.2015.00038

[52] K. Maity, S. Garain, K. Henkel, D. Schmeißer, D. Mandal, Self-powered human-health monitoring through aligned PVDF nanofibers interfaced skin-interactive piezoelectric sensor. ACS Appl. Polym. Mater. 2 (2020) 862-878. https://doi.org/10.1021/acsapm.9b00846

[53] L. Su, Z. Jiang, Z. Tian, H. Wang, H. Wang, Y. Zi, Self-powered, ultrasensitive, and high-resolution visualized flexible pressure sensor based on color-tunable triboelectrification-induced electroluminescence. Nano Energy 79 (2021) 105431. https://doi.org/10.1016/j.nanoen.2020.105431

[54] K. Lee, S. Jang, K.L. Kim, M. Koo, C. Park, S. Lee, J. Lee, G. Wang, C. Park, Artificially intelligent tactile ferroelectric skin. Adv. Sci. (2020) 2001662. https://doi.org/10.1002/advs.202001662

[55] B. Zhao, J. Hu, W. Ren, F. Xu, X. Wu, P. Shi, Z.G. Ye, A new biosensor based on PVDF film for detection of nucleic acids. Ceram. Int. 41 (2015) S602-S606. https://doi.org/10.1016/j.ceramint.2015.03.253

[56] S.K. Hwang, H.Y. Hwang, Development of a tactile sensing system using piezoelectric robot skin materials. Smart Mater. Struct. 22 (2013) 055004. https://doi.org/10.1088/0964-1726/22/5/055004

[57] Jing Zhang, Zhong MA, Sheng Li, Lijia Pan, Yi Shi, Recent research progress in biomimetic tactile sensors. SCIENTIA SINICA Technologica, 50 (2020) 1-16. https://doi.org/10.1360/SST-2019-0204

[58] Encyclopedia Brtitannica.

[59] B. İlik, A. Koyuncuoğlu, H. Uluşan, S. Chamanian, D. Işık, Ö. Şardan-Sukas, H. Külah In: Proceedings of Thin Film PZT Acoustic Sensor for Fully Implantable Cochlear Implants 1 (2017) 366. https://doi.org/10.3390/proceedings1040366

[60] J. Chung, Y. Jung, S. Hur, J. H. Kim, S. J. Kim, W. D. Kim, Y.H. Choung, S.H. Oh, Development and characterization of a biomimetic totally implantable artificial basilar membrane system, Front. Bioeng. Biotechnol., (2021), https://doi.org/10.3389/fbioe.2021.693849. https://doi.org/10.3389/fbioe.2021.693849

[61] A.J. Maniglia, H. Abbass, T. Azar, M. Kane, P. Amantia, S. Garverick, et al. The middle bioelectronic microphone for a totally implantable cochlear hearing device for profound and total hearing loss. Otol. Neurotol. 20 (1999) 602-611.

[62] J.H. Kwak, Y. Jung, K. Song, S. Hur, Fabrication of Si3N4-based artificial basilar membrane with ZnO nanopillar using MEMS process. J. Sensor. (2017), https://doi.org/10.1155/2017/1308217. https://doi.org/10.1155/2017/1308217

[63] J. Jang, J.H. Jang, H. Choi, Biomimetic artificial basilar membranes. Adv. Healthc. Mater. 6 (2017) 1700674. https://doi.org/10.1002/adhm.201700674

[64] P. Hennet, Piezoelectric bone surgery: A review of the literature and potential applications in veterinary oromaxillofacial surgery. Front. Vet. Sci. 2 (2015) 8. https://doi.org/10.3389/fvets.2015.00008

[65] M. Thomas, U. Akula, K. K. R. Ealla, N. Gajjada, Piezosurgery: A boon for modern periodontics. J. Int. Soc. Prev. Community Dent. 7 (2017) 1-7. https://doi.org/10.4103/2231-0762.200709

[66] N. Vyas, E. Pecheva, H. Dehghani, R.L. Sammons, Q.X. Wang, D.M. Leppinen, et al. High speed imaging of cavitation around dental ultrasonic Scaler tips. PLoS One. 11 (2016) e0149804. https://doi.org/10.1371/journal.pone.0149804

[67] T.K. Corcoran, R. Venkataramanan, R.M. Hoffman, M.P. George, A. Petrov, T. Richards, et al. Systemic delivery of atropine sulfate by the microdose dry-powder inhaler. J. Aerosol Medi. Pulmo. Drug Deliv. 26 (2013) 46-55. https://doi.org/10.1089/jamp.2011.0948

[68] H. Wang, T. Zhang, M. Zhao. Micro-dosing of fine cohesive powders actuated by pulse inertia force. Micromach. 9 (2018) 73. https://doi.org/10.3390/mi9020073

[69] S. Kar, S. McWhorter, S.M. Ford, S.A. Soper. Piezoelectric mechanical pump with nanoliter per minute pulse-free flow delivery for pressure pumping in micro-channels. The Analyst. 123 (1998) 1435-1441. https://doi.org/10.1039/a800052b

[70] M. Wahbah, M. Alhawari, B. Mohammad, H. Saleh, M. Ismail, Characterization of human body-based thermal and vibration energy harvesting for wearable devices. IEEE J. Emer. Select. Topic. Circu. Syst. 4 (2014) 354-363. https://doi.org/10.1109/JETCAS.2014.2337195

[71] M.W. Shafer, E. Garcia, The power and efficiency limits of piezoelectric energy harvesting. J. Vibrat. Acous. 136 (2014) 021007. https://doi.org/10.1115/1.4025996

[72] S.K. Ghosh, D. Mandal, Sustainable energy generation from piezoelectric biomaterial for noninvasive physiological signal monitoring. ACS Sustain. Chem. Eng. 5 (2017) 8836. https://doi.org/10.1021/acssuschemeng.7b01617

[73] S. Li, M. Alam, R. U. Ahmed, H. Zhong, Ultrasound-driven piezoelectric current activates spinal cord neurocircuits and restores locomotion in rats with spinal cord injury. Bioelectron Med., 2020, https://doi.org/10.1186/s42234-020-00048-2. https://doi.org/10.1186/s42234-020-00048-2

[74] J. Kim, A.S. Campbell, B.E.F. de Ávila, J. Wang, Wearable biosensors for healthcare monitoring. Nat. Biotechnol. 37 (2019) 389-406. https://doi.org/10.1038/s41587-019-0045-y

[75] D. Rodrigues, A.I. Barbosa, R. Rebelo, I.K. Kwon, R.L. Reis, V.M. Correlo, Skin-Integrated Wearable Systems and Implantable Biosensors: A Comprehensive Review. Biosensors 10 (2020) 79. https://doi.org/10.3390/bios10070079

[76] S. Sharafkhani, M. Kokabi, Ultrathin-shell PVDF/CNT nanocomposite aligned hollow fibers as a sensor/actuator single element. Comp. Sci. Technol. 200 (2020) 108425. https://doi.org/10.1016/j.compscitech.2020.108425

[77] N. Bourasseau, E. Moulin, C. Delebarre, P. Bonniau, Radome health monitoring with Lambwaves: experimental approach. NDT E. Int. 33 (2000) 393-400. https://doi.org/10.1016/S0963-8695(00)00007-4

[78] T. Ghosh, T. Kundu, P. Karpur, Efficient use of lamb modes for detecting defects in large plates. Ultrasonics 36 (1998) 791-801. https://doi.org/10.1016/S0041-624X(98)00012-2

[79] K.S. Tan, N. Guo, B.S. Wong, C.G. Tui, Comparison of lamb waves and pulse echo in detection of near-surface defects in laminate plates. NDT E. Int. 28 (1995) 215-223. https://doi.org/10.1016/0963-8695(95)00023-Q

[80] O.S. Salawu, Detection of structural damage through changes in frequency: A review. Eng. Struct. 19 (1997) 718-723. https://doi.org/10.1016/S0141-0296(96)00149-6

[81] C.K. Lee, F.C. Moon, Laminated piezopolymer plates for torsion and bending sensors and actuators. J. Acoust. Soc. Am. 85 (1989) 2432-2439. https://doi.org/10.1121/1.397792

[82] C.T. Sun, N.C. Cheng, Piezoelectric waves on a layered cylinder. J. Appl. Phys. 45 (1974) 4288-4294. https://doi.org/10.1063/1.1663048

[83] Q. Wang, S.T. Quek, Enhancing flutter and buckling capacity of column by piezoelectric layers. Int. J. Solids Struct. 39 (2002) 4167-4180. https://doi.org/10.1016/S0020-7683(02)00334-7

[84] W. H. Duan, Q. Wang, S. T. Quek, Applications of piezoelectric materials in structural health monitoring and repair: selected research examples. Mater. 3 (2010) 5169-5194. https://doi.org/10.3390/ma3125169

[85] Q. Wang, S.T. Quek, Repair of delaminated beams via piezoelectric patches. Smart Mater. Struct. 13 (2004)1222-1229. https://doi.org/10.1088/0964-1726/13/5/026

Advanced Functional Piezoelectric Materials and Applications
Materials Research Foundations 131 (2022) 165-185

Materials Research Forum LLC
https://doi.org/10.21741/978164490209-6

Chapter 6

Piezoelectric Materials for Biomedical and Energy Harvesting Applications

Tanzeel Munawar[1], Nadia Akram[1]*, Khawaja Taimoor Rashid[1], Asim Mansha[1], Akbar Ali[1]

[1]Department of Chemistry, Government College University Faisalabad, Faisalabad-38000, Pakistan

*nadiaakram@gcuf.edu.pk

Abstract

Researchers explore alternative energy harvesting technologies because nonrenewable energy sources cause environmental pollution and energy crises. Mechanical energy can be turned into usable electricity, which may help to fulfill energy demand without environmental issues because it is the most common form of energy. Piezoelectric behavior is the distended energy harvesting mechanism based on high electromechanical connection influence and piezoelectric influence. Recent research in mineral, polymer, natural, and advance functional piezoelectric materials (AFPM) with biological effects is discussed. Piezoelectric power harvesting at different scales (nano, micro, etc.) has been discussed in various fields like transport, biomedical uses, wearing and inserting electronic devices, and tissue redevelopment. Piezocomposite and piezoelectric energy harvesting technology are examined along with their developments, limitations, and possible enhancements. This study covers a wide range of piezoelectric materials that may provide power to wireless devices in various applications.

Keywords

Advance Functional Piezoelectric Materials (AFPM), Piezoelectric Energy Harvesting, Biomedical uses of Piezocomposite, Applications of Piezoelectric Materials

Contents

1.   Introduction

Piezoelectric materials have been utilized in various applications, including healthcare, energy harvesting, and electrochemistry. Piezoelectric materials have been efficiently working in different fields such as medical, energy harvesting, and electrochemical. Based on widely usage, piezoelectric materials have attained the great attention of researchers to enhance the piezoelectric behavior of functional piezoelectric materials (F.P.M.). A range of organic or inorganic materials (natural or synthetic) are commonly utilized as reinforcement (Filler) in F.P.M. operating system to prepare advanced functional piezoelectric materials (AFPM) or Piezoelectric Composite materials (P.C.M.).

Even though the components (which are generally inexpensive) were first included as extensions to lower the material's price, their capacity to modify certain desirable attributes rapidly became apparent, and they became multifunctional fillers [1-4]. Fillers may increase workability or change the tensile properties, thermal properties, life duration, and other properties of P.F.M.S in terms of cost control. Simultaneously, the produced materials may have various drawbacks, such as being brittle or opaque. The amount of filler is very important to determine the extent and degree of changes, but thickness, shapes, and topography are important factors. Fillers may also increase workability or change the P.F.M.'s thermodynamic, tensile, electric, magnet, and other properties, in terms of cost control [5].

As compared to single-phase materials, composite materials have superior characteristics and novel uses. This chapter will look at the concepts of composite effects and their usual applications, with piezo composites as an example. Due to their high tolerability, these materials are made of piezoelectrically active ceramic and polymer. On the base of conductivity of each phase (1, 2, or 3 dimensions), the shape of two blended materials can be categorized into ten structures: 0-0, 0-1, 0-2, 0-3, 1-1, 1-2, 1-3, 2-2, 2-3, and 3-3. The most useful piezo compound is regarded to be a 1-3 piezo compound [6].

1.1 Types of advance piezoelectric functional materials

Like semiconductors, many piezoelectrically active inorganic materials include (ZnO, ZnS, GaN, etc.), [7-9] as well as ceramics (P.Z.T., $BaTiO_3$, $NaNbO_3$, $KNbO_3$, K.N.N., $ZnSnO_3$, etc.) [10-16] polymers (PVDF polyamides, parylene -C, etc.) [3, 5, 17] and bio-substance (bone-based materials, hairs, proteins, bagasse, etc.) [18-20]. Numerous piezoelectric materials have also been developed by combining active ingredients in specified amounts to tailor the electrical properties of the resulting composites to specific purposes.

There are many types of piezoelectric materials based on the matrix and filler nature. However, The polymeric piezo composite is more common and widely used in different fields.

1.1.1 Polymer piezocomposite

The employment of piezo-polymers in the industry is limited by their low piezo sensitivity and severe effectiveness loss at high heating rates. The additives has been used to boost the piezoelectrical properties of the polymers [21]. There are many possible combinations of organic, inorganic, and biological materials in conducting polymers to synthesize advanced piezoelectric functional composite materials with unusual properties which can be used for harvesting energy, medical applications, and to enhance the efficiency of electrochemical devices [5, 17, 22-24]. A list of some conducting polymers with their electrical constant [D_{31} (pC/N)] is given in table 1. According to Judeinstein and Sanchez1 [25], additives are chemically bonded and linked together through intermolecular forces. Understanding and evaluating the characteristics of hybrid materials requires understanding the chemical bonding between the organic and inorganic phases [26-28].

Here, some research-based examples for polymeric piezo composite have been presented in table 1. Various organic fillers have improved the piezoelectric performance of conductive polymers. P-toluene sulphonic acid (PTSA), dodecylbenzene sulphonic acid (DBSA), naphthalene, lauric Acid (L.A.), and amphiphilic fillers from the sulphonic acid family have all been investigated as fillers for polypyrrole and polyaniline [33-36].

Table 1: List of Conducting polymer [29-32]

Polymer name	T_{max} (°C)	D_{31} (pC/N)
PVDF	100	20
Vinylidene Cyanide-co-Vinyl Acetate	170	7
Polyurea	200	0.3
Polyimides	250	10
Polyamides	150	-6

Beleze and. Zarbin reported the synthesis of the polymeric piezo composite by blending the inorganic acid with conducting polymers (polypyrrole and polyaniline). The addition of antumonic acid ($H_2Sb_2O_6.nH_2O$) has confirmed the proton's high conductance due to reactions in which proton was exchanged. Such materials with high protonic conductivities are much more usable for numerous devices [27]. Mihail Iacob *et al.* Have reported the synthesis of polymeric piezo composite using the natural bentonite into polydimethylsiloxane-a,u-diol (PDMS). The prepared polymeric piezo composites have improved dielectric permittivity and piezoresponse and improved other mechanical properties such as increasing Young's modulus and decreasing break elongation [2].

Efforts to incorporate bioactive fillers, on the other hand, has shown great potential and can be used to alter derived hybrid materials for the clinical field. For example, Hanninen *et al.* reported the fabrication of cellulosic nanofiber and chitosan into Polyvinylidene Fluoride (PVDF) to develop innovative functional piezoelectric material using the stir casting technique. It has been discovered that biological fillers with greater symmetry have larger piezoelectric properties for films [37-39]. Some type of polymer piezoelectric composite along with the thier propeties have been given in Figure 1.

Bio-waste-based piezo composite compounds have been reported for energy harvesters, according to Kumar *et al.* The weakly piezo fish scale was employed to create a hybrid polymer piezo composite using the PVDF. The solution approach produced a homogenous distribution of fish scale particles. As shown by additional FTIR spectra and larger melting points in D.S.C. and XRD tests, fish scale corresponding results in the piezoelectric-phase-in matrix PVDF. [40].

Advanced Functional Piezoelectric Materials and Applications   Materials Research Forum LLC
Materials Research Foundations 131 (2022) 165-185   https://doi.org/10.21741/978164490209-6

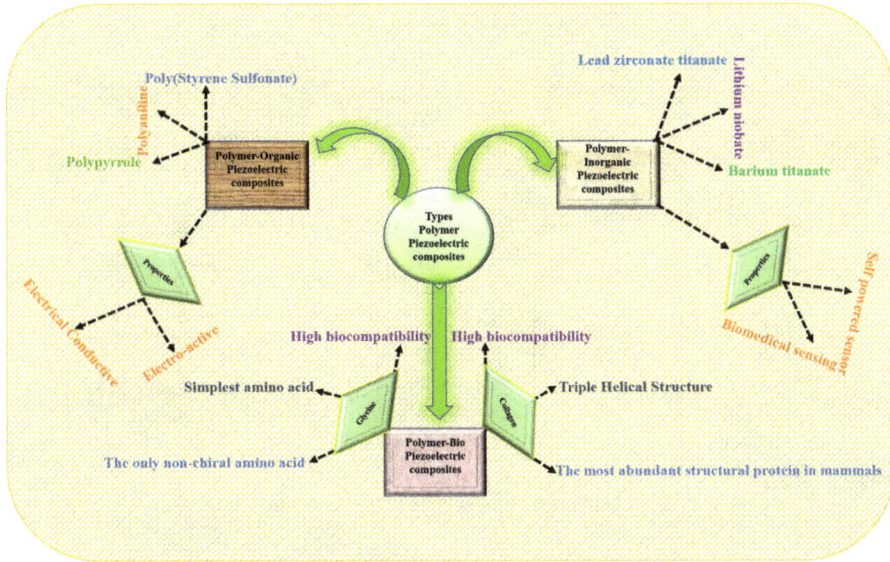

Figure 1: Types of polymer piezoelectric composite along with their properties.

1.1.2 Ceramics piezocomposite

Lead-free fillers are preferred in hybrids because lead-based fillers have environmental consequences. Sodium potassium niobate based fillers, barium titanate-based fillers, bismuth layer-based fillers, and molybdenum sulfide-based fillers are most often used lead-free ceramic fillers as given in Figure 2. Ceramic fillers are frequently treated with additional metals to create blended fillers that improve the composite's fundamental piezo capabilities. The large dielectric constant reported in $BaTiO_3$ owing to its ferroelectricity observed due to its ferroelectricity led to the development of piezoceramics [41].

A barium titanate (B.T.) based system is an example of a lead-free piezoelectric system. Its dielectric constant confirmed electrical behavior, good electromechanical connections, and low dielectric loss; that's why it is usable for piezoelectrical materials. The piezoelectrical efficiency can be improved by adding other piezo-active substances [42].

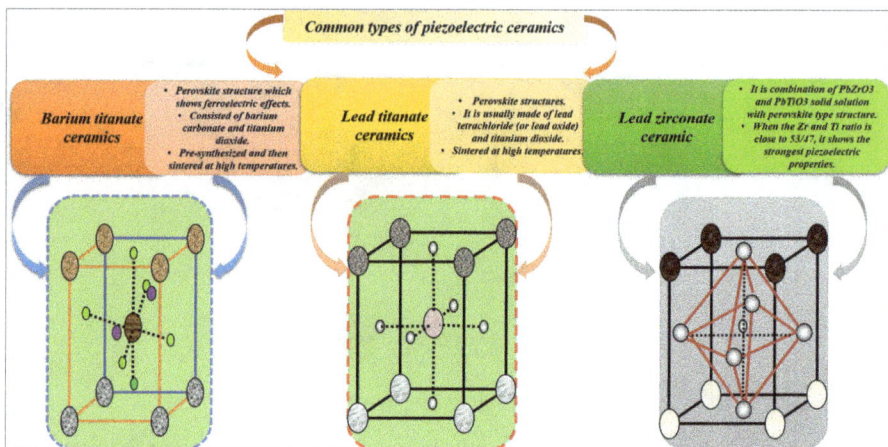

Figure 2: Common type of piezoelectric ceramics materials

1.1.3 Polymer ceramics piezocomposite

Polymer ceramic composites have emerged in popularity because they can incorporate the outstanding pyroelectric and piezo properties of ceramics with the strength, versatility, processing facility, reasonably high dielectric properties, and malfunction power of polymer that a single-phase piezoelectric material cannot achieve [43-46]. Composites comprised of electro-active ceramics and a ferroelectric polymer are extremely appealing for applications because of these features, and some polymer ceramics piezo composites are given in Figure 3. They have a low absorption coefficient that matches aqueous solvent and epidermis, and their characteristics may be modified to meet various needs [45, 47]. Hu et. al [48], Pauer [49], and Lu et. al [50] were among the first to try to make such composites. They used P.Z.T. as an additive substance with connective polyurethane (0– 3). The dielectric constant of the polymeric matrix controls the electromagnetic current that operates on each isolated sphere piezo grains in a 0–3 composite. The majority of the applied field can travel through the lower dielectric constkant stage since most polymers have a smaller dielectric constant than piezoelectric ceramic materials [48].

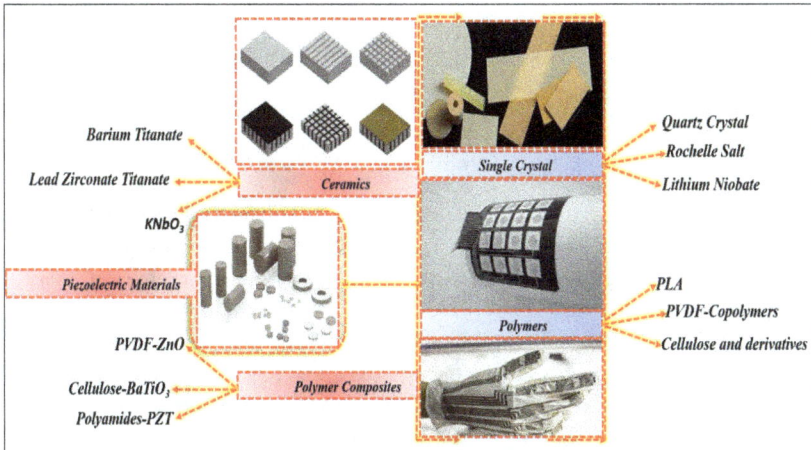

Figure 3: Some polymer ceramics piezo composite

The polymers' low toughness and elasticity inspired the researchers to utilize piezoceramics as an additive for piezo-active uses such as sensing devices. BaTiO, P.Z.T., ZnO, etc., are examples of piezoelectric ceramics, which can easily be used as an additive for piezo composites. Due to their great piezoelectric capability, P.Z.T. and its nanostructures are often combined with polymeric materials like PVDF, epoxy, and polyamides. It's important to remember that a high P.Z.T. concentration in the composites may operate against the intended tensile properties. Furthermore, lead-based ceramics are not recommendable for biological purposes and have major environmental implications. Many novel lead substitutes are now being explored, including LiNbO and KnaLiNbO.

2.   Applications

2.1   Microelectromechanical system (MEMS) devices

Traditional renewable energy often takes enormous machinery, elevated turbine, and electromagnetic devices. It is thus prohibitively costly, making it impossible to gather and exploit minor mechanical energy sources such as wind, flow of water, and animal body motions. On the other hand, scaled-down handheld gadgets and micro sensors often use very little electricity and are relatively cheaper than batteries. As a result, a new category of compensators as nanogenerators have been created to effectively capture the mechanical force and convert it into electricity. Prof. Zhonglin Wang [51] was the first person to suggest the notion of a nanogenerator. Unlike electric current, the nanogenerator is a system that uses the piezo effect to transform small amounts of mechanical energy into electricity. The piezo nanogenerator (PENG) [52-54] and the tribo nanogenerator (TENG) are two frequently used names for these two devices [54-56].

2.2 MEMS generators for energy harvesting

Because the piezoelectric effect depends on the material's underlying polarization and does not require an additional power supply, magnetic field, or interaction with some other substance. It is a very efficient framework for collecting acoustic physical energy and converting it into electrical energy [24, 57]. Compared to other energy harvesting technologies, piezoelectric generators are more robust, dependable, and responsive to small stresses, with a 3–5-fold greater energy harvest and greater voltage yield [40, 58, 59]. As a result, piezoelectric transmission has been the most potential external energy harvesting technique, with applications in radio electronics, MEMS (microelectromechanical systems), wear implantation bioelectronics, and other domains. A summary of piezoelectirc materials for energy harvasting appalication given in table 2. Furthermore, piezoelectric generators can be made in tiny sizes and shapes, and they can be readily incorporated into microelectromechanical systems. Furthermore, climatic conditions like moisture have little effect on them [42, 60, 61].

Over the past decade, several articles have attempted to build different nanogenerators using unique device architectures and sophisticated components [54, 56, 69]. Piezo biomaterials have played a significant role as potential functional applications in humans, including implanted and attachable energy harvesters because Piezo biomaterials are biodegradable and biocompatible as given in Figure 4. Vivekananthan et al. developed a piezo collagen nanofibers film in which mechanical energy can be converted into electric energy and serves as a moisture sensor [70].

Hansen et al. prepared the hybrid Nanogenerator using the PVDF and Bio-filler. They utilized it to power a single U.V. sensor made of nanowires [71].

Aside from in situ forms of energy, biomechanical energy is produced by sports exercise such as footsteps, arm whacks, touch pressing, etc., which can be used to generate electricity in a range of technological equipment such as led bulbs, wearable electronics, cellphones, and surgically implanted medical technology by employing piezo power generators on different body parts like feet, legs, arms, etc. [23, 24, 71, 72].

Advanced Functional Piezoelectric Materials and Applications   Materials Research Forum LLC
Materials Research Foundations 131 (2022) 165-185   https://doi.org/10.21741/978164490209-6

Table 2: Application of piezoelectric materials with mechanical energy sources.

Piezoelectric marterials	Energy harvester	Mechanical Energy sources	Refe
Piezo Ceramics (PZT)	Shoe's sole	Runing	[62, 63]
PiezoPolymer (PVDF)	Shoe's sole	Runing	[63]
	Air generator	Wind flow	[63, 64]
	Rains fall harvester	Rain drop	[60]
	Roadways energy harvester	Vehicular vibration	[63]
Microfiber Composite (NCMF-PDMS)	Piezoelectric generator	Mechanical vibration	[65]
Nanofiber PiezoComposite (PZT+PDMS)	Piezo-Nano-Generator		[66]
Cymbal	Vibrational harvester		[67]
Piezopolymer (3D).	Biomedical, Vibrational harvester	Movement	[68]
Polymer ceramic piezocomposite (PZT+PVDF)	Piezo-Nano-Generator	Body Movements	[55]

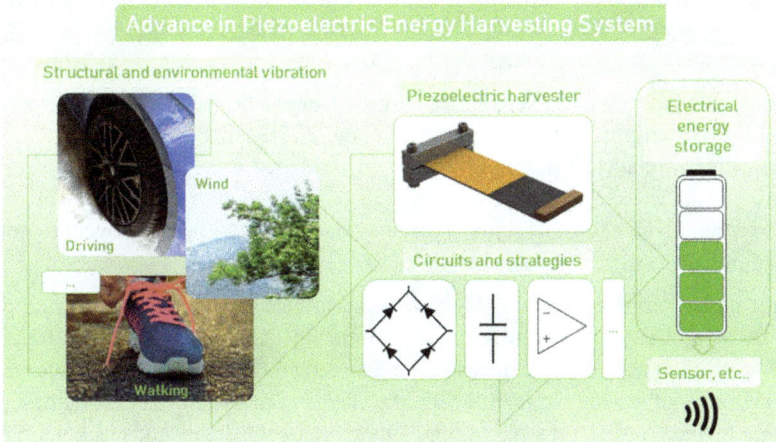

Figure 4: Different piezoelectric energy harvesting system

2.3 MEMS sensor

2.3.1 Pressure sensor

Snyder *et al.* employed a piezo generator installed in automobile tires to activate tire pressure sensors. The tire's vibration drove the generator while tires moved, and unusual tire compression is used for drivers via a less-energy radiation connection. Energy collection in road transport is piezoelectric sensors' wireless system transforming rolling stock vibration into electrical energy to forecast railway roller failure [73, 74].

Bodkhe *et al.* developed Nano piezo composite using the 10% of $BaTiO_3$ nanoparticle with PVDF as a pressure sensor for 3D printing techniques [75].

Curry *et al.* developed a piezoelectric pressure sensing device by encapsulating degrading Poly (L-lactic Acid) and Mo-electrodes. This gadget could measure a wide range of pressures (0 to 18 kilo Pascal) [76]. Furthermore importantly, the sensing device was destroyed during 56 days at 74°C, showing that this disintegrating sensor has therapeutic potential. Several papers have used nature-inspired piezoelectric sensors to produce patient physiological monitors and electronics skins [77-80].

2.3.2 Healthcare sensor

As proven by Liu *et al.,* the portable piezo-sensor may be used as a health tracking device to detect respiratory rates, activity, and voice cord movements [81].

Park *et al.* fabricated PVDF using graphene oxide to create a versatile artificial epidermis that can sense many stimuli, such as heat and pressure, with high sensitivity for concurrent arterial pulse rates and temperature sensors [82].

Cheng *et al.* created a piezoelectric self-powered implanted health tracking system for detecting hypertension levels using polarization PVDF film. The proposed device's favorable linear relationship among maximum produced energy and blood circulation pressure and its excellent precision, maximum output energy, and exceptional stability during repeated operating sessions indicated that it might be employed in medical applications [83].

Park and colleagues constructed a self-powered heartbeat sensor to monitor spiral pulses. The Sol-Gel process was used to create a high-quality P.Z.T. thin film, then transferred onto a 4.8 m thick plastic sheet. It was possible to directly interact with the device and the rough skin's complicated texture, allowing the device to monitor even the slightest pulses variations on the skin's surface. Accordingly, the device's sensibility and reaction time were 0.018 kPa and 60 milliseconds. The device's endurance was proved by putting it through 5000 pressing sessions [84].

Advanced Functional Piezoelectric Materials and Applications   Materials Research Forum LLC
Materials Research Foundations 131 (2022) 165-185   https://doi.org/10.21741/978164490209-6

2.3.3 Cell and tisusse regenration

The bioelectric microenvironment is critical in promoting tissue regeneration [85-87]. Several alterations are made to its molecular components. Nonetheless, contemporary artificial matter scaffolds (known as static scaffolds) are inactive and do not react to outer stimuli. Static scaffolds obstruct natural signaling pathways by preventing bioelectric impulses from being sent. Stimuli-responsive scaffolds are highly wanted because they allow for the creation and transmission of bioelectric signals like native tissues, allowing for the maintenance of correct physiological processes [88]. Piezoelectric materials may produce electrical impulses in response to apply stress, induced by cell adhesion, motility, and body movements [87].

Hoop *et al.* discovered that radio signals aided in boosting the prospective in piezo PVDF, improving neurite formation in cells, using an ultrasonic approach [89]. Similarly, the piezoelectric action of the PVDF–TrFE fibril scaffolds stimulated neural cell development, neurite expansion, and neuronal differentiation [90].

As a result, piezoelectric materials might be key components of dynamic scaffolding that permit electric impulses with no need for external power [91-93].

Conclusion

Nanofiber, thin-film, and layer types of mineral, organic, and composite piezoelectric materials have been studied. The piezoelectric efficiency of inorganic materials is better. However, their brittleness and hazardous lead levels make them unsuitable for use. In most applications, piezoelectric generators' integrability is determined by their flexibility. The most studied flexible piezopolymers are PVDF and its copolymers. Several successful ways are discovered to improve the poor piezoelectric efficiency of these polymers. Dispersing ceramic nanoparticles into the polymer matrices is one way. This innovation drew the most interest because it allowed composite generators to merge the benefits of piezoelectric material with the elasticity of polymer.

In this studies, bio-inspired natural fibers with naturally desired material properties including biocompatibility, good biocompatibility, elasticity, and longevity have potential piezoelectric performance. However, this field of study is still in its infancy, with plenty of room for discovering novel biomaterials. Piezoelectricity is being studied for auto monitoring, energy production, and activation in a variety of applications. Despite encouraging research findings, only a few piezoelectric items have been marketed, with the rest still in the research and development phases.

Furthermore, new materials with improved qualities are projected to find use in various industries in the future. Finding high-temperature materials, for example, will allow piezoelectric power collectors to be used in thermal applications requiring fluid motion.

Piezoelectric generators may be used to power wearing and implanted biomedical equipment such as self-powered detectors for healthcare management and growth factors for research regenerating brain and skeletal tissues. The current tendancy in electronic device development is to minimize device volume, decrease energy requirements, and increase device elasticity and integration. Micro and nano assembling of piezoelectric generators, on the other hand, is made possible by materials and manufacturing processes that increase the reliability, integration, and power output efficiency. As a result, most wireless electronic gadgets should be able to be powered by a piezoelectric generator in the future.

Reference

[1] T. Siponkoski, M. Nelo, N. Ilonen, J. Juuti, H. Jantunen, High performance piezoelectric composite fabricated at ultra low temperature, Compos. B. Eng. 229 (2022) 109486. https://doi.org/10.1016/j.compositesb.2021.109486

[2] M. Iacob, V. Tiron, G.-T. Stiubianu, M. Dascalu, L. Hernandez, C.-D. Varganici, C. Tugui, M. Cazacu, Bentonite as an active natural filler for silicone leading to piezoelectric-like response material, J. Mater. Res. Technol. 17 (2022) 79-94. https://doi.org/10.1016/j.jmrt.2021.12.125

[3] Q.T. Nguyen, D.G. Baird, Preparation of polymer-clay nanocomposites and their properties, Adv. Polym. Technol. 25 (2006) 270-285. https://doi.org/10.1002/adv.20079

[4] M. Iacob, C. Tugui, V. Tiron, A. Bele, S. Vlad, T. Vasiliu, M. Cazacu, A.-L. Vasiliu, C. Racles, Iron oxide nanoparticles as dielectric and piezoelectric enhancers for silicone elastomers, Smart Mater. Struct. 26 (2017) 105046. https://doi.org/10.1088/1361-665X/aa867c

[5] L. Ruan, X. Yao, Y. Chang, L. Zhou, G. Qin, X. Zhang. Properties and applications of the beta phase poly (vinylidene fluoride). Polymers. 10 (2018) 228. https://doi.org/10.3390/polym10030228

[6] Z. Cui, N.T. Hassankiadeh, Y. Zhuang, E. Drioli, Y.M. Lee. Crystalline polymorphism in poly(vinylidenefluoride) membranes. Prog. Polym. Sci. 51 (2015) 94-126. https://doi.org/10.1016/j.progpolymsci.2015.07.007

[7] H.K. Park, K.Y. Lee, J.S. Seo, J.A. Jeong, H.K. Kim, D. Choi, S.W. Kim, Charge-generating mode control in high-performance transparent flexible piezoelectric nanogenerators, Adv. Funct. Mater. 21 (2011) 1187-1193. https://doi.org/10.1002/adfm.201002099

[8] C.-Y. Chen, G. Zhu, Y. Hu, J.-W. Yu, J. Song, K.-Y. Cheng, L.-H. Peng, L.-J. Chou, Z.L. Wang, Gallium nitride nanowire based nanogenerators and light-emitting diodes, ACS. nano. 6 (2012) 5687-5692. https://doi.org/10.1021/nn301814w

[9] J.M. Wu, C.C. Kao, Self-powered pendulum and micro-force active sensors based on a ZnS nanogenerator, RSC. Advances. 4 (2014) 13882-13887. https://doi.org/10.1039/C3RA47435F

[10] P.G. Kang, B.K. Yun, K.D. Sung, T.K. Lee, M. Lee, N. Lee, S.H. Oh, W. Jo, H.J. Seog, C.W. Ahn, Piezoelectric power generation of vertically aligned lead-free (K, Na) NbO 3 nanorod arrays, RSC. Advances. 4 (2014) 29799-29805. https://doi.org/10.1039/C4RA02921F

[11] Z.-H. Lin, Y. Yang, J.M. Wu, Y. Liu, F. Zhang, Z.L. Wang, BaTiO3 nanotubes-based flexible and transparent nanogenerators, J. Phys. Chem. Lett. 3 (2012) 3599-3604. https://doi.org/10.1021/jz301805f

[12] J.H. Jung, M. Lee, J.-I. Hong, Y. Ding, C.-Y. Chen, L.-J. Chou, Z.L. Wang, Lead-free NaNbO3 nanowires for a high output piezoelectric nanogenerator, ACS. nano. 5 (2011) 10041-10046. https://doi.org/10.1021/nn2039033

[13] M.-R. Joung, H. Xu, I.-T. Seo, D.-H. Kim, J. Hur, S. Nahm, C.-Y. Kang, S.-J. Yoon, H.-M. Park, Piezoelectric nanogenerators synthesized using KNbO 3 nanowires with various crystal structures, J. Mater. Chem. A 2 (2014) 18547-18553. https://doi.org/10.1039/C4TA03551H

[14] C.K. Jeong, K.I. Park, J. Ryu, G.T. Hwang, K.J. Lee, Large-area and flexible lead-free nanocomposite generator using alkaline niobate particles and metal nanorod filler, Adv. Funct. Mater. 24 (2014) 2620-2629. https://doi.org/10.1002/adfm.201303484

[15] Q.-t. Xue, Z. Wang, H. Tian, Y. Huan, Q.-Y. Xie, Y. Yang, D. Xie, C. Li, Y. Shu, X.-H. Wang, A record flexible piezoelectric KNN ultrafine-grained nanopowder-based nanogenerator, AIP. Adv. 5 (2015) 017102. https://doi.org/10.1063/1.4905698

[16] M.M. Alam, S.K. Ghosh, A. Sultana, D. Mandal, Lead-free ZnSnO3/MWCNTs-based self-poled flexible hybrid nanogenerator for piezoelectric power generation, Nanotechnology. 26 (2015) 165403. https://doi.org/10.1088/0957-4484/26/16/165403

[17] K.S. Ramadan, D. Sameoto, S. Evoy, A review of piezoelectric polymers as functional materials for electromechanical transducers, Smart Mater. Struct. 23 (2014) 033001. https://doi.org/10.1088/0964-1726/23/3/033001

[18] B.Y. Lee, J. Zhang, C. Zueger, W.-J. Chung, S.Y. Yoo, E. Wang, J. Meyer, R. Ramesh, S.-W. Lee, Virus-based piezoelectric energy generation, Nat. Nanotechnol.7 (2012) 351-356. https://doi.org/10.1038/nnano.2012.69

[19] J. Yang, Q.S. Chen, F. Xu, H. Jiang, W. Liu, X. Q. Zhang, Z.X. Jiang, G.D. Zhu. Epitaxy enhancement of piezoelectric properties in P(VDF-TrFE) copolymer films and applications in sensing and energy harvesting. Adv. Electron. Mater. 6 (2020) 2000578.. https://doi.org/10.1002/aelm.202000578

[20] J.S. Andrew, D.R. Clarke Effect of electrospinning on the ferroelectric phase content of polyvinylidene difluoride fibers. Langmuir. 24 (2008) 24, 670-672. https://doi.org/10.1021/la7035407

[21] C. Baur, Y. Zhou, J. Sipes, S. Priya, W. Voit, Organic, flexible, polymer composites for high-temperature piezoelectric applications, Energy Harvest. Syst. 1 (2014) 167-177. https://doi.org/10.1515/ehs-2013-0015

[22] A. Palshikar, N. Sharma, Review on Piezoelectric Materials as Thin Films with their Applications. Mater. Sci. Res. India. 12 (2015) 79-84. https://doi.org/10.13005/msri/120113

[23] S.P. Beeby, M.J. Tudor, N. White, Energy harvesting vibration sources for microsystems applications, Meas. Sci. Technol. 17 (2006) R175. https://doi.org/10.1088/0957-0233/17/12/R01

[24] Q. Zheng, B. Shi, Z. Li, Z.L. Wang, Recent progress on piezoelectric and triboelectric energy harvesters in biomedical systems, Adv. Sci. 4 (2017) 1700029. https://doi.org/10.1002/advs.201700029

[25] P. Judeinstein, C. Sanchez, Hybrid organic-inorganic materials: a land of multidisciplinarity, J. Mater. Chem. 6 (1996) 511-525. https://doi.org/10.1039/JM9960600511

[26] A.B. Gonçalves, A.S. Mangrich, A.J.G. Zarbin, Polymerization of pyrrole between the layers of α-Tin (IV) Bis (hydrogenphosphate), Synth. Met. 114 (2000) 119-124. https://doi.org/10.1016/S0379-6779(00)00227-7

[27] F.A. Beleze, A.J. Zarbin, Synthesis and characterization of organic-inorganic hybrids formed between conducting polymers and crystalline antimonic acid, J. Braz. Chem. Soc. 12 (2001) 542-547. https://doi.org/10.1590/S0103-50532001000400017

[28] M.M. Chamakh, D. Ponnamma, M.A.A. Al-Maadeed, Vapor sensing performances of PVDF nanocomposites containing titanium dioxide nanotubes decorated multi-walled carbon nanotubes, J. Mater. Sci. Mater. Electron. 29 (2018) 4402-4412. https://doi.org/10.1007/s10854-017-8387-z

[29] M. Silva, C.M. Costa, V. Sencadas, A. Paleo, S. Lanceros-Méndez, Degradation of the dielectric and piezoelectric response of β-poly (vinylidene fluoride) after temperature annealing, J. Polym. Res. 18 (2011) 1451-1457. https://doi.org/10.1007/s10965-010-9550-x

[30] E. Fukada, New piezoelectric polymers, Jpn. J. Appl. Phys. 37 (1998) 2775. https://doi.org/10.1143/JJAP.37.2775

[31] S. B. Lang, S. Muensit. Review of some lesser-known applications of piezoelectric and pyroelectric polymers. Appl. Phys. A . 85 (2006) 125-134. https://doi.org/10.1007/s00339-006-3688-8

[32] S. Mathur, J. Scheinbeim, B. Newman, Piezoelectric properties and ferroelectric hysteresis effects in uniaxially stretched nylon-11 films, J. Appl. Phys. 56 (1984) 2419-2425. https://doi.org/10.1063/1.334294

[33] S. Sinha, S. Bhadra, D. Khastgir, Effect of dopant type on the properties of polyaniline, J. Appl. Polym. Sci. 112 (2009) 3135-3140. https://doi.org/10.1002/app.29708

[34] V.L. Reena, J.D. Sudha, C. Pavithran, Role of amphiphilic dopants on the shape and properties of electrically conducting polyaniline-clay nanocomposite, J. Appl. Polym. Sci. 113 (2009) 4066-4076. https://doi.org/10.1002/app.30525

[35] J. Liu, X. Wang, X. Hu, A. Xiao, M. Wan, Studies of influence of naphthalene mono/disulfonic acid dopant on thermal stability of polypyrrole, J. Appl. Polym. Sci. 109 (2008) 997-1001. https://doi.org/10.1002/app.28051

[36] M. Liu, K. Tzou, R. Gregory, Influence of the doping conditions on the surface energies of conducting polymers, Synth. Met. 63 (1994) 67-71. https://doi.org/10.1016/0379-6779(94)90251-8

[37] A. Hänninen, E. Sarlin, I. Lyyra, T. Salpavaara, M. Kellomäki, S. Tuukkanen, Nanocellulose and chitosan based films as low cost, green piezoelectric materials, Carbohydr. Polym. 202 (2018) 418-424. https://doi.org/10.1016/j.carbpol.2018.09.001

[38] L. Csoka, I.C. Hoeger, P. Peralta, I. Peszlen, O.J. Rojas, Dielectrophoresis of cellulose nanocrystals and alignment in ultrathin films by electric field-assisted shear

assembly, J. Colloid. Interface. Sci. 363 (2011) 206-212.
https://doi.org/10.1016/j.jcis.2011.07.045

[39] L. Csoka, I.C. Hoeger, O.J. Rojas, I. Peszlen, J.J. Pawlak, P.N. Peralta, Piezoelectric effect of cellulose nanocrystals thin films, ACS. Macro. Lett. 1 (2012) 867-870. https://doi.org/10.1021/mz300234a

[40] C. Kumar, A. Gaur, S. Tiwari, A. Biswas, S.K. Rai, P. Maiti, Bio-waste polymer hybrid as induced piezoelectric material with high energy harvesting efficiency, Compos. Commun.11 (2019) 56-61. https://doi.org/10.1016/j.coco.2018.11.004

[41] S. Banerjee, S. Bairagi, S.W. Ali, A critical review on lead-free hybrid materials for next generation piezoelectric energy harvesting and conversion, Ceram. Int. 47 (2021) 16402-16421. https://doi.org/10.1016/j.ceramint.2021.03.054

[42] M. Jaafar, Development of hybrid fillers/polymer nanocomposites for electronic applications, Book: Hybrid Nanomaterials: Advances in Energy, Environment and Polymer Nanocomposites, John Wiley & Sons, Inc., Hoboken, NJ, USA (2017) 349-369. https://doi.org/10.1002/9781119160380.ch7

[43] Z.M. Dang, Y.Q. Lin, H.P. Xu, C.Y. Shi, S.T. Li, J. Bai, Fabrication and dielectric characterization of advanced BaTiO3/polyimide nanocomposite films with high thermal stability, Adv. Funct. Mater. 18 (2008) 1509-1517. https://doi.org/10.1002/adfm.200701077

[44] J. Li, Seok SIl Chu B., Dogan F., Zhang Q., Wang Q. Nanocomposites of Ferroelectric Polymers with TiO2 Nanoparticles Exhibiting Significantly Enhanced Electrical Energy Density, Adv. Mater. 21 (2009) 217-221. https://doi.org/10.1002/adma.200801106

[45] P. Kim, S.C. Jones, P.J. Hotchkiss, J.N. Haddock, B. Kippelen, S.R. Marder, J.W. Perry, Phosphonic acid-modified barium titanate polymer nanocomposites with high permittivity and dielectric strength, Adv. Mater. 19 (2007) 1001-1005. https://doi.org/10.1002/adma.200602422

[46] F.R. Fan, W. Tang, Z.L. Wang, Flexible nanogenerators for energy harvesting and self-powered electronics. Adv. Mater. 28 (2016) 4283-4305. https://doi.org/10.1002/adma.201504299

[47] C. Dias, D. Das-Gupta, Inorganic ceramic/polymer ferroelectric composite electrets, IEEE Trans. Dielectr. Electr. Insul. 3 (1996) 706-734. https://doi.org/10.1109/94.544188

[48] D.W. Hu, M.G. Yao, Y. Fan, C.R. Ma, M.J. Fan, M. Liu. Strategies to Achieve High Performance Piezoelectric Nanogenerators. Nano. Energy. 55 (2019) 288-304 https://doi.org/10.1016/j.nanoen.2018.10.053

[49] L. Pauer, Flexible piezoelectric material, IEEE Int. Conv. Rec. 21 (1973) 1-3.

[50] L.J. Lu, W. Q. Ding, J.Q. Liu, B. Yang. Flexible PVDF based piezoelectric nanogenerators. Nano. Energy 78 (2020) 78, 105251. https://doi.org/10.1016/j.nanoen.2020.105251

[51] J.Z. Gui, Y.Z. Zhu, L.L. Zhang, X. Shu, W. Liu, S.S. Cuo, X.Z. Zhao, Enhanced output-performance of piezoelectric poly(vinylidene fluoride trifluoroethylene) fibers-based nanogenerator with interdigital electrodes and well-ordered cylindrical cavities. Appl. Phys. Lett. 112 (2018) 072902. https://doi.org/10.1063/1.5019319

[52] M. Lee, C.Y. Chen, S. Wang, S.N. Cha, Y.J. Park, J.M. Kim, L.J. Chou, Z.L. Wang, A hybrid piezoelectric structure for wearable nanogenerators, Adv. Mater. 24 (2012) 1759-1764. https://doi.org/10.1002/adma.201200150

[53] J.Z. Gui, Y.Z. Zhu, L.L. Zhang, X.Shu, W. Liu, S. S. Cuo, X. Z. Zhao Enhanced output-performance of piezoelectric poly(vinylidene fluoride trifluoroethylene) fibers-based nanogenerator with interdigital electrodes and well-ordered cylindrical cavities. Appl. Phys. Lett. 112 (2018) 072902. https://doi.org/10.1063/1.5019319

[54] X. Li, L. Zhang, Y. Feng, X. Zhang, D. Wang, F. Zhou, Solid-liquid triboelectrification control and antistatic materials design based on interface wettability control, Adv. Funct. Mater. 29 (2019) 1903587. https://doi.org/10.1002/adfm.201903587

[55] Y. Yang, L. Lin, Y. Zhang, Q. Jing, T.-C. Hou, Z.L. Wang, Self-powered magnetic sensor based on a triboelectric nanogenerator, ACS. nano. 6 (2012) 10378-10383. https://doi.org/10.1021/nn304374m

[56] L. Zhang, X. Li, Y. Zhang, Y. Feng, F. Zhou, D. Wang, Regulation and influence factors of triboelectricity at the solid-liquid interface, Nano. Energy. 78 (2020) 105370. https://doi.org/10.1016/j.nanoen.2020.105370

[57] X. Xie, Q. Wang, Energy harvesting from a vehicle suspension system, Energy. 86 (2015) 385-392. https://doi.org/10.1016/j.energy.2015.04.009

[58] K. Shi, B. Sun, X. Huang, P. Jiang, Synergistic effect of graphene nanosheet and BaTiO3 nanoparticles on performance enhancement of electrospun PVDF nanofiber mat for flexible piezoelectric nanogenerators, Nano. Energy. 52 (2018) 153-162. https://doi.org/10.1016/j.nanoen.2018.07.053

[59] H. Madinei, H.H. Khodaparast, S. Adhikari, M. Friswell, Design of MEMS piezoelectric harvesters with electrostatically adjustable resonance frequency, Mech. Syst. Signal Process. 81 (2016) 360-374. https://doi.org/10.1016/j.ymssp.2016.03.023

[60] M. Shirvanimoghaddam, K. Shirvanimoghaddam, M.M. Abolhasani, M. Farhangi, V.Z. Barsari, H. Liu, M. Dohler, M. Naebe, Towards a green and self-powered Internet of Things using piezoelectric energy harvesting, IEEE. Access. 7 (2019) 94533-94556. https://doi.org/10.1109/ACCESS.2019.2928523

[61] Y. Sun, J. Chen, X. Li, Y. Lu, S. Zhang, Z. Cheng, Flexible piezoelectric energy harvester/sensor with high voltage output over wide temperature range, Nano. Energy. 61 (2019) 337-345. https://doi.org/10.1016/j.nanoen.2019.04.055

[62] A.C. Turkmen, C. Celik, Energy harvesting with the piezoelectric material integrated shoe, Energy. 150 (2018) 556-564. https://doi.org/10.1016/j.energy.2017.12.159

[63] L.J. Lu, W.Q. Ding, J.Q. Liu, B. Yang. Flexible PVDF based piezoelectric nanogenerators. Nano. Energy 78 (2020) 105251. https://doi.org/10.1016/j.nanoen.2020.105251

[64] Z.C. He, F. Rault, M. Lewandowski, E. Mohsenzadeh, F. Salaün. Electrospun PVDF nanofibers for piezoelectric applications: A review of the influence of electrospinning parameters on the phase and crystallinity enhancement. Polymers. 13 (2021) 174. https://doi.org/10.3390/polym13020174

[65] C. Dagdeviren, B.D. Yang, Y. Su, P.L. Tran, P. Joe, E. Anderson, J. Xia, V. Doraiswamy, B. Dehdashti, X. Feng, Conformal piezoelectric energy harvesting and storage from motions of the heart, lung, and diaphragm, Proc. Natl. Acad. Sci. 111 (2014) 1927-1932. https://doi.org/10.1073/pnas.1317233111

[66] M. Háková, L.C. Havlíková, J. Chvojka, J. Erben, P. Solich, F. Švec, D. Šatínský, A comparison study of nanofiber, microfiber, and new composite nano/microfiber polymers used as sorbents for on-line solid phase extraction in chromatography system, Anal. Chim. Acta. 1023 (2018) 44-52. https://doi.org/10.1016/j.aca.2018.04.023

[67] B. Ren, S.W. Or, X. Zhao, H. Luo, Energy harvesting using a modified rectangular cymbal transducer based on 0.71 Pb (Mg 1/3 Nb 2/3) O 3-0.29 PbTiO 3 single crystal, J. Appl. Phys. 107 (2010) 034501. https://doi.org/10.1063/1.3296156

[68] M. Han, H. Wang, Y. Yang, C. Liang, W. Bai, Z. Yan, H. Li, Y. Xue, X. Wang, B. Akar, Three-dimensional piezoelectric polymer microsystems for vibrational energy

harvesting, robotic interfaces and biomedical implants, Nat. Electron. 2 (2019) 26-35. https://doi.org/10.1038/s41928-018-0189-7

[69] D.-M. Shin, H.J. Han, W.-G. Kim, E. Kim, C. Kim, S.W. Hong, H.K. Kim, J.-W. Oh, Y.-H. Hwang, Bioinspired piezoelectric nanogenerators based on vertically aligned phage nanopillars, Energy. Environ. Sci. 8 (2015) 3198-3203. https://doi.org/10.1039/C5EE02611C

[70] V. Vivekananthan, N.R. Alluri, Y. Purusothaman, A. Chandrasekhar, S. Selvarajan, S.-J. Kim, Biocompatible collagen nanofibrils: an approach for sustainable energy harvesting and battery-free humidity sensor applications, ACS. Appl. Mater. Interfaces. 10 (2018) 18650-18656. https://doi.org/10.1021/acsami.8b02915

[71] B.J. Hansen, Y. Liu, R. Yang, Z.L. Wang, Hybrid nanogenerator for concurrently harvesting biomechanical and biochemical energy, ACS. nano. 4 (2010) 3647-3652. https://doi.org/10.1021/nn100845b

[72] G. Zhu, Z.-H. Lin, Q. Jing, P. Bai, C. Pan, Y. Yang, Y. Zhou, Z.L. Wang, Toward large-scale energy harvesting by a nanoparticle-enhanced triboelectric nanogenerator, Nano. Lett. 13 (2013) 847-853. https://doi.org/10.1021/nl4001053

[73] A. Baji, Y. W. Mai, Q. Li, Y. Liu. Electrospinning induced ferroelectricity in poly(vinylidene fluoride) fibers. Nanoscale. 3 (2011) 3068-3071. https://doi.org/10.1039/c1nr10467e

[74] C.X. Zhao, J. Niu, Y.Y. Zhang, C. Li, P. H. Hu. Coaxially aligned MWCNTs improve performance of electrospun P(VDF-TrFE)- based fibrous membrane applied in wearable piezoelectric nanogenerator. Compos. B Eng. 178 (2019) 107447. https://doi.org/10.1016/j.compositesb.2019.107447

[75] S. Bodkhe, G. Turcot, F.P. Gosselin, D. Therriault, One-step solvent evaporation-assisted 3D printing of piezoelectric PVDF nanocomposite structures, ACS. Appl. Mater. Interfaces. 9 (2017) 20833-20842. https://doi.org/10.1021/acsami.7b04095

[76] E.J. Curry, K. Ke, M.T. Chorsi, K.S. Wrobel, A.N. Miller, A. Patel, I. Kim, J. Feng, L. Yue, Q. Wu, Biodegradable piezoelectric force sensor, Proc. Natl. Acad. Sci. 115 (2018) 909-914. https://doi.org/10.1073/pnas.1710874115

[77] J. Joseph, S.G. Singh, S.R.K. Vanjari, Leveraging innate piezoelectricity of ultra-smooth silk thin films for flexible and wearable sensor applications, IEEE Sens. J. 17 (2017) 8306-8313. https://doi.org/10.1109/JSEN.2017.2766163

[78] K. Shi, M. Ren, I. Zhitomirsky, Activated carbon-coated carbon nanotubes for energy storage in supercapacitors and capacitive water purification, ACS Sustain. Chem. Eng. 2 (2014) 1289-1298. https://doi.org/10.1021/sc500118r

[79] X. Wang, Y. Gu, Z. Xiong, Z. Cui, T. Zhang, Silk-molded flexible, ultrasensitive, and highly stable electronic skin for monitoring human physiological signals, Adv. Mater. 26 (2014) 1336-1342. https://doi.org/10.1002/adma.201304248

[80] D.-M. Shin, S.W. Hong, Y.-H. Hwang, Recent advances in organic piezoelectric biomaterials for energy and biomedical applications, Nanomater. 10 (2020) 123. https://doi.org/10.3390/nano10010123

[81] Z. Liu, S. Zhang, Y. Jin, H. Ouyang, Y. Zou, X. Wang, L. Xie, Z. Li, Flexible piezoelectric nanogenerator in wearable self-powered active sensor for respiration and healthcare monitoring, Semicond. Sci. Technol. 32 (2017) 064004. https://doi.org/10.1088/1361-6641/aa68d1

[82] J. Park, M. Kim, Y. Lee, H.S. Lee, H. Ko, Fingertip skin-inspired microstructured ferroelectric skins discriminate static/dynamic pressure and temperature stimuli, Sci. Adv. 1 (2015) e1500661. https://doi.org/10.1126/sciadv.1500661

[83] X. Cheng, X. Xue, Y. Ma, M. Han, W. Zhang, Z. Xu, H. Zhang, H. Zhang, Implantable and self-powered blood pressure monitoring based on a piezoelectric thinfilm: Simulated, in vitro and in vivo studies, Nano. Energy. 22 (2016) 453-460. https://doi.org/10.1016/j.nanoen.2016.02.037

[84] D.Y. Park, D.J. Joe, D.H. Kim, H. Park, J.H. Han, C.K. Jeong, H. Park, J.G. Park, B. Joung, K.J. Lee, Self-powered real-time arterial pulse monitoring using ultrathin epidermal piezoelectric sensors, Adv. Mater. 29 (2017) 1702308. https://doi.org/10.1002/adma.201702308

[85] K. Kapat, Q.T. Shubhra, M. Zhou, S. Leeuwenburgh, Piezoelectric nano-biomaterials for biomedicine and tissue regeneration, Adv. Funct. Mater. 30 (2020) 1909045. https://doi.org/10.1002/adfm.201909045

[86] C. Shuai, G. Liu, Y. Yang, W. Yang, C. He, G. Wang, Z. Liu, F. Qi, S. Peng, Functionalized BaTiO3 enhances piezoelectric effect towards cell response of bone scaffold, Colloids Surf. B: Biointerfaces.185 (2020) 110587. https://doi.org/10.1016/j.colsurfb.2019.110587

[87] A. Zaszczyńska, A. Gradys, P. Sajkiewicz, Progress in the applications of smart piezoelectric materials for medical devices, Polymers. 12 (2020) 2754. https://doi.org/10.3390/polym12112754

[88] S.M. Damaraju, Y. Shen, E. Elele, B. Khusid, A. Eshghinejad, J. Li, M. Jaffe, T.L. Arinzeh, Three-dimensional piezoelectric fibrous scaffolds selectively promote mesenchymal stem cell differentiation, Biomaterials. 149 (2017) 51-62. https://doi.org/10.1016/j.biomaterials.2017.09.024

[89] M. Hoop, X.-Z. Chen, A. Ferrari, F. Mushtaq, G. Ghazaryan, T. Tervoort, D. Poulikakos, B. Nelson, S. Pané, Ultrasound-mediated piezoelectric differentiation of neuron-like PC12 cells on PVDF membranes, Sci. Rep. 7 (2017) 1-8. https://doi.org/10.1038/s41598-017-03992-3

[90] Y.-S.L.a.T.L. Arinzeh, The Influence of Piezoelectric Scaffolds on Neural Differentiation of Human Neural Stem/Progenitor Cells, Tissue Eng. 18 (2012) 19-20. https://doi.org/10.1089/ten.tea.2011.0540

[91] J. Jacob, N. More, K. Kalia, G. Kapusetti, Piezoelectric smart biomaterials for bone and cartilage tissue engineering, Inflamm. Regen. 38 (2018) 1-11. https://doi.org/10.1186/s41232-018-0059-8

[92] A.H. Rajabi, M. Jaffe, T.L. Arinzeh, Piezoelectric materials for tissue regeneration: A review, Acta Biomater. 24 (2015) 12-23. https://doi.org/10.1016/j.actbio.2015.07.010

[93] F. Ali, W. Raza, X. Li, H. Gul, K.-H. Kim, Piezoelectric energy harvesters for biomedical applications, Nano. Energy. 57 (2019) 879-902. https://doi.org/10.1016/j.nanoen.2019.01.012

Advanced Functional Piezoelectric Materials and Applications   Materials Research Forum LLC
Materials Research Foundations 131 (2022) 186- 221   https://doi.org/10.21741/978164490209-7

Chapter 7

Piezoelectric Thin Films and their Applications

M. Rizwan[1*], A. Ayub[2], I. Ilyas[2], F. Seeart[2], T. Noor[3]

[1]School of Physical Sciences, University of the Punjab, Lahore, Pakistan

[2]Department of Physics, University of the Punjab, Lahore, Pakistan

[3]Department of Physics, University of Gujrat, Gujrat, Pakistan

*rizwan.sps@pu.edu.pk

Abstract

Lead free piezoelectric thin films are very pivotal in technological applications. There has been a tremendous amount of research in the lead free piezoelectric thin films. KNN and ZnO piezoelectric thin films are very crucial in the miniaturization of piezoelectric devices. Characterization methods for piezoelectric thin films such as resonant spectrum method and pneumatic loading method are briefly deliberated here. Piezoelectric thin films provide several advantages in various applications such as highly sensitive sensors, large displacements and low voltage actuators. All applications of piezoelectric materials exhibit a high piezoelectric coefficient, insight into future outlook and integration for real-world applications is deliberated here.

Keywords

Thin Films, ZnO, Actuators, Sensors, Piezoelectric Properties, BAW Sensors

Contents

1.   Piezoelectric thin films

Bulk and thin film piezoelectric materials have garnered a lot of attraction due to applications in energy harvesting in the light of the energy crisis. The main principal behind these piezoelectric materials is that a stress caused by an external stimulus produces response in the form of potential difference [1]. This response is termed a piezoelectric effect. Piezoelectric materials and their applications are characterized by parameters such as piezoelectric coefficient, electromechanically coupling, curie temperature, dielectric constant, depolarization temperature, remanent polarization, spontaneous polarization. Lead zirconate titante materials also known as PZT were the first lead based piezoelectric materials that were extensively studied. Later on other lead dependent piezoelectric materials like lead niobate, lead titante and modifications of these solid solutions were explored for piezoelectric applications. PZT materials became the mainstream constituents for piezoelectric uses because of their amazing piezoelectric, optical, ferroelectric, dielectric and elastic materials. PZT materials ruled the ceramic industry but the use of lead was very alarming since its use in the field lead to continuous release of these toxins in the environment. This lead to a novel research of Pb free piezoelectric elements including organic and inorganic lead free piezo ceramics [2]. These alternative lead free piezoelectric materials had comparable and ever superior piezoelectric properties to PZT materials. With time there have been extensive research in the lead free piezoceramic industry and new materials, their processing and characterization and implementation in the practical world is continuously explored [3].

Piezoelectric thin films have applications as actuators, sensors, generators and as well in microelectromechanical system (MEMS). The pyroelectric properties of piezoelectric thin films are utilized in infrared detectors. integrated optical systems that use optical switches have been explored with high dielectric piezoelectric thin films. $PbZr_{1-x}Ti_xO_3$ or PZT) thin films have applications as access memory (NVRAM) and it has superior properties to current semiconductor based devices. PZT thin film layers are very promising because of great dielectric coefficients, lower coercive field and high curie temperature and can easily be integrated into semiconductor technology [4].

PZT films are a solid solution of lead zirconate and lead titanate perovskites. The PZT thin films have perovskite structure, wherever the A position is positioned through lead and B is positioned by zirconium or titanium and oxygen is at face centered. The characteristics of PZT thin films based on concentration of lead titante and temperature. There is a boundary that is called phase boundary (MPB) in which where ferroelectric phase changes to antiferroelectric [5].

PZT epitaxial thin films can be deposited with RF sputtering and metal oxide chemical vapor deposition (MOCVD). These films are very crucial for the integration of piezoelectric devices. Piezoelectric thin films exist in perovskite or wurtzite structure. PZT thin films have dominated piezoelectric devices but new alternative lead free thin films such as ZnO, AlN thin films will be incorporated in innovative piezoelectric devices. PZT thin films are important for piezoelectric transduction [6]. Even though PZT thin films have superior properties, their growth is complicated and thus lags behind bulk piezoelectric materials. PZT thin films are very attractive for high piezoelectric coefficient applications. Fig. 1 shows a PZT thin films based micro actuator

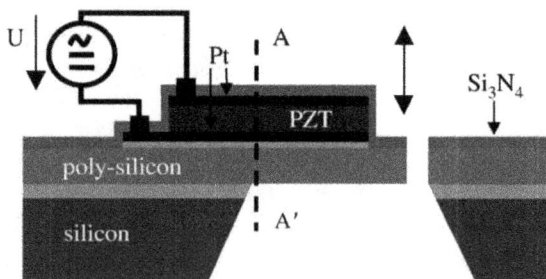

Figure 1. A PZT thin film based micro actuator [7].

Advanced Functional Piezoelectric Materials and Applications   Materials Research Forum LLC
Materials Research Foundations 131 (2022) 186- 221   https://doi.org/10.21741/978164490209-7

2.   Lead free piezoelectric thin films

Lead based piezoceramics have dominated the ceramic industry for more than five decades because of their high piezoelectric and electrochemical properties and low cost processing courtesy to existence of morphotropic phase boundary. Although PZT possess superior properties their lead toxicity is a big concern for commercial applications. The transition to alternative Pb free piezoceramics is still in the works, Pb-free replacements need to have similar or superior characteristics to replace PZT thin films. Some films like these are perovskite based thin films such as barium titanate, alkali niobate thin films, bismuth based thin films and potassium sodium niobate, an aluminum nitride and zinc oxide thin films [8].

2.1   AlN thin films

These thin films have been implemented in energy harvesting, transducers, resonators and actuators because of its wide band gap, high thermal stability and high piezoelectric characteristics [9]. AlN is very suitable for CMOS processing in comparison to PZT thin films [10]. There are many methods how AlN thin films can be deposited like RF and Dc sputtering, metal organic chemical vapor deposition (MOCVD), pulsed laser deposition (PLD), and hydride vapor phase deposition (HVPD), microbeam epitaxy. These optimized techniques allow us to alter the characteristics of these thin films. AlN exist in rock salt, zinc-blende, and wurtzite structure [11]. Out of all these structure wurtzite is the most stable AlN thin film structure. Rock salt structure is also found to be stable in scandium doped AlN thin films. Only wurtzite structure is important since its shows piezoelectric properties. Wurtzite structure is a compatible structure because of the presence of natural polarization. In the wurtzite assembly of AlN thin film, there is a tetrahedral structure in which aluminum is bounded by four nitrogen atoms. Spontaneous polarization in AlN thin films is entirely due to its crystal structure and not due to external electric field or strain [12]. Polarization of these films could be characterized through wet etching process. AlN thin film synthesis needs a substrate or three or six-fold structure the piezoelectric response of bulk counterpart is greater than the film. Silicon oxide is used as a s substrate for deposition of good quality AlN thin films. These films can be synthesized through MOCVD, PLD, MBE, REF and Dc sputtering. In all of these growth methods the growth is along the preferred direction along the direction in which surface energy is lowest. The polarity of these films is measured by the polarity of substrates. The most mutual technique for these film installation is low temperature sputtering [13].

Advanced Functional Piezoelectric Materials and Applications   Materials Research Forum LLC
Materials Research Foundations 131 (2022) 186- 221   https://doi.org/10.21741/978164490209-7

2.2   ZnO thin films

A piezoelectric thin film is pivotal for the miniaturization of electronic devices and in comprehending many piezoelectric devices. The main structure in piezoelectric thin films are perovskite type and wurtzite. ZnO and AlN piezoelectric thin films are non-ferroelectric and have wurtzite structure and have applications as resonator [14]. Zinc oxide is a very promising material due to its large band gap of 3.37eV [15] and high piezoelectric coefficient for its wurtzite structure [16]. These astonishing properties make ZnO a good contender for applications in optoelectronics, sensors, transducers and in energy harvesting [15, 17]. ZnO is an n-type semiconductor, its properties scan be modified via doping of suitable elements or during the process of annealing. Piezoelectric characteristics of ZnO thin films based on the deposition techniques also as well as on the growth parameters [18].

These non-ferroelectric piezoelectric thin films have low coupling coefficients and thus have limiting applications as actuators. The crystallinity of substrate controls crystallinity of ZnO thin films. ZnO piezoelectric thin films have applications as actuators, sensors and as bulk acoustic wave (BAW) resonator and surface acoustic (SAW) filter. PZT/ZnO based actuator and a nano tip integrated actuator performance is explored in literature [19-21]. The rising application of thin films as differential liquid flow sensor is reported in [22]. When ZnO thin films are annealed at 300ºC, they give superior performance as vibrational sensor. Piezoelectric ZnO thin films lean on the methods and conditions of deposition techniques. Atomic layer deposition is implemented for determining structural and optoelectronic properties [23], on the other hand another technique such a pulse laser deposition is implemented in aluminium, boron and gallium doped ZnO thin films for solar cell applications [24].

ZnO thin films have prospective applications in energy harvesting. The characterization of ZnO thin films is performed for the intention of embedment of materials to convert mechanical energy into electrical energy. ZnO thin films characterization also makes nanoelectromechanical system self-powered [24]. A ZnO based power generator is built by integrating ZnO piezoelectric transducer within the stainless harden substrate [25].

ZnO thin films is a promising candidate for BAW resonator devices because of its great k^2, but its temperature coefficient of frequency needs to be decreased for ZnO thin films to be used as BAW resonator [26].

Table 1, gives various ZnO thin films deposited by different methods of different thickness and piezoelectric properties.

Table 1. Piezoelectric coefficient of ZnO films of various thicknesses.

Deposition technique	Thin Film thickness (nm)	Piezoelectric Coefficient d_{33} (pmV^{-1})
Pulse Laser Deposition	200	12 [18]
Pulse Laser Deposition	50	25 [27]
Pulse Laser Deposition	80	49.7 [28]
RF Magnetron Sputtering	285	5 [29]
RF Magnetron Sputtering	710	5.3 [29]
DC Sputtering	210	110 [29]

ZnO thin films operate in the lower frequency region and is very suitable for medical application along with other applications aforementioned due to it being biosafe [30]. When alternating frequency or radio frequency is provided to a piezoelectric material it generates acoustic waves either perpendicular to into the medium (bulk acoustic wave(BAW)) or parallel to the surface (surface acoustic waves(SAW)). SAW devices that have high operational frequency can be realized via ZnO thin films [31]

Zinc oxide thin films and its employment in many applications is due to its versatile properties. ZnO is a very important material for application as transducer due to its high electromechanical coupling coefficient [32]. ZnO thin films show maximum piezo response when the axis is perpendicular to substrate because of its strong piezoelectricity about the c-axis. The main advantages of ZnO is that during thin film deposition via pulsed laser deposition or sputtering it naturally grows in this direction, thus giving maximum piezo activity [33].

ZnO can be implemented as resonator in micro electrical mechanical system (MEMS) technologies. MEMS utilize both direct and inverse piezoelectric transduction to convert electrical and mechanical signal. For direct piezoelectric effect, the ZnO resonator will act as a sensor, on the other hand for reverse piezoelectric effect it as an actuator. Cantilever beam resonator made with ZnO is made of ZnO layer sandwiched between two electrodes [33]. Made of Al or Pt. the electrodes are a necessity for the generation of strain to detect voltage. The frequency of applied voltage is similar to resonance frequency of cantilever to produce resonant vibration, that produce large deflection in the beam causing a stronger output voltage [34].

Advanced Functional Piezoelectric Materials and Applications   Materials Research Forum LLC
Materials Research Foundations 131 (2022) 186- 221   https://doi.org/10.21741/978164490209-7

2.2.1 Synthesis of ZnO thin films

ZnO thin films can be deposited on a substrate in multiple ways such as through sputtering and pulsed laser deposition (PLD). In thin film deposition, parameters like temperature of substrate , energy and chamber pressure play a pivot role in the thin film growth and can greatly control the properties of the film, and as well as the deposition rate [35]. In PLD, laser hits the target material and creates a plasma, that then hits the substrate positioned right in front of it. The ions in the plasma form the thin film upon hitting the substrate [36]. To find the apt deposition technique many deposition techniques are tested and the best suited technique is opted for deposition. Schematic diagram of deposition of ZnO thin films by PLD is given in Fig. 2 [37].

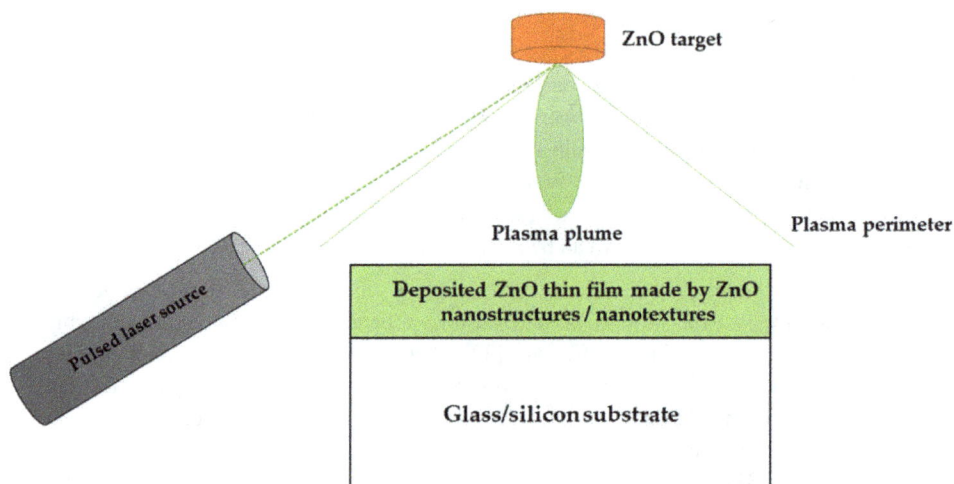

Figure 2. Schematic diagram of PLD of ZnO [37].

The chamber is evacuated to avoid contamination using roughing pump to achieve a very low pressure. The deposition can take place in vacuum or via a reactive gas such as with oxygen in case of oxide thin films. The addition of reactive gas such as oxygen should not affect the partial pressure and mean free path, because that will decrease deposition rate. The temperature of substrate should also be just right because discrepancies in it will affect the crystallinity of the film [38]. The more close the substrate is to the target, the higher is the deposition rate thus deposition rate also depend on the distance between substrate and target [36].

Materials Research Forum LLC
https://doi.org/10.21741/978164490209-7

Sputtering is very similar to PLD, in this plasma plume is formed with a radio frequency power source instead of laser source. The plasma ions are imparted on the target to cause sputtering. The deposition rate is based on the number of ions in the plasma, since ions are causing the sputtering in the target material. Just as discussed previously deposition rate is controlled by substrate temperature, pressure of reactive gas, distance between substrate and target and RF power. Oxygen is added for ZnO deposition, the rate of oxygen to argon is very crucial in ZnO deposition [38]. The schematic diagram of RF sputtering for Al doped ZnO thin films is given in Fig. 3 [39].

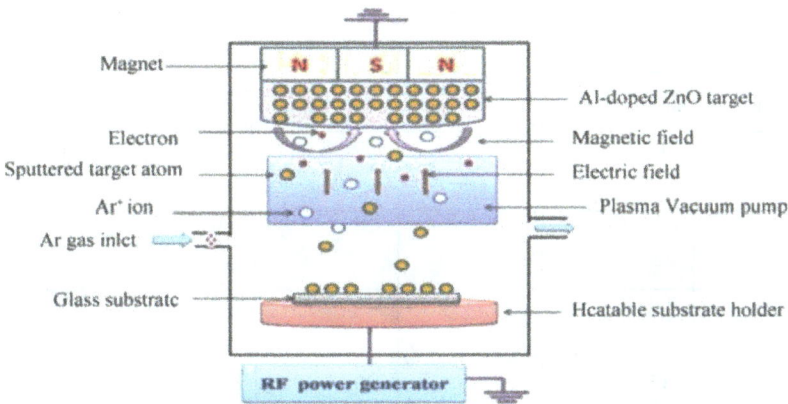

Figure 3. Schematic diagram of Al-doped ZnO deposition via RF sputtering [39] with permission from Copyright Engineered Science Publisher.

2.3   KNN thin films

Potassium sodium niobate thin films are another very important piezoelectric thin films that have garnered attention a as lead free alternative of PZT thin films [40]. KNN thin films have application a microsensors, actuators and as energy generators due to their dielectric properties and electromechanically coupling constants. These devices are fabricated via etching and photolithography [41]. The deposition of KNN thin films by RF sputtering will give high piezoelectric properties. KNN is nontoxic is thus safe for medical application such as a cardio mechanical electrical sensors(CMES) [42].

Ferroelectricity in $KNbO_3$ was discovered in the 1950s. KNN thin films show great piezoelectric response, which can be further modified through addition of systems like $LiTaO_3$ $LiSbO_3$. The modification in piezoelectric response due the addition of these systems caused a huge uproar and boosted up research in KNN thin films. The reason behind enhancement in piezoelectric response is due to polymorphic phase transition between the tetragonal and orthorhombic phase of KNN. Lithium and thallium addition improves piezoelectric coefficient to a value greater than 300pC/N in modified KNN compositions, the improved piezoelectric properties are owed to room temperature instability.

Table 2 gives some KNN compositions and their improved piezoelectric properties.

Table 2. Piezoelectric properties of some modifies KNN compositions.

Material	d_{33} (pC/N)	k_p	T_c	$\dfrac{\varepsilon}{\varepsilon_0}$	Reference
$(K_{0.5}Na_{0.5})NbO_3$	80	0.35	420	290	[43]
0.93KNN-0.07LiNbO$_3$	240	0.45	460	950	[44]
0.82KNN-0.07AgNbO$_3$	175	0.42	360	500	[45]
0.95KNN-0.05LiTaO$_3$	200	0.36	430	570	[46]
KNN-LT	215	0.35	320	1210	[47]
KNN-LiSbO$_3$	265	0.5	392	1380	[48]
KNN-LT-LS	340	0.48	264	1650	[49]

2.3.1 Synthesis of KNN thin films

KNN thin films permittivity has a stable dependence on frequency. Thin films can be fabricated via chemical solution deposition that gives control over homogeneity and composition of thin films. KNN thin films have been fabricated by chemical solution deposition process [50]. Other methods to obtain KNN thin films are PLD, RF-magnetron sputtering and CVD [51-53]. A schematic diagram for the procedure of CVD for KNN thin film is give in

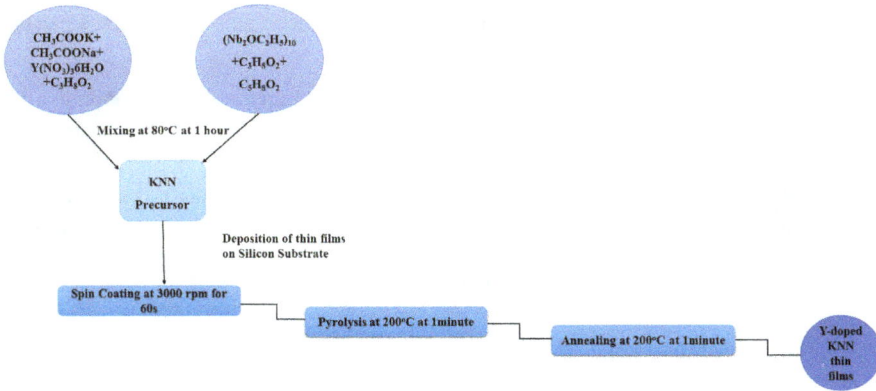

Figure 4. Schematic diagram of process of chemical solution deposition of KNN thin films [54].

Deposition of KNN thin films via physical vapor deposition is complicated due to instability of potassium and sodium. Therefore, sol-gel method is implemented which offers fabrication at a much lower temperature, control over chemical composition and homogeneity [55].

3. Characterization techniques for piezoelectric thin film

Piezoelectric thin films can be fabricated through many techniques; some of the common methods are deliberated here.

3.1 Resonance spectrum method

The resonance frequency is the most precise measurement for characterizing acoustic characteristics of piezoelectric thin films. The piezoelectric film has to be deposited on the solid substrate because at high frequencies it's very thin to be self-reliant, and its properties rely on substrate properties. The characterization of a piezoelectric film that is coated on the anisotropic substrate for forming a two-layers thickness composite resonator have been studied [56]. If we know the resonant spectrum of composite resonators we can determine three piezoelectric film parameters, i.e., the elastic constant, coefficient of electromechanical coupling, and density.

There are two groups of explicit approximate formulas on which resonant spectrum method principals is based. The density, $\hat{\rho}$, and the relation between three (spacing of parallel resonance frequencies) SPRF distinctive values, and elastic constant, $\hat{C}_{33}^D$, piezoelectric

Advanced Functional Piezoelectric Materials and Applications   Materials Research Forum LLC
Materials Research Foundations 131 (2022) 186- 221   https://doi.org/10.21741/978164490209-7

film and the relation between the effective coupling factor of a specific mode $k^2_{eff}(m)$, and k^2_t of the piezoelectric film.

Similarities between the modified model frequency spacing method and distribution of the SPRF method has been observed [57, 58], and it has certain relation with constraints of the piezoelectric film [59, 60]. The consequences are concise by way of:

A) When $\gamma \approx n\pi$ (n=0,1,2,...), at the center of fixed regions, we can derive SPRF values from the following equation:

$$\tan \gamma + z_b \tan \gamma_b = 0, \qquad \text{Eq.1}$$

and is given approximately

$$\Delta f_N \approx \Delta f_0 \cdot \left(1 + \frac{2.(\hat{\rho}l)}{\rho_b b}\right)^{\frac{-1}{2}}, \qquad \text{Eq.2}$$

Where:

$$\Delta f_0 = V_b / 2b , \qquad \text{Eq.3}$$

is the fundamental modes, mechanical resonance frequency, or spacing of the plain plate modes? It is supposed that totally the substrate constraints are identified and equation (2) represents that Δf_N depends only on $\hat{\rho}l$. By the evaluation of Δf_N, from SPRF data the value of $\hat{\rho}l$ could be calculated.

B) When $\gamma \approx (n + \frac{1}{2})\pi$ (n=0,1,2,...), at the shift states center, the SPRF value can be consequent from the equation (1) and is given as

$$\Delta f_T \approx \Delta f_0 \cdot (1 + \frac{\rho_b V_b^2}{b} \cdot \frac{l}{\hat{C}_{33}})^{-1}, \qquad \text{Eq.4}$$

It shows that Δf_T just depend on $\frac{l}{\hat{C}_{33}}$. By the evaluation of Δf_T, from SPRF data the value $\frac{l}{\hat{C}_{33}}$ of can be calculated.

C) The SPRF distribution, the period is taken from (1) and given as

$$\Delta f_C = \frac{\hat{V}}{2l} = \frac{1}{2} \cdot \sqrt{\frac{(\hat{C}_{33}/l)}{(\hat{\rho}l)}} , \qquad \text{Eq.5}$$

SPRF period depend just on $\hat{V}$. By the evaluation of Δf_C and $\hat{V}$ can be find out directly. Either $(\hat{\rho}l)$ or $(\hat{C}_{33}/l)$ is known, others can be found out from this formula.

The equations (2), (4) and (5) gives the approximate relationship between three distinct values $\Delta f_N, \Delta f_T, and \ \Delta f_C$ and the material & geometrical parameters of the two layers and. These three distinct values could be estimated from the dissemination of SPRF, that is obtained from experimental data. These three equations contain the following parameters, bulk modulus,

$(\lambda_1 + 2\mu_1 = \rho_b V_b^2)$, densities, $\hat{\rho}$ and ρ_b, and $\hat{C}_{33}$, and thickness, l and b of two layers, individually [61].

By utilizing the meaning of operative pairing factor given by

$$\Delta f_P(m) = f_P(m + 1) - f_P(m), \qquad \text{Eq.6}$$

and resonance frequency equations (1) and

$$\tan \gamma + z_b \tan \gamma_b = \frac{k_t^2}{\gamma} [\tan (\gamma/2) + z_b \tan \gamma_b] \tan \gamma, \qquad \text{Eq.7}$$

We can stem two formulas that relate k_t^2 to "effective coupling factor" for specific mode at core of the normal area. $k_{eff}^2(m_N)$ and at the 1st transition area $k_{eff}^2(m_T)$, respectively.

$$k_{eff}^2(m_N) = \frac{\pi^2}{4} \cdot \frac{f_s(m)}{f_P(m)} \cdot [1 - \frac{f_s(m)}{f_P(m)}], \qquad \text{Eq.8}$$

In equation (6) $f_P(m + 1), and \ f_P(m)$ are the parallel resonance frequencies of $(m + 1) - th \ and \ m - th$ order HBAR.

At the 1st normal region; $\tan \gamma \approx 0$, we can write down $\gamma = \pi + \varepsilon$ and $\gamma_b = m_N \pi + \delta$ and by ignoring the term that incorporates $\varepsilon. \delta$ to acquire the formula;

$$k_t^2 = \left(1 + \frac{\rho_b b}{\hat{\rho} l}\right) \cdot (k_{eff}^2(m_N)), \qquad \text{Eq.9}$$

Where $(m_N) = round \left(\frac{\hat{V} b}{V_b l}\right) + 1$, it's the order of mode at the core of the region.

At the 1st transition region; $\tan \gamma \to \infty$, we can show that $\gamma = \left(\frac{1}{2}\right). \pi + \varepsilon$; and

$\gamma_b = \left(m_T + \frac{1}{2}\right) \pi + \delta$, we acquire the formula, after a few algebraic process

$$k_t^2 = [1 + \left(\frac{b}{l}\right) \cdot (C_{33}^D/\rho_b \cdot V_b^2)] \cdot \frac{1}{\Gamma} \cdot (k_{eff}^2(m_N)), \qquad \text{Eq. 10}$$

Where $(m_T) = round[\left(\frac{\hat{V} b}{2V_b l}\right) + \frac{1}{2}]$ and it's the order of mode at core of the region.

$$\Gamma = 1 + 2 . \frac{\nabla \hat{\rho}}{\rho_b v_b} . (1 + \varepsilon) . \delta, \qquad\qquad \text{Eq.11}$$

Is the correction factor and not surely near to unity when m_T is not very big.

Equation (9) and (10) shows the measuring methods of k_t^2 value of film can be found out from $k_{eff}^2(m_N)$, and ration of masses of the two layers from $k_{eff}^2(m_T)$, and elasticity and thickness of two layers. Its alluring to account that in equation (9) the relationship between k_t^2 and k_{eff}^2 is engrossed by vibrational kinetic energy distribution, and in (10) relation is engrossed by the distribution of elastic P.E. By making use of resonance frequency spectra the effective coupling factors $k_{eff}^2(m_T)$, and $k_{eff}^2(m_N)$, are measured.

By taking two assumptions, we derive the two groups of explicit formulas,

     a) The lossless materials, so the series & parallel resonance frequency formulas are specified through equation (1) and (7).

     b) The electrode properties are neglected.

There are always some harms in constituents and resonance frequencies are measured from

$$Z_{in} = \frac{V}{I} = \frac{1}{j\omega C_0}\left[1 - \frac{k_t^2}{\gamma} . \frac{2\tan\left(\frac{\gamma}{2}\right) + Z_b \tan\gamma_b}{1 + \frac{Z_b \tan\gamma_b}{\tan\gamma}}\right], \qquad\qquad \text{Eq.12}$$

directly.

To identify the validity of this method we are presenting numerical calculation consequences and a few experimental consequences to demonstrate the validation of this technique and electrodes belongings on the correctness of this method.

3.2   Pneumatic loading method and normal loading method

The piezoelectric coefficient is the essential parameter to develop (microelectromechanical systems) MEMS that consist of these thin films. Actually, for the assessment technique of piezoelectric thin films, there are two trends. The 1st one is to make usage of straight piezoelectric result like normal loading technique, and the second one is to apply opposite piezoelectric result such as interferometer. Before preceding to high field poling process, the antecedent can tell information to us. Because after applying voltage for poling it leads to substantial reorientation of domains in as-deposited film. The method piezoelectric effect is inescapable to study phenomena of piezoelectric in as-deposited films. To compute clamped films, the use of direct effects is less fallacious than the one by using converse effects as supported by the mathematical manipulation of Lefki's [62]. From a practical

Advanced Functional Piezoelectric Materials and Applications   Materials Research Forum LLC
Materials Research Foundations 131 (2022) 186- 221   https://doi.org/10.21741/978164490209-7

point of view, there is no other procedure for the dimension like polishing of back-side and gluing that essentially needed for the application of the converse effect method.

Lead zirconate-titanate (PZT) thin films that have the configuration of MPB and 400nm thickness were assembled by metal-oxide deposition (MOD0 process on silicon substrates which are platinized. The thickness of the normal silicon substrate and bottom electrode of platinum was 0.5mm and 100nm respectively. Dong-Guk Kim and Ho-Gi Kim [63] depicted Polarization hysteresis and XRD pattern of PZT film which are fast thermal processed for crystallization respectively at 650^0C. It was found that well-developed thin films of PZT had small chunk of pyrochlore phase and no favorable orientation. Polarization hysteresis loops were computed at an extreme voltage of 1V to 14V by RT66A.

Experimental set-up of inflated filling technique has been shown by a study performed by Dong-Guk Kim and Ho-Gi Kim [63]. This arrangement was boosted by insistent probe to deliver pressure on the surface of the film, to check normal operation pressure gauge is used, for the cause of inflatable power vacuum pump is used, to measure electrical charge from the piezoelectric film the charge amplifier is used and few valves are present for pneumatic control. The thin-film capacitor with a platinum topmost electrode having a diameter of a small number of millimeters was strained alternately through 1 atm pressure, and at the same time, the piezoelectric charge induced from the thin film was detected by the charge amplifier. We can calculate the piezoelectric d-coefficient from its thermodynamic definition. To examine loading nature, we make use of four types of metal tips equally tip with diameter 1 mm & 2 mm and the surface of tip refined and un-polished. Before characterization, totally cells are propelled at 10 V for 10 minutes.

The charge induced from the this film was computed as burden enlarges, for the usual charging with four distinct types of instructions as displayed in Dong-Guk Kim and Ho-Gi Kim [63]. It results in four distinct sloped lines rather than a single line. It implies that the usual filling technique must have to acknowledge with planes of contact. The piezoelectric coefficient computed from each slope ranges from 479pC/N for the tip of unpolished 1 mm diameter to 352pC/N for the polished tip of 2 mm-diameter.

To describe this fact AFM analysis was reported in [63] for the surface exposed to piezoelectric film and tip, and load bias experiments was also studied for the sophisticated and unsophisticated tip. From AFM analysis irregularity data is concluded in the Table 3. The microscopic images of the surfaces express us the interaction conditions at loading time.

Advanced Functional Piezoelectric Materials and Applications   Materials Research Forum LLC
Materials Research Foundations 131 (2022) 186- 221   https://doi.org/10.21741/978164490209-7

Table 3. Roughness conclusion of AFM analysis (Units: Scan area, Angstrom, 10 micron by micron) [63].

Surface	Rms Roughness	Peak to Valley	Average Roughness
Thin film before loading	**141**	**18**	**15**
Polished Tip	**4508**	**343**	**21**
Thin film after loading	**250**	**33**	**26**

The film surface is roughened by rough tips. That's not the conformal contact between films and tips and the distribution of pressure mostly relies on location. Some of the areas does not press and some regions withstand large pressure which may leads to damage to the surface. This section of pressure distribution shows that the piezoelectric coefficient depends on the size of the tip and polishing state. With the increase in load the contact becomes closer between tips and films. The study performed in [63] shows that with the load bias the difference in interaction circumstances decrease between unsophisticated and sophisticated tip. It implies that for authentic calculations the uniform pressure circulation on film exterior is an essential state.

In [63], the typical outputs for piezoelectric cells of diameter 3 mm, for several computing circumstances of time, driving direction and voltage were shown. There is a direct relation between peak of each line and the piezoelectric constant of films experiencing a definite history. Dense lines are said to be for the similar cell as computed through the normal loading technique. The outcome of the inflated filling technique is 333pC/N for similar cells, a slightly smaller value than normal loading process. Identical filling by the inflatable method can acquire additional true values without harming the surface of the film.

The non-uniformity of pressure distribution leads to the usual filling technique is exorbitantly sensitive to the conditions of a tip. In addition to it, the inflated filling technique has been suggested and it illustrates a more consistent method for characterizing piezoelectric thin film.

3.3   Characterizations using capacitance measurements

Here we discuss variation in dimensions of the piezoelectric thin film of Aluminum Nitride when we apply dc bias. There is no need for a polling process in AlN because of its oriented composition. For the characterization of piezoelectric film parameters, the measurements in the variation of transversal and longitudinal directions are the key, namely, its charge constants and strain.

The equation shows the capacitance of a dielectric film [64]

$$C_0 = \frac{\epsilon_0 \epsilon A}{d} \, , \qquad \text{Eq.13}$$

Here ϵ is the dielectric constant or relative permittivity, d is the thickness of film, A is the capacitor area and ϵ_0 is the permittivity of free space and. Eq(13) can be utilized to compute electronic dielectric constant if capacitance of AlN is measured.

It was shown in [64] that AlN piezoelectric material domains will contract when it is driven by applying dc field **E**, therefore its film thickness increased by Δd and area reduced by ΔA [64]. Both the variation in ΔA and vertical extension Δd of AlN material are corresponds with the magnitudes of both transvers d_{31} and longitudinal d_{33} charge constants as given [65]:

$$\Delta d = V d_{33} \, , \qquad \text{Eq.14}$$

$$\Delta A = -V d_{31} A/d \, . \qquad \text{Eq.15}$$

Moreover

$$d_{33} = -x \times d_{31}, \qquad \text{Eq.16}$$

Where $|x|$ ranges from 2 to 2.5 [66]. Thus

$$\frac{\Delta d}{d} \approx -x \frac{\Delta A}{A} \, , \qquad \text{Eq.17}$$

The alteration in material shape is linked with the change in capacitance value, that is computed approximately from the formula of capacitance of parallel plate capacitor:

$$C_V = \epsilon_0 \epsilon \frac{A - \Delta A}{d + \Delta d} \, , \qquad \text{Eq.18}$$

$$C_V = \epsilon_0 \epsilon \frac{A(1 - \frac{\Delta A}{A})}{d(1 + \frac{\Delta d}{d})} \, , \qquad \text{Eq.19}$$

$$C_V = C_0 \frac{(1 - \frac{\Delta A}{A})}{(1 + \frac{\Delta d}{d})} \, , \qquad \text{Eq.20}$$

Putting (17) into (20) and solving for ΔA and Δd respectively;

$$\Delta A = A \left(\frac{C_r - 1}{C_r + x} \right) \, , \qquad \text{Eq.21}$$

$$\Delta d = \frac{d(1-C_r)}{(x^{-1}+C_r)},$$
Eq.22

Here, C_r is the ratio between C_V and C_0. By applying the dc field piezoelectric material is activated and because of change in geometrical dimensions, the capacitance will decrease as predicted by Eq(2) and (3). Hence by knowing the thin-film capacitance ratio and thickness we can determine the vertical or horizontal extension of piezoelectric material.

A parallel plate capacitor was fabricated containing AIN piezoelectric in [67]. The dielectric material of the capacitor is made up of AIN thin film having a thickness of $0.6\ \mu m$. The electrodes are made up of Molybdenum of thickness $0.1\ \mu m$. The area of the capacitor is $700\times200\ \mu m^2$. HP4294A impedance analyzer is used to measure the capacitance of fabricated device with and without dc bias [67]. There is a smooth and stable behavior over the frequency. Because of the converse piezoelectric effect, the application of dc bias results in the expansion of the material. Thus becomes the reason to decrease capacitance results from this structure. When we apply dc bias the change in capacitance takes place about 3.5% from the unbiased value.

This method eludes the preparation of complex sample preparation and the utilization of complicated test sets. When the ratio of capacitance is taken into account then the interfacial problems between the AIN film and metallization are automatically implanted. As compared to other methods this method consequence in a higher value. The real value of the coefficient lies in the calculated range. The estimated error will be 16% if we assume the calculated mid-range value to be the real parameter of material.

4.   Applications

Piezoelectric thin films (PZT) having width less than 5 µm have widely been utilized in several microelectromechanical systems (MEMS) applications. PZT materials are one of the greatest materials due to their outstanding transformation of energy from mechanical to electrical domains even at a small scale. PZT films are more attractive in those devices in which piezoelectric coefficients are firstly concerned and their high values are required [42, 68]. PZT based ferroelectric thin films are also an important class and are widely used in piezoelectric transduction. Their piezoelectricity is also employed in actuators, sensors, and transducers. In MEMS devices PZT thin films are being applied in several electronic devices such as actuators, biosensors, resonators, filters, etc. Early, many problems were suffered for integration of PZT thin films but now recently many advances have been carried out for fabrication of no of devices increasing its applications. Here we will discuss some presentations of thin films in plans at micro and nano scales. In microdevices to grow identifying and motivating functionalities deposition of PZT thin films upon substrates in

Advanced Functional Piezoelectric Materials and Applications   Materials Research Forum LLC
Materials Research Foundations 131 (2022) 186- 221   https://doi.org/10.21741/978164490209-7

front of manufacturing is an well-organized technique [42]. Gadgets that associated electronics with other devices like actuators, transducers and sensors have also gained attention. Energy harvesting, actuators, electronics, MEMS, and acoustic biosensors are some promising applications of PZT films that are explained simultaneously. All these offer many advantages compared with competitor technologies.

4.1 Energy harvesting

For supplying low-power mobile electronics, implantable devices, and wireless microsystems and sensors which are important for sensing automotive applications and for monitoring the environment; PZT energy harvesters are promising transducers. These also gained much attention as a key technology in autonomous wireless sensors network (WSN) in which changing batteries is not practical, and in e-health, etc. Piezoelectric materials that can generate enough power even from minute biomechanical excitations at very small frequencies and at non-resonant conditions are required by WSN applications. New advances in energy harvesting are paving a new era of applications for the realization of replacing traditional batteries with sustainable, rechargeable, and maintenance-free batteries. Many piezoelectric materials now are used for energy harvesting applications but Aluminum nitride (AlN), zinc oxide (ZnO), lead zirconate titanate (PZT), and polyvinylidene fluoride (PVDF) are most commonly used. Presentation of altered constituents for piezoelectric energy harvesting presentations is associated that is termed as energy harvesting figure of merit (FOM) and is expressed by the following relation [69]:

$$FOM = \frac{e_{31}^2}{\varepsilon_o \varepsilon_r} \qquad Eq.23$$

Here e_{31} and ε_r denotes piezoelectric coefficient and dielectric permittivity of material respectively. In usually utilized PZT elements high values of piezoelectric coefficient in between -8 to -12$^C/_{m^2}$ and large values of dielectric permittivity reduce its efficiency for energy harvesting devices. AlN thin films having a low piezoelectric coefficient ($\approx$ 1 $^C/_{m^2}$) and small dielectric permittivity (≈ 10) are also considered a noble applicant for energy gathering applications [69]. Conversion of mechanical energy exhibits many techniques and effects but energy harvesting based upon PZT thin films is an outstanding class for the conversion of electrical energy into mechanical energy. This transfer of energy is because of the capability of piezoelectric constituents to transform useful strain energy into electrical energy utilizing some micromachining fabrication techniques. Some challenges are faced during mechanical energy harvesting like weak excitation causes with low frequency ($< 100Hz$) [70]. Conversion of mechanical form into electrical form is always proportional to piezoelectric coupling (k^2) that does not represent the material constant of piezoelectric for MEMS structure but rigidities of elastic materials evolved in deformation of piezoelectric material [42]. PZT materials are used in 33 modes showing both voltages and stress are acting in 3-direction. However, PZT materials can also be controlled in 31 modes showing voltages in 3-direction while mechanical stressing in 1-direction. Bimorph is the most common type of 31 modes. In it, energy is fluctuating between potential energy and kinetic energy form. The detention of energy through an involved circuit moisture the oscillator. Other parasitic damping mechanisms compete with PZT energy harvesting. It is originating that the damping constant gained through energy gathering necessity be equivalent to or greater than the dependent checking constant. Cantilever-involved materials such as PZT exhibit better k^2 standards of 5% or more can be supportive in two means when utilized for energy harvesting. First, these provide superior matching with damping coefficient when operated in air. Second, these deliver much energy in the electrical tanks for harvesting. Furthermore, one more issue of corresponding of operative conflict of the harvesting circuit with an interior resistance of PZT also affects [71]. High combination PZT procedures possess a large frequency break between resonance frequency in which voltages are highest and internal resistances are lowest and in antiresonance frequency in which both voltages and internal resistances are highest. A working point does not necessarily exist near resonance as often taken but is found near to antiresonance. Typically, those PZT thin films that are employed for energy harvesting applications possessed wurtzite or perovskite crystal structure increasing its demand for further applications. Generally, resonant PZT energy harvesters consisting of firm substrates have a beam shape and manifest important resonant frequency as [70]:

$$\omega_n = A_n \sqrt{\frac{K_{eff}}{M}} \qquad \text{Eq.24}$$

Here K_{eff} represents effective spring constant taking into account thickness and stiffness constants of considered material, A_n is a parameter describing geometrical dimensions such as width, cross-sectional area and length of a vibrating system. Nowadays however most energy collecting devices are dependent upon piezoelectric thin films yet many challenges are faced such as inadequacy of easy fabrication techniques, high use of toxic components such as lead (Pb) that limit their use for flexible, and wearable devices. Choice of electrodes is also an important point in thin-film piezoelectric harvesters. Piezoelectric thin film parallel plate capacitors possessed more capacitance but low output voltages. Vibration collecting devices have AC at output therefore rectification is done for energy storage in batteries that need at least 500 mV even when charge transfer switches are utilized for voltage multiplication. Finally, it is more preferable to use interdigitated electrodes for energy harvesting with high permittivity piezoelectric [71]. Despite all, now significant investigation and many energies are being completed on Pb-free high-quality PZT thin films to generate high power output for their potential use as energy harvesters to supply sustainable and implantable electronics. High power levels can be obtained via a micropower generator if a frequency of greater than 100Hz and acceleration of 0.1 to 1g is utilized in vibration sources.

4.2   Actuators

The properties of the piezoelectric thin films differ from piezoelectric bulk materials. Clamping of films with an elastic body (substrate) is the main difference between piezo films and bulk piezoelectric. The film is composed of a composite structure that is clamped in a film frame but moves freely in an off-plane direction, and the result of calculations of formal within two piezoelectric along numerous elastic constants is required for the rigorous treatment of this problem. Elastic coefficients are not known exactly up-to-now therefore most effective method is to use those piezoelectric constants of film that are compressed with a firm substrate [72]. The main hindrance during thin film measurement is only the existence of substrate as for determining mechanical characteristics of sample both PZT and substrate are evolved. Then measured piezoelectric coefficients will be functions of standard piezoelectric coefficients. For actuators in-plane stresses σ_1 and σ_2 and off-plane strain s_3 transformed such as a function of a practical field E_3 as:

$$\sigma_{1,2} = -e_{31,f} \cdot E_3$$

Eq.25

$$s_3 = d_{33,f} \cdot E_3 \text{sa}$$

$d_{33,f}$ and $e_{31,f}$ are effective piezoelectric coefficients that can be determined from bulk tensor properties. These effective coefficients are linked with bulk coefficient by following relations as [73]:

$$e_{31,f} = \frac{d_{31}}{s_{11}^E + s_{12}^E} = e_{31} - \frac{c_{13}^E}{c_{33}^E} e_{33} \qquad |e_{31,f}| > |e_{31}|$$

Eq.26

$$d_{33,f} = \frac{e_{33}}{c_{33}^E} = d_{33} - \frac{2s_{13}^E}{s_{11}^E + s_{12}^E} d_{31} \qquad < d_{33}$$

Here s_{11}^E and s_{12}^E are known as elastic compliance of PZT at the constant electric field that is very important in evaluating piezoelectric characteristics of thin films. $d_{33,f}$ is also known as longitudinal effect and used to explain a change in thickness with the applied field. $d_{33,f}$ can be obtained by measuring strain s_3 as a function of an electric field by using precisely a double-beam interferometer or by bending of substrate and then accumulating developed charges linked with in-plane stresses as [74]:

$$D_3 = e_{31,f} \ (x_1 + x_2) \qquad\qquad \text{Eq.27}$$

Advanced Functional Piezoelectric Materials and Applications   Materials Research Forum LLC
Materials Research Foundations 131 (2022) 186- 221   https://doi.org/10.21741/978164490209-7

Despite these many other methods can also be used to measure the thickness of PZT; however, each method has pros and cons. So, the method of measurement should be nominated according to the purpose of capacity. Piezoelectric charge (D) uses as a function of in-plane stress σ_1 can be used to determine $e_{31,f}$ coefficient that is a transverse effect. In a precise manner, this can be achieved by creating a bending in cantilever or by frequency change in resonator originated by piezoelectric stresses. From Equation (24) it is clear that $e_{31,f}$ is greater than e_{31} as if the model is permitted to move along the longitudinal path, then more piezoelectric strains can be evolved in transverse instructions. Transverse coefficient $e_{31,f}$ has many potential applications, especially in actuators. PZT solutions become more suitable for actuator applications when a current signal, force, and power outputs are demanded. High piezoelectric and electromechanical coupling coefficients play a critical role in such cases.

In cantilever structures, piezoelectric in-plane pressures produce bending moments in the membrane concerning the neutral level.

$$M_1 = dt_p e_{31,f} E_3 \qquad\qquad \text{Eq.28}$$

Below piezoelectric film thickness of the passive part (h) must be close to the width of the piezoelectric film (t_p). k is a very important coupling coefficient for ultrasonic applications whose value k^2 evaluates the transfer of energy from an electrical to a machine-driven arrangement. The combination constant (k) based upon the concerned material and geometry and its value in nearly modest models can be designed systematically. For its calculation consider a free-standing plate for minor bends that radii of curvatures are equal to moments M_1 and M_2 respectively.

$$\frac{1}{R_1} + \frac{1}{R_2} = \frac{1}{N}(M_1 + M_2) \qquad\qquad \text{Eq.29}$$

Here N denotes rigidity for plates. Both directions can be the same for the isotropic system as considered here and moments are distributed homogeneously in such a way that the whole plate is covered by one top electrode. For this simple system, the whole expression is written in 2 × 2 matrices where curvature ($\frac{1}{R}$) and D as a function of the applied voltages and moment are represented [75]. For symmetrical matrix moment and curvature are expressed as:

$$\frac{1}{R} = \frac{\frac{1}{R_1} + \frac{1}{R_2}}{\sqrt{2}}, \qquad M = \frac{(M_1 + M_2)}{\sqrt{2}} \qquad\qquad \text{Eq.30}$$

Then constitutive relation for the matrix is:

$$\begin{bmatrix} R^{-1} \\ D \end{bmatrix} = \begin{bmatrix} \dfrac{1}{N} & \dfrac{\sqrt{2}e_{31,f}d}{N} \\ \dfrac{\sqrt{2}e_{31,f}d}{N} & \dfrac{\varepsilon_0\varepsilon_{33,f}}{t_p} \end{bmatrix} \begin{bmatrix} M \\ V \end{bmatrix} \qquad \text{Eq.31}$$

The square of the coupling constant (k^2) is equivalent to the ratio of the square without diagonal elements to the multiplication of both diagonal components. For case $t_p << h$ values of coefficients N and d in terms of Poisson's ratio (υ) and bulk modulus (Y) of silicon can be expressed as:

$$N \cong \frac{Yh^3}{12(1-\upsilon)}, \qquad d = \frac{h}{2} \qquad \text{Eq.32}$$

Then putting values of coefficient N and d final expression of k^2 can be transcribed as given below [73]:

$$k^2 = 3k_p^2 \frac{\varepsilon_{33}^\sigma}{\varepsilon_{33,f}} \cdot \frac{(s_{11} \cdot (1-\upsilon))_{Si}}{(s_{11}^E \cdot (1-\upsilon_{12}))_{PZT}} \cdot G \cdot \alpha \left(\frac{h}{t_p}\right) \cdot \left(\frac{t_p}{n}\right) \qquad \text{Eq.33}$$

Here k_p and G denotes planer coupling constant of geometrical factor and piezoelectric material followed from type and shape of the energetic assembly and electrodes, and symmetry of vibration modes respectively. α being a function of $\left(\frac{h}{t_p}\right)$ is zero if $h = 0$ and 1 for $h >> t_p$. Coupling coefficient can also be calculated using energy method or can be derived considering clamped disk micro motor or ultrasonic stator. For each schematic representation, a FOM for the substantial can be developed. For force yield PZT is suitable material however gained power efficiency having the relation of $\frac{k^2}{\tan\delta} \propto \frac{(e_{31,f})^2}{\varepsilon_{33,f}.\tan\delta}$ for ultrasonic applications is not always suitable with piezoelectric thin films.

Here $\tan\delta$ represents dielectric losses that are very high (generally 2 to 4% at 1kHz) for PZT thin films than ZnO and AlN thin films [53]. Therefore, for higher efficiency, AlN and ZnO thin films are competitive for PZT. Furthermore, further improvements must be done to reduce losses for PZT thin films.

4.3 Electronics

Reduced size and weight are significantly needed in modern electronic devices to evolve the electronic modulus of increasing capacity and functionality into large areas of applications, especially in future mobile platforms. Integration of downscaled electronic systems with highly acquired characteristics into several technologies such as nano or microelectromechanical systems (NEMS and MEMS) is the main hindrance for the next generation of electronics. As fabrication of these small-scale technologies required complicated and multistep processes. Also, the fabrication cost of these small-scale technologies is very high because of the utilization of not only expensive electrodes but also noble metal electrodes and complicated processing devices. In addition, various significant degradation processes (for example, delamination. Cracking, warping, etc.) are also incorporated during the etching processes of device miniaturization. Therefore, although the performance of MEMS and NEMS-based devices is sufficient yet cost still matters for their large production [76]. Piezoelectric thin films are promising candidates for the miniaturization of electronic components. PZT thin films put forward revolutionary advances in microelectronic devices comprising sensors, actuators, resonators, filters, and transducers. PZT thin films provide precise displacements at increased integration densities, and integrated control electronics with small size and complexity of circuits give a fast response at the output. Also, with the use of PZT wireless electronics those high-frequency systems have become able to produce that are reliable, miniaturized, and having low cost with good performance. To face all challenges during the fabrication of electronics mostly passive elements are also preferred as these occupy up to 60% of the total area of devices. Integration of these passive elements on a substrate can be more beneficial in both cost and size reduction. For MEMS applications PZT thin films have been used in many kinds of electronic devices. A zone which has acknowledged considerable consideration is that of rf controls. These switches can be capacitive or resistive in which associations are made through cantilever actuators utilizing either another metal or insulator at top of the electrode. MEMS technologists also observed that it is not always necessary for some standard materials like Si, and silicon oxide to provide efficient solutions for the integration of components. Because of such reasons PZT thin films now are introduced in electronics.

There are main 3 sets of rf-MEMS modules in which polar tinny films are utilized that are given as (1) Components based upon bulk acoustic wave (BAW) that use piezoelectric effect control characteristics of the circuit (2) tunable capacitors in which capacitance of the circuit is modified by utilizing permittivity of ferroelectric materials depending upon electric field (3) in capacitive switches replacing dielectric layer with a material having a high dielectric constant [77]. Many studies on BAW devices suggest that these devices have achieved both miniaturization and excellent electric characteristics. PZT performs good characteristics and crystallinity for BAW filters and resonators. Having excellent piezoelectricity ZnO and AlN thin films are suitable piezoelectric materials for BAW applications. For ZnO not only does lattice fitting between ZnO and substrate but also surface smoothness of substrate strongly affect and force c-axis orientation of ZnO thin films. AlN being polar, non-ferroelectric material of wurtzite structure, having high

Advanced Functional Piezoelectric Materials and Applications   Materials Research Forum LLC
Materials Research Foundations 131 (2022) 186- 221   https://doi.org/10.21741/978164490209-7

frequency for high sound velocity, and large Q is more preferred for fabrication of gigahertz BAW resonators. One more attractive feature of AlN is its chance of showing extremely accomplishment films at a low treating temperature of 200°C. Ferroelectric thin films can also be used for tunable microwave applications. The performance of these devices depends upon the composition of the film, defects, fabrication methods, strain, design, etc. These ferroelectric devices have many advantages such as high tuning speed, dc control power, small leakage of currents, and high breakdown fields. Despite many advantages, these devices suffer some difficulties like hysteresis, and requirement of some limitations on temperature showing some further investigations on these devices. In addition, BAW thin films are less luxurious because their engineering is dependent upon group IC skill. Also, thin-film gadgets have capabilities of good power handling, small size, small noise due to pyroelectric current, and low-temperature drift. Polar thin film-based devices are growing fast commercially. There is a chance of capturing a large portion of the market of microwave filters by thin film-based filters for handheld devices.

4.4 Acoustic biosensors

Precision cell positioning has become a critical issue of research in the realm of precision medicine, as proper classification of cellular behaviour can only be done through investigating separate cell reactions. To move micro particles and cells remotely, dielectrophoretic [78], magnetic, optical, and acoustic [79] forces are utilized. Single-beam acoustic tweezers (SBAT) have gained a lot of attention among acoustic devices, and its viability was practically and theoretically established some years previously [3]. SBAT uses the acoustic radiation force to catch tiny particles using a greatly motivated ultrasound beam, similar near optical tweezers.

In the Mie regime, the ultrasound beam width methodology that of a particular cell to accomplish specific cell mobilization. As a result, a low focal length-number with a high-frequency ultrasonic transducer is required. Piezoelectric film technology [80] is preferred for achieving high frequencies. AlN films (particularly Sc-doped AlN) have been shown to have good thermal and chemical stability, a high longitudinal wave velocity and a low dielectric constant, allowing them to be employed on larger-frequency uses. Without contacting each other, 10 m epidermoid carcinoma cell and 10 m polystyrene microspheres were independently operated. The transverse radiation force was predicted to have a maximum value of 10 nN [81].

4.5 Surface acoustic wave (SAW) biosensors

Cellphones will leverage substantially greater data transfer speeds and additional frequency spectrums in next-generation communication systems. A phone with more than forty filters is expected to be released. Different materials, cutting orientations, and/or methods may be used to create filters for different frequency bands. This would result in significantly greater interference throughout the assimilation of multiple policies, particularly in GHz region, as well as an increase in the RF front-end volume [69]. Extremely high sound waves velocity of AlN films suggests that AlN-based SAW filters are expected to become a very challenging future technology. The large lithography line width with an equal center working frequency enables the increase in yield, fabrication tolerance, and allows for a more advanced feasible employed frequency. In SAW technologies, a set of comb-like sensors are positioned on the surface of a piezoelectric material. The frequency is calculated using the given formula [82]:

$$f = \frac{1}{2W} \sqrt{E}/\rho \qquad \text{Eq.34}$$

The resonant frequency, Young's modulus, field of two adjacent probes and piezo-electric substance density are all represented as f, E, W, respectively. Using a mask with multiple pitches, procedures with altered frequencies can be fabricated in only lithography method. The parameters of the equivalent circuit can be determined by new geometrical measurements such as n (reflector finger number) and N (comb finger pairs). These variables are usually chosen to improved fit the 50-ohm arrangement. An external sound wave is produced and spreads beside the film apparent when opposed electric aptitudes are supplied to electrodes placed in a discontinuous fashion [83].

AlN SAW devices have a piezoelectric coupling coefficient K_t^2 of less than 2%. Because the filters' bandwidth is proportional to K_t^2, they can only be used as transitional filters. Though, the greatest recorded piezoelectric pairing constant of AlN sheets was 6.5 percent, representing that AlN SAW device performance has a lot of room for improvement [84]. When sputtering polycrystalline AlN films, a metal seed layer is commonly utilized. This can improve the device's film crystallinity as well as its coupling coefficient.

AlN films have transverse and longitudinal velocities of roughly 6000 and 11000 m/s, which are substantially higher than additional materials like quartz. The Rayleigh-mode wave, with speed about 5000 m/s, is generally employed for SAW devices; this value is higher than some others, but not high enough. Different substrates can be used to optimize the thin film's velocity. For this objective, the installation of AlN films on SiC and diamond substrates was examined, with the speed of AlN upon diamond being around 12000 m/s [85].

Surface acoustic wave-based microfluidic gadgets have also fascinated researchers' curiosity, particularly in the biological area for "lab-on-a-chip" campaigns. Biotic research is always carried out in a liquid environment. The auditory energy spreads into the liquefied body when a droplet of fluid is placed in the way of these waves. The sound wave alteration, also called the Rayleigh angle, $\theta_R = \sin^{-1}(\frac{c_l}{c_s})$, where c_l and c_s are the liquid and solid velocity, respectively, determines the diffraction angle. The diffraction angle for water is almost 22 degrees [86].

SAW can combine many actions on a single chip, such as concentration, translation, mixing, pumping, collecting, separation and droplet collection. LOC gadgets are now mostly depending on the piezo-electric measurable 128° Rayleigh wave and YX-LiNbO$_3$ constructions, while these haven't been extensively explored. Furthermore, AlN has a lower electro-mechanical combination constant and a lower piezoelectric reaction than LiNbO$_3$. The modifying impact of these devices is influenced by the film depth [87]. Rayleigh waves are dispersive, and increasing layer thickness increases the electromechanical coupling coefficient.

The electromechanical combination is the electro-auditory transformation proficiency, that is directly related to the largeness of the surface element that determines the wave's force. As a result, a denser film was needed to transfer a droplet at a faster rate. These are cost-effective and versatile, making it easy to integrate microfluidic and microelectronic devices into a LOC method. Furthermore, the Sc-doped AlN film has a substantially greater coupling coefficient than un-doped AlN, resulting in improved manipulation abilities [88]. However, more device wave mode and structural improvement is necessary.

5.   Recent developments in piezoelectric thin film devices

Piezoceramic materials that are devoid of lead Ceramic manufacturers and, particularly in Asia, final device manufacturers pushed the improvement of sintered ceramic components and lead-free powders. The latter's accomplishments were discussed in previous parts, and the advancements of piezoceramic manufacturing businesses are discussed next. PI Ceramic GmbH (Germany) is a pioneer in the improvement of novel Pb-free piezo-ceramics in Europe [16]. The investigation of these works has been around since the 2000s. with financing from both national and European governments. The PIC700 was the 1st new piezoceramic lead free product that's accessible on the market in 2014, was the result of the efforts.

This BNT-based material has a high thickness coupling coefficient ($k_t = 0.40$), a low density ($\rho = 5.6$ g/cm^3), and a temperature range of equal to 200°C, it's a good option for MHz-range ultrasonic transducers and sonar/hydrophone applications. SONOX P1 LF, a BaTiO$_3$-based piezoelectric material with a thickness coupling coefficient (k_t) of 0.45, piezoelectric coefficient (d_{33}) of 135 pC/N, and a temperature coefficient (T_C) about 115°C, was created by Ceram Tec GmbH (Germany) [89]. Meggitt A/S (Denmark)

Advanced Functional Piezoelectric Materials and Applications   Materials Research Forum LLC
Materials Research Foundations 131 (2022) 186- 221   https://doi.org/10.21741/978164490209-7

commercialized a KNN-based lead-free composition and Ferro perm Piezoelectric Pz61. Because of the piezoceramic's low density, it has a low frequency response of 24.6 MRayl, which is important for improved resonant frequency in nondestructive testing and medical imaging.

The maximum working temperature is $180°C$ and with a piezoelectric coefficient (d_{33}) of 80 pC/N. The material is also available as a thick film for medical use such as in vivo skin imaging [18]. Linear-array transducers made of field-engineered Pb-free $KNbO_3$ a polymer and single crystals achieved even greater picture resolution [90]. Meggitt A/S established a non-perovskite material depends on Bi-titanate for high-temperature applications. The advantage of this piezoceramic is its extraordinarily high curie temperature about $650°C$, but the disadvantage is its low d_{33} of roughly 15–20 pC/N. The firm uses an environmentally approachable spray-pyrolysis method and can produce 100 kg batches [20].

Conclusion

This chapter reviews piezoelectric thin films and lead free piezoelectric thin films properties, synthesis and application. The lead free piezoelectric thin films such as ZnO and KNN thin films are deliberated here. ZnO, KNN thin films properties, synthesis process and applications are discussed. Characterization techniques for deposition of piezoelectric thin films and pneumatic method are suggested to be reliable. Piezoelectric thin films can be used for several applications of MEMS like energy harvesting, actuators, and electronics. Piezoelectric energy harvesters are promising transducers that can generate high power output and are widely used in autonomous wireless sensor applications. Piezoelectric films are more attractive in those devices in which piezoelectric coefficients are firstly concerned. An analytical model for measuring the piezoelectric coefficient relevant to actuators is derived. For all applications, Piezoelectric allows increasing force and power output compared to other films. Dielectric losses are very high for piezoelectric thin films that must be taken into account for ultrasonic applications. The reduction of dielectric losses in Piezoelectric films is desirable to enhance their efficiency. Further improvements must be taken to low dielectric losses of piezoelectric thin films. In this chapter the applications of lead free piezoelectric thin films were described. How these are used in acoustic biosensors and in surface acoustic wave biosensors were explained in detail. Recent developments in piezoelectric thin film devices were also explained here.

References

[1] A.M. Abyzov, Aluminium oxide and alumina ceramics (review). Part 2. Foreign manufacturers. Technologies and research in the field of alumina ceramics,Refract. Ind. Ceram. 2 (2019) 13-22. https://doi.org/10.17073/1683-4518-2019-2-13-22

[2] M.A. Ahmad, R. Planas, Piezoelectric coefficients of thin film aluminum nitride characterizations using capacitance measurements. IEEE Microw. Wirel. Compon. Lett. 19(3). (2009) 140-142. https://doi.org/10.1109/LMWC.2009.2013682

[3] A.S. Bhalla, R. Guo, The perovskite structure-a review of its role in ceramic science and technology. Materials research innovations, Mater. Res. Innov. 4(1) (2000) 3-26. https://doi.org/10.1007/s100190000062

[4] P.K. Bhujbal , H.M. Pathan, Temperature dependent studies on radio frequency sputtered al doped zno thin film, Eng. 10(2020) 58-67. https://doi.org/10.30919/es8d1003

[5] J.Böhlmark, Fundamentals of high power impulse magnetron sputtering, kemi och biologi,Institutionen för fysik, (2006)

[6] M.A. Caro, S. Zhang, T. Riekkinen, M. Ylilammi, M.A. , Lopez-Acevedo, O., Molarius,J. T. Laurila, Piezoelectric coefficients and spontaneous polarization of scaln, J. Condens. Matter Phys. 27(24) (2015) p. 245901. https://doi.org/10.1088/0953-8984/27/24/245901

[7] Chelliah, C.R. John, R. Swaminathan, Pulsed laser deposited hexagonal wurzite zno thin-film nanostructures/nanotextures for nanophotonics applications, J. 12(1) (2018) p. 016013. https://doi.org/10.1117/1.JNP.12.016013

[8] Z. Chen, Y. Wu, Y. Yang, J. Li, , B. Xie, , X. Li, S. Lei, J. Ou-Yang, X. Yang, Q. Zhou, Multilayered carbon nanotube yarn based optoacoustic transducer with high energy conversion efficiency for ultrasound application, Nano Energy. 46 (2018) p. 314-321. https://doi.org/10.1016/j.nanoen.2018.02.006

[9] N. Chidambaram, A. Mazzalai,P. Muralt, Measurement of effective piezoelectric coefficients of pzt thin films for energy harvesting application with interdigitated electrodes, IEEE Trans. Ultrason. Ferroelectr. Freq. Controll. 59(8) (2012) p. 1624-1631. https://doi.org/10.1109/TUFFC.2012.2368

[10] C.R. Cho, C-axis oriented (na, k) nbo3 thin films on si substrates using metalorganic chemical vapor deposition, Mater. Lett. 57(4) (2002) p. 781-786. https://doi.org/10.1016/S0167-577X(02)00872-8

[11] D. D'Agostino, C. Di Giorgio,A. Di Trolio, A. Guarino,A.M. Cucolo, A. Vecchione,F. Bobba, Piezoelectricity and charge trapping in zno and co-doped zno thin films, AIP. Adv. 7(5) (2017) p. 055010. https://doi.org/10.1063/1.4983474

[12] Directive, 96/ec of the european parliament and of the council of 27 january 2003 on waste electrical and electronic equipment (weee), Official Journal of the European Union , 37 (2002) p. 24-38.

[13] R.E. Eitel, C.A. Randall, T.R. Shrout, P.W. Rehrig, W. Hackenberger,S.E. Park, New high temperature morphotropic phase boundary piezoelectrics based on bi (me) o3-pbtio3 ceramics, J. Appl. Phys.40(10R) (2001) p. 5999. https://doi.org/10.1143/JJAP.40.5999

[14] R. Elfrink, T.M.Kamel,M. Goedbloed, S. Matova, D. Hohlfeld, Y. Van Andel,R. Van Schaijk, Vibration energy harvesting with aluminum nitride-based piezoelectric devices, J. Micromech. Microeng. 19(9) (2009)p. 094005. https://doi.org/10.1088/0960-1317/19/9/094005

[15] R. Elfrink, M. Renaud, T.M. Kamel, C. de Nooijer, M. Jambunathan, M. Goedbloed, D. Hohlfeld, S. Matova, V. Pop, L. Caballero, R. van Schaijk , Vacuum-packaged piezoelectric vibration energy harvesters: Damping contributions and autonomy for a wireless sensor system, J. Micromech. Microeng. 20(10) (2010) p. 104001. https://doi.org/10.1088/0960-1317/20/10/104001

[16] L. Fan, S.Y. Zhang, H. Ge, HH. Zhang, Theoretical investigation of acoustic wave devices based on different piezoelectric films deposited on silicon carbide, J. App.PhyS. 114(2) (2013) p. 024504. https://doi.org/10.1063/1.4813491

[17] C. Fei, X. Liu, B. Zhu, D. Li, X. Yang, Y. Yang, Q. Zhou, Aln piezoelectric thin films for energy harvesting and acoustic devices, Nano Energy, 51 (2018)p. 146-161. https://doi.org/10.1016/j.nanoen.2018.06.062

[18] M.H. Frey, D.A. Payne, Nanocrystalline barium titanate: Evidence for the absence of ferroelectricity in sol-gel derived thin layer capacitors, App. Phys Lett. 63(20) (1993)p. 2753-2755. https://doi.org/10.1063/1.110324

[19] S. Fujii, T. Odawara, T. Omori, K.Y. Hashimoto, H. Torii, H. Umezawa,S. Shikata. Low propagation loss in a one-port resonator fabricated on single-crystal diamond. in 2011 IEEE International Ultrasonics Symposium, IEEE. (2011) https://doi.org/10.1109/ULTSYM.2011.0134

[20] H. Gullapalli, V.S. Vemuru, A. Kumar, A. Botello Mendez, R. Vajtai, M. Terrones, S. Nagarajaiah, P.M. Ajayan, Flexible piezoelectric zno-paper nanocomposite strain sensor, Small. 6(15) (2010)p. 1641-1646. https://doi.org/10.1002/smll.201000254

[21] Y. Guo, K.I. Kakimoto, H. Ohsato,. Phase transitional behavior and piezoelectric properties of $(Na_{0.5}K_{0.5})NbO_3$-$LiNbO_3$ ceramics, App. Phys. Lett. 85(18) (2004) pp.4121-4123. https://doi.org/10.1063/1.1813636

[22] M.A.M. Hatta, M.W.A. Rashid, U.A.A.H. Azlan, N.A. Azmi, M.A. Azam, T. Moriga, Influence of yttrium dopant on the structure and electrical conductivity of potassium sodium niobate thin films, Mater. Res. 19 (2016) p. 1417-1422. https://doi.org/10.1590/1980-5373-mr-2016-0076

[23] S. Hirsch, J. Guo, R. Reiter, S. Papazoglou, T. Kroencke, J. Braun, I. Sack, Mr elastography of the liver and the spleen using a piezoelectric driver, single shot wave field acquisition, and multifrequency dual parameter reconstruction, Magn. Reson. Med. 71(1 (2014) p. 267-277. https://doi.org/10.1002/mrm.24674

[24] E. Hollenstein, M. Davis, M. Damjanovic, N. Setter, Piezoelectric properties of li-and ta-modified $(k_{0.5}Na_{0.5})NbO_3$ ceramics, App. Phys. Lett. 87(18) (2005)p. 182905. https://doi.org/10.1063/1.2123387

[25] Y. Hou, M. Zhang, G. Han, C. Si, Y. Zhao, J. Ning, A review: Aluminum nitride mems contour-mode resonator, J. Semicond. 37(10) (2016)p. 101001. https://doi.org/10.1088/1674-4926/37/10/101001

[26] H.W. Huang, W.C. Chang, S.J. Lin, Y.L. Chueh, Growth of controllable zno film by atomic layer deposition technique via inductively coupled plasma treatment, J. App. Phys. 112(12) (2012)p. 124102. https://doi.org/10.1063/1.4768839

[27] S. Humad,R. Abdolvand, G.K. Ho, G. Piazza, F. Ayazi, High frequency micromechanical piezo-on-silicon block resonators. in IEEE International Electron Devices Meeting 2003, IEEE (2003) pp.39-3

[28] A. Illiberi, F. Roozeboom, P. Poodt, Spatial atomic layer deposition of zinc oxide thin films, AACS Appl. Mater. Interfaces, 4(1) (2012) p. 268-272. https://doi.org/10.1021/am2013097

[29] Y. Isogai, M. Miyayama,H. Yanagida, Mechanical response to reducing gases in pzt/zno actuator, J. Ceram. Soc. JAPAN, 103(1193) (1995) p. 96-98. https://doi.org/10.2109/jcersj.103.96

[30] Y. Isogai, M. Miyayama, H. Yanagida, Pzt/zno actuator responding to reducing gases, Sensors and Actuators B: Chemical, 30(1) (1996) p. 47-53. https://doi.org/10.1016/0925-4005(95)01749-L

[31] R.E. Jaeger , L. Egerton, Hot pressing of potassium sodium niobates, J. Am. Ceram. Soc. 45(5) (1962) p. 209-213. https://doi.org/10.1111/j.1151-2916.1962.tb11127.x

[32] R.L. Johnson, Characterization of piesoelectric zno thin films and the fabrication of piezoelectric micro-cantilevers, Ames Lab., Ames, IA (United States) (2005) https://doi.org/10.2172/850081

[33] I. Kanno, Piezoelectric pzt thin films: Deposition, evaluation and their applications. in 2019 20th International Conference on Solid-State Sensors, Actuators and Microsystems & Eurosensors XXXIII (TRANSDUCERS & EUROSENSORS XXXIII), IEEE. (2019) https://doi.org/10.1109/TRANSDUCERS.2019.8808525

[34] A. Khan, Z. Abas,H.S. Kim, I.K. Oh, Piezoelectric thin films: An integrated review of transducers and energy harvesting, Smart Mater. Struct. 25(5) (2016) p. 053002. https://doi.org/10.1088/0964-1726/25/5/053002

[35] D.G. Kim, H.G. Kim, A new characterization of piezoelectric thin films. in ISAF 1998. Proceedings of the Eleventh IEEE International Symposium on Applications of Ferroelectrics (Cat. No. 98CH36245), IEEE. (1998)

[36] Kubo, Kawabata,Kobayashi, Electronic properties of small particles. Annual Review of Materials Science, 1984. 14(1) p. 49-66. https://doi.org/10.1146/annurev.ms.14.080184.000405

[37] H. Kueppers, T. Leuerer, U. Schnakenberg, W. Mokwa,M. Hoffmann, T. Schneller, U. Boettger,R. Waser, PZT thin films for piezoelectric microactuator applications, Sens. Actuator A Phys. 97 (2002) p. 680-684. https://doi.org/10.1016/S0924-4247(01)00850-0

[38] A. Kuoni, R. Holzherr, M. Boillat, N.F. de Rooij, Polyimide membrane with zno piezoelectric thin film pressure transducers as a differential pressure liquid flow sensor, J. Micromech. Microeng. 13(4) (2003) p. S103. https://doi.org/10.1088/0960-1317/13/4/317

[39] F. Kurokawa, R. Yokokawa, H.Kotera, F. Horikiri, K. Shibata, T. Mishima, M. Sato,I. Kanno, Micro fabrication of lead-free (K,Na) NbO3 piezoelectric thin films by dry etching, Micro Nano Lett. 7(12) (2012)p. 1223-1225. https://doi.org/10.1049/mnl.2012.0570

Advanced Functional Piezoelectric Materials and Applications   Materials Research Forum LLC
Materials Research Foundations 131 (2022) 186- 221   https://doi.org/10.21741/978164490209-7

[40] K.W. Kwok, H.L.W.Chan, C.L. Choy, Evaluation of the material parameters of piezoelectric materials by various methods, IEEE Trans. Ultrason. Ferroelectr. Freq. Control. 44(4) (1997) p. 733-742. https://doi.org/10.1109/58.655188

[41] M. Laurenti, S. Stassi, M. Lorenzoni, M. Fontana, G. Canavese, V. Cauda, C.F.Pirri, Evaluation of the piezoelectric properties and voltage generation of flexible zinc oxide thin film,. Nanotechnology. 26(21) (2015) p. 215704. https://doi.org/10.1088/0957-4484/26/21/215704

[42] H.J. Lee, I.W. Kim, J.S. Kim, C.W. Ahn, B.H. Park, Ferroelectric and piezoelectric properties of $Na_{0.52}K_{0.48}NbO_3$ thin films prepared by radio frequency magnetron sputtering, App.Phys. Lett. 94(9) (2009) p. 092902. https://doi.org/10.1063/1.3095500

[43] S. H. Lee, S.S. Lee, J.J. Choi, J.U. Jeon, K. Ro, Fabrication of a high-aspect-ratio nano tip integrated micro cantilever with a zno piezoelectric actuator. in Key Engineering Materials, Trans Tech Publ.270 (2004) pp. 1095-1100 https://doi.org/10.4028/www.scientific.net/KEM.270-273.1095

[44] K. Lefki, G.J.M. Dormans, Measurement of piezoelectric coefficients of ferroelectric thin films, J.App.Phys. 76(3) (1994) p. 1764-1767. https://doi.org/10.1063/1.357693

[45] C. Lei, Z.G. Ye, Lead-free piezoelectric ceramics derived from the $K_{0.5}Na_{0.5}NbO_3$-$AgNbO_3$ solid solution system, App. Phys. Lett. 93(4) (2008) p. 042901. https://doi.org/10.1063/1.2956410

[46] F. Levassort, J.M. Grégoire, K. Lethiecq, K. Astafiev, L. Nielsen, R. Lou-Moeller, W.W. Wolny, High frequency single element transducer based on pad-printed lead-free piezoelectric thick films. in 2011 IEEE International Ultrasonics Symposium, IEEE. (2011) pp. 848-851 https://doi.org/10.1109/ULTSYM.2011.0207

[47] W. Li, Z. Xu, R. Chu, P. Fu, G. Zang, Large piezoelectric coefficient in $(Ba_{1-x}Ca_x)(Ti_{0.96}Sn_{0.04})O_3$ lead free ceramics, J. Amer.Ceram.Soc. 94(12) (2011) p. 4131-4133. https://doi.org/10.1111/j.1551-2916.2011.04888.x

[48] Y. Lu, N.W. Emanetoglu, Y. Chen, Zno piezoelectric devices, in Zinc oxide bulk, thin films and nanostructures, Elsevier (2006)p. 443-489. https://doi.org/10.1016/B978-008044722-3/50013-0

[49] M. Lukacs, M. Sayer,S. Foster, Single element high frequency (< 50 mhz) pzt sol gel composite ultrasound transducers, IEEE Trans. Ultrason. Ferroelectr. Freq. Control. 47(1)(2000)p. 148-159. https://doi.org/10.1109/58.818757

[50] N. Marandian Hagh, B. Jadidian, A. Safari, Property-processing relationship in lead-free (k, na, li) nbo3-solid solution system, J. Electroceramics. 18(3) (2007) p. 339-346. https://doi.org/10.1007/s10832-007-9171-x

[51] K. Matsubara, P. Fons, K. Iwata, A. Yamada, K. Sakurai, H. Tampo, S. Niki, Zno transparent conducting films deposited by pulsed laser deposition for solar cell application,Thin Solid Films. 431 (2003) p. 369-372. https://doi.org/10.1016/S0040-6090(03)00243-8

[52] P. Muralt, Pzt thin films for microsensors and actuators: Where do we stand? , I IEEE Trans. Ultrason. Ferroelectr. Freq. Control. 47(4) (2000) p. 903-915. https://doi.org/10.1109/58.852073

[53] P. Muralt, Piezoelectric thin films for mems, Integr. Ferroelectr. 17(1-4) (1997) p. 297-307. https://doi.org/10.1080/10584589708013004

[54] P. Muralt, Ferroelectric thin films for micro-sensors and actuators: A review, J. Micromech. Microeng. 10(2) (2000) p. 136. https://doi.org/10.1088/0960-1317/10/2/307

[55] P. Muralt, R.G. Polcawich,S. Trolier-McKinstry, Piezoelectric thin films for sensors, actuators, and energy harvesting, MRS Bull. 34(9) (2009)p. 658-664. https://doi.org/10.1557/mrs2009.177

[56] Y. Nakashima, W. Sakamoto, T. Shimura, T. Yogo, Chemical processing and characterization of ferroelectric (K, Na)NbO$_3$ thin films, Jpn. J. Appl. Phys. 46(10S) (2007) p. 6971. https://doi.org/10.1143/JJAP.46.6971

[57] M. Nemoz, R. Dagher, S. Matta,A. Michon, P. Vennéguès, J. Brault, Dislocation densities reduction in mbe-grown aln thin films by high-temperature annealing, J. Cryst. Growth. 461 (2017)p. 10-15. https://doi.org/10.1016/j.jcrysgro.2016.12.089

[58] P.K. Panda, Environmental friendly lead-free piezoelectric materials, J. Mater.Sci. 44(19) (2009) p. 5049-5062. https://doi.org/10.1007/s10853-009-3643-0

[59] S.Y. Pao, M.C. Chao, Z. Wang, C.H. Chiu, K.C. Lan, Z.N. Huang, L.R. Shih, C.L. Wang, Analysis and experiment of hbar frequency spectra and applications to characterize the piezoelectric thin film and to hbar design. in Proceedings of the 2002 IEEE International Frequency Control Symposium and PDA Exhibition (Cat. No. 02CH37234), IEEE. (2002) pp. 27-35

[60] K. Park, D.K. Lee, B.S. , H. Jeon, N.E.Lee, D.Whang, H.J. Lee, Y.J.Kim, J.H. Ahn, Stretchable, transparent zinc oxide thin film transistors. Adv.Func.Mater. 20(20) (2010) p. 3577-3582. https://doi.org/10.1002/adfm.201001107

[61] B. Pickarski, D. DeVoe, M. Dubey, R. Kaul, J. Conrad, Surface micromachined piezoelectric resonant beam filters, Sens. Actuator A Phys. 91(3) (2001)p. 313-320. https://doi.org/10.1016/S0924-4247(01)00601-X

[62] A. Qi, L.Y. Yeo, J.R. Friend, Interfacial destabilization and atomization driven by surface acoustic waves, Phys. Fluids. 20(7) (2008) p. 074103. https://doi.org/10.1063/1.2953537

[63] W. Qin, T. Li, Y. Li, J. Qiu, X. Ma, X. Chen, X. Hu, W. Zhang, A high power zno thin film piezoelectric generator, Appl. Surf. Sci. 364 (2016) p. 670-675. https://doi.org/10.1016/j.apsusc.2015.12.178

[64] N.N. Rogacheva, The theory of piezoelectric shells and plates, CRC press. (2020) https://doi.org/10.1201/9781003068129

[65] R. Ruby. 11e-2 review and comparison of bulk acoustic wave fbar, smr technology. in 2007 IEEE Ultrasonics Symposium Proceedings. IEEE. (2007) pp. 1029-1040 https://doi.org/10.1109/ULTSYM.2007.262

[66] R.C. Ruby, P. Bradley, Y. Oshmyansky, A. Chien, J.D. Larson, Thin film bulk wave acoustic resonators (fbar) for wireless applications. in 2001 IEEE Ultrasonics Symposium. Proceedings. An International Symposium (Cat. No. 01CH37263). IEEE.1 (2001) pp. 813-821

[67] N. Setter, D. Damjanovic, L. Eng, G. Fox, S. Gevorgian, S.Hong, A. Kingon, H. Kohlstedt, N.Y. Park, G.B. Stephenson, Ferroelectric thin films: Review of materials, properties, and applications, J. of App. Phys. 100(5) (2006) p. 051606. https://doi.org/10.1063/1.2336999

[68] T.R. Shrout , S.J. Zhang, Lead-free piezoelectric ceramics: Alternatives for pzt? ,J. Electroceram. 19(1) (2007) p. 113-126. https://doi.org/10.1007/s10832-007-9047-0

[69] L. Shu, B. Peng, C. Li, D. Gong, Z. Yang, X.Liu, W. Zhang, The characterization of surface acoustic wave devices based on aln-metal structures, Sensors. 16(4) (2016) p. 526. https://doi.org/10.3390/s16040526

[70] M.T. Todaro, F. Guido, L. Algieri, V.M. Mastronardi, D. Desmaële, G. Epifani, M. De Vittorio, Biocompatible, flexible, and compliant energy harvesters based on piezoelectric thin films, IIEEE Trans Nanotechnol , 17(2) (2018) p. 220-230. https://doi.org/10.1109/TNANO.2017.2789300

[71] S. Vicknesh, S. Tripathy, V.K. Lin, S.J. Chua, B. Kumar, H. Gong, Surface micromachined freestanding zno microbridge and cantilever structures on Si (111) substrates, App. Phys. Lett., 90(9) (2007) p. 091913. https://doi.org/10.1063/1.2642619

[72] N.V. Viet, X.D. Xie, K.M. Liew, N. Banthia, Q. Wang, Energy harvesting from ocean waves by a floating energy harvester, Energy. 112 (2016)p. 1219-1226. https://doi.org/10.1016/j.energy.2016.07.019

[73] Y.Y. Villanueva, D.R. Liu, P.T. Cheng, Pulsed laser deposition of zinc oxide, Thin Solid Films. 501(1-2) (2006) p. 366-369. https://doi.org/10.1016/j.tsf.2005.07.152

[74] J. Voldman, Electrical forces for microscale cell manipulation, Annu. Rev. Biomed. Eng. 8 (2006)p. 425-454. https://doi.org/10.1146/annurev.bioeng.8.061505.095739

[75] Wadell, transmission line design handbook, artech house inc. (1991) p. 291-293.

[76] Z. Wang, J.D.N. Cheeke, Characterizing unpoled piezoelectric ceramic film by lamb-waves, IEEE Trans. Ultrason. Ferroelectr. Freq. Control. 46(5) (1999) p. 1094-1100. https://doi.org/10.1109/58.796115

[77] Y. Wang, D. Damjanovic, N. Klein, E. Hollenstein,N. Setter, Compositional inhomogeneity in Li and Ta modified (K, Na)NbO$_3$ ceramics, J. Amer. Ceram. Soc. 90(11) (2007) p. 3485-3489. https://doi.org/10.1111/j.1551-2916.2007.01962.x

[78] W.B. Wang, Y.Q. Fu, J.J. Chen, W.P. Xuan, J.K.Chen, X.Z. Wang, P. Mayrhofer, P.F. Duan, A. Bittner, U. Schmid, Alscn thin film based surface acoustic wave devices with enhanced microfluidic performance, J. Micromech. Microeng. 26(7) (2016) p. 075006. https://doi.org/10.1088/0960-1317/26/7/075006

[79] JZ. Wang, X. Li, J.D.N. Cheeke, A modified modal frequency spacing method for coating characterization, J. Acoust. Soc. Am. 104(5) (1998) p. 3119-3122. https://doi.org/10.1121/1.423904

[80] Z. L. Wang, J. Song, Piezoelectric nanogenerators based on zinc oxide nanowire arrays, Science. 312(5771) (2006) p. 242-246. https://doi.org/10.1126/science.1124005

[81] Z. Wang, Y. Zhang, J.D.N. Cheeke, control, Characterization of electromechanical coupling coefficients of piezoelectric films using composite resonators, IEEE Trans. Ultrason. Ferroelectr. Freq. Control. 46(5) (1999) p. 1327-1330. https://doi.org/10.1109/58.796138

[82] S.S. Won, H. Seo, M. Kawahara, S. Glinsek, J. Lee, Y. Kim, C.K. Jeong, A.I. Kingon, S.H. Kim, Flexible vibrational energy harvesting devices using strain-engineered perovskite piezoelectric thin films, Nano Energy. 55(2019) p. 182-192. https://doi.org/10.1016/j.nanoen.2018.10.068

[83] J. Wu, J. Wang, Phase transitions and electrical behavior of lead-free (K$_{0.50}$Na$_{0.50}$)NbO$_3$ thin film, AIP. (2009)

Advanced Functional Piezoelectric Materials and Applications   Materials Research Forum LLC
Materials Research Foundations 131 (2022) 186- 221   https://doi.org/10.21741/978164490209-7

[84] X. Yan, W. Ren, X. Wu, P. Shi, X. Yao, Lead-free (K, Na)NbO$_3$ ferroelectric thin films: Preparation, structure and electrical properties, J. Alloys Compd. 508(1) (2010) p. 129-132. https://doi.org/10.1016/j.jallcom.2010.08.025

[85]Y. Yoshino, Piezoelectric thin films and their applications for electronics, J. App. Phys. 105(6) (2009) p. 061623. https://doi.org/10.1063/1.3072691

[86]Y. Zhang, Z. Wang, J.D.N. Cheeke, Simulation of electromechanical coupling coefficient by modified modal frequency spectrum method including the electrode effect, Ultrasonic, 38(1-8) (2000) p. 114-117. https://doi.org/10.1016/S0041-624X(99)00172-9

[87] Y.Zhang, Z. Wang, J.D.N.Cheeke, F.S. Hickernell, Direct characterization of zno films in composite resonators by the resonance spectrum method. in 1999 IEEE Ultrasonics Symposium. Proceedings. International Symposium (Cat. No. 99CH37027), IEEE.Vol. 2 (1999) pp. 991-994

[88] S. Zhang, R. Xia,T.R. Shrout, G. Zang, J. Wang, Piezoelectric properties in perovskite $_{0.948}$(K$_{0.5}$Na$_{0.5}$)NbO$_3$-$_{0.052}$LiSbO$_3$ lead-free ceramics. J. App. Phys. 100(10) (2006)p. 104108. https://doi.org/10.1063/1.2382348

[89] J. Zhou, M. DeMiguel-Ramos, L. Garcia-Gancedo, E. Iborra, J. Olivares, H. Jin, J.K. Luo, A.S. Elhady, S.R. Dong, D.M. Wang, Characterization of aluminum nitride films and surface acoustic wave devices for microfluidic applications, Sens. Actuators B Chem. 202 (2014) p. 984-992. https://doi.org/10.1016/j.snb.2014.05.066

[90] B. Zhu, J. Xu, Y. Li, T. Wang, K. Xiong, C. Lee, X. Yang,M. Shiiba, S. Takeuchi, Q. Zhou, Micro-particle manipulation by single beam acoustic tweezers based on hydrothermal pzt thick film, AIP Adv. 6(3) (2016) p. 035102. https://doi.org/10.1063/1.4943492

Advanced Functional Piezoelectric Materials and Applications   Materials Research Forum LLC
Materials Research Foundations 131 (2022) 222-258   https://doi.org/10.21741/978164490209-8

Chapter 8

Bulk Lead-Free Piezoelectric Perovskites and their Applications

M. Rizwan[1*], A. Ayub[2], S. Fatima[2], F. Aleena[2], I. Ilyas[2], A. Shoukat[3]

[1]School of Physical Sciences, University of the Punjab, Lahore, Pakistan

[2]Department of Physics, University of the Punjab, Lahore, Pakistan

[3]Department of Physics, University of Gujrat, Gujrat, Pakistan

*rizwan.sps@pu.edu.pk

Abstract

Perovskites are an interesting class of materials that have tremendous applications as actuators, sensors and in photovoltaics. Lead based perovskites exhibit piezoelectricity and other interesting properties and thus have conquered the ceramic industry for a long time. Lead free piezoelectric perovskites are the need of the hour because lead based piezoceramics are toxic and a danger to the environment. There are various contenders of lead free alternatives of lead zirocnate-titnate (PZT) based ceramics including potassium-sodium niobate, barium titanate, bismuth based perovskites that exhibit similar piezoelectric and ferroelectric properties in comparison to PZT ceramics. These lead free piezoceramics and their important properties and respective applications such as sensors, transducers and actuators, is briefly explored in this chapter.

Keywords

Piezoelectricity, Ceramics, BNT, Phase Boundary, Sensors, Actuators

Contents

1.   Perovskites

Perovskites are minerals possessing the chemical formula of ABX$_3$, in this formula, A and B are acting as metallic cations of varying size and X act as a halide. Perovskites originally located by Gustave rose and have chemical formula of CaTiO$_3$ [1]. Perovskites are very exciting material and the field of perovskites have grown many times over the past years due to tremendous amount of research dedicated to implementing these fascinating materials in multiple applications. Perovskites are polymorphous and can exist in multiple crystalline structures, the characteristics of perovskites can very easily be altered via replacement of foreign entities on the A site and B site [2]. Most commonly cubic structure of perovskite is considered ideal since it gives most stability according to Goldschmidt criteria.

Polymorphic nature of perovskites accommodates small and large sized cations and thus perovskites can exist in many space groups [3]. The distortion in the structure produces vacancies and accommodates any adjustments that lead to a flexible perovskite structure that we can alter depending on our requirements. The bandgap of perovskites is very versatile and these materials can be semiconductors, insulators an conductor [4]. Perovskites have a colossal impact on the solar cell technology due to the impressive perovskites solar cells that are lead free and provide a new source for energy production. Perovskites are hugely popular in piezoceramics industry. The cogitation of perovskites is very crucial since they make up to 90% of earth lower mantle.

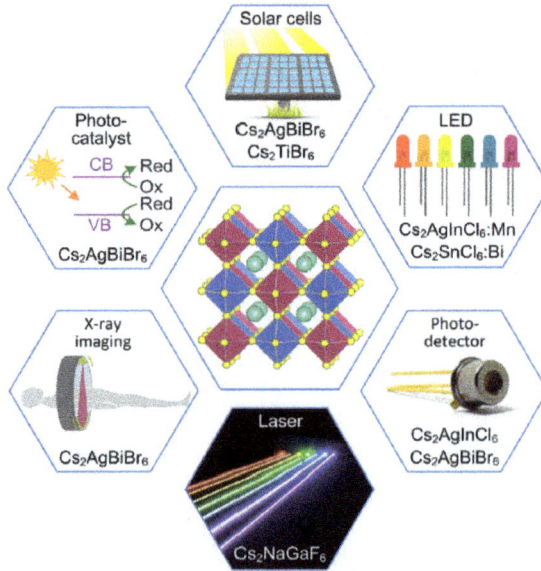

Figure 1. Applications of perovskites [9].

Oxide based perovskites have been dominant in the industry, but fluoroperovskites are also emerging in many applications. In most perovskites, B cation is a transition metal and A cation belongs to group I or II of the periodic table. Oxygen in perovskites forms an octahedral cage, whose tilting due to introduction of impurities, or stress greatly influence its physical properties [5].

Perovskites have extensive application from solar cells to dielectric materials [5]. The vast horizon of applications of perovskites are due to its superior properties as high temperature superconductors [6]. Perovskites have applications as relaxers, ferroelectric and piezoelectric materials. Ferroelectric perovskites is applied in RAM cells which have exceptional properties such as consumption of less power, more charge storage capacity in comparison to DRAM cell [7, 8]. Some applications of perovskites are given in Fig. 1 [9].

2.   Lead free perovskites

Ceramics industry is dominated by perovskites and especially by lead dependent piezoelectric perovskites such as $(Pb, Zr)TiO_3(PZT)$ perovskites [10]. These piezoceramics have exceptional ferroelectric and dielectric properties and thus have dominated the piezoceramics industry for a long time. But one drawback of these ceramics is the poisonousness of the lead and the predominant problem of continuous emission of lead in the environment [11]. The toxicity of these materials posed a serious danger to human safety and the future of the environment, this caused a stir for the search of new superior or compatible lead-free alternative of these PZT ceramics. Piezoelectricity was first observed in 1880s by Jacques and pieree Curie, these materials produce a voltage signal upon stress. Perovskites like $BaTiO_3$ and $PbTiO_3$ revolutionized the ceramic properties [11]. Bismuth based perovskites, $BaTiO_3$, and other perovskites became good contender for alternatives of PZT ceramics [12].

These new alternatives are now being used commercially instead of the toxic PZT ceramics and new efforts are continually made to find better alternatives of these ceramics and to enhance their properties. PZT materials have very huge reach as actuators, sensors and transducer in ceramic industry. Lead free perovskites are implemented in these applications and efforts to achieve same performance is still in the works. PZT ceramics had good piezoelectric properties due to lead ferroelectric character [13]. The polarization is driven by lead in these materials, but despite these properties, they are toxic. The main goal of Lead free piezoceramic industry is not only to find alternative, but to also ensure that they give the same or superior performance in comparison to PZT ceramics, which has been a challenging task, but scientists and researchers continuous hard work has led to a great deal of progress in this field. Fig. 2, lays out the complete layout of the development of lead free perovskites [14].

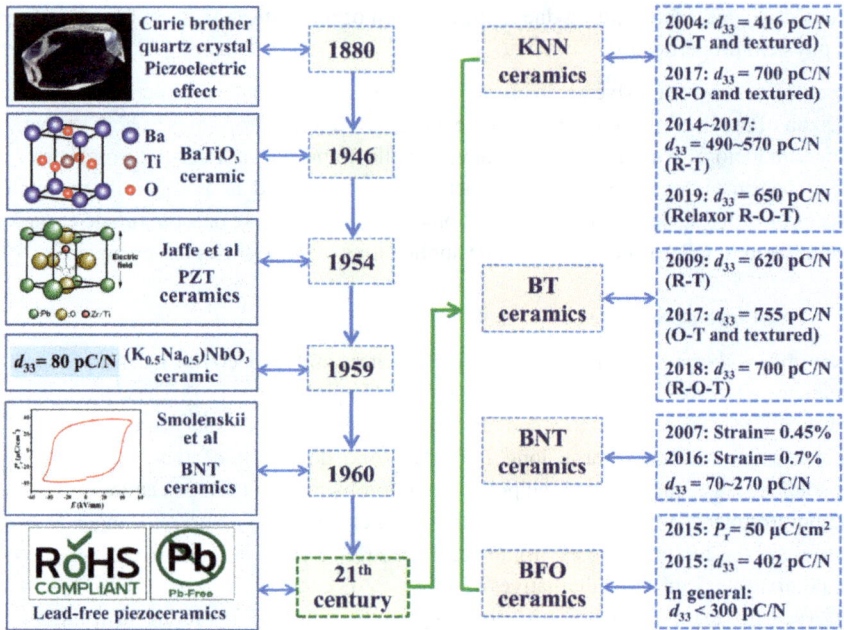

Figure 2. Development of lead free piezoceramics [14].

Barium titanate and its composition were quite easy to achieve and promoted future applications as transducers. These ceramics have large coupling coefficient, but their temperature was quite high due to phase transition. PZT ceramics had superior electromechanical coupling around the morphotropic phase boundary than BT ceramics [15].

3.   Processing of lead-free perovskites

These piezoceramics are fabricated through sintering or solid-state synthesis of metal oxide and carbonate solutions. Other sintering processes such as pressure or electric field based sintering are also being considered along with mechanochemical synthesis [16].

For potassium sodium niobate and sodium niobate titanate-based formulations, typical synthesis temperatures range from 700 to 900°C [17]. Potassium sodium niobate is known for its sensitivity to humidity of alkaline reagents, particularly K_2CO_3, and has a small sintering range near to 1140°C, hard microstructure control, easy presence of secondary phases, and chemical heterogeneity, particularly when chemical modification is involved

Advanced Functional Piezoelectric Materials and Applications   Materials Research Forum LLC
Materials Research Foundations 131 (2022) 222-258   https://doi.org/10.21741/978164490209-8

[18, 19]. Densification of NBT necessitates sintering temperatures of around 1100°C. Bismuth evaporation, which results in oxygen vacancies, a coarse microstructure can be remedied by addition of more bismuth oxide. It's important to memo that the vapor pressure of bismuth over $K_{0.5}Bi_{0.5}TiO_3$ is many times greater than that of Na and K over their respective niobates [19, 20]. When sintering and BKT- NBT and NBT–BT solid solutions at 1050 and 1200°C, respectively, they attain high relative densities. BZT–BCT solid state synthesis occurs at temperatures ranging from 1200–1300°C, while sintering occurs at temperatures ranging from 1300-1500°C [21]. To evade the development of harmful secondary phases, precise control of stoichiometry is required, just as it is in the case of pure BT. To lower the high sintering temperature, substituent or impurities such as CuO, CeO_2, Y_2O_3, or Bi_2O_3 were utilized [22-25].

4.   Piezoelectricity in lead free perovskite

Due to the unique properties such as coupling of mechanical and electrical displacement, the piezoelectric materials are employed in advanced technologies [26]. For actuator devices, the piezoelectric devices provide a high ratio of pressure per density a significant environmental and chemical stability as well as ability to operate at high temperature and pressure as compared to other electromechanical technologies. Buzzers, ultrasound, and Nano positioners in scanning microscopes, SONAR, diesel engine fuel injectors are all examples of piezoelectric materials in use.

PZT are the most frequently used piezoelectric ceramic. PZT has a significant disadvantage in that it comprises of more than 60% lead by weight. This high Pb- concentration poses a risk during processing, restricts uses, and is possibly hazardous to the environment upon disposal. Regulatory bodies across the globe have begun to establish tight limitations on the utilization of lead in recent years, with the exemption of the electronics sector, which has yet to find a viable alternative for PZT. No single composition has been presented with qualities equivalent to PZT, therefore apt lead-free piezoelectric materials are still being investigated [26].

4.1   Fundamentals of piezoelectricity

For a material to exhibit the characteristics of a piezoelectric material, it needs to be non-centro-symmetric. Aluminum nitride and quartz are some non-centro symmetric piezoelectric material used in piezoelectric devices. There is spontaneous polarization in piezoelectric materials owed to the separation of + and - e charges centers. Perovskite that has the general structure of ABO_3 undergoes spontaneous polarization and thus are very pivot as piezoelectric materials. In perovskites phase transition occurs from centrosymmetric cubic structure at high temperature to rhombohedral, tetragonal, orthorhombic, or monoclinic at low temperatures. Phase transition results in a non-centrosymmetric structure that is due to non-uniform octahedral tilting and alteration of the B cation with respect to A cation. The phase transition from cubic phase to a first non-centrosymmetric structure occurs at Curie temperature. When a temperature material lowered to curie temperature during phase transition, different parts of a material opts different lower symmetry crystallographic orientations. Ferroelectricity is material's ability to alters its spontaneous polarization in the influence of an applied electric field. The reorientation of spontaneous polarization requires domain walls to move which are basically regions that separate two domains and, is provided by cohesive force. In a polycrystalline material there are many local cohesive fields due to presence of compositional changes in various grains. Macroscopically observable coercive field (E_c) is quantity that is needed to make the polarization zero in order to compensate positive and negative local polarization. Since polycrystalline materials are made of equal number of domain orientations that are directed opposite therefore, in polycrystalline ceramics there is no macroscopic polarization nor any piezoelectricity at the macroscopic level in the presence of high processing temperature. The oppositely directed domains cancel out the polarization and results in no net polarization. To have piezoelectricity at macroscopic length scale, electric field must be applied, as a result domains become parallel with the applied electric field, this process is called polling. The poling process will produce a polarization parallel to electric field and therefore produce piezoelectricity parallel to the field direction at macroscopic length scale. The strain introduced in a piezo-material in reaction to an introduced electric field is described by the reversed piezoelectric effect [27]. This reversed piezoelectric effect is written as follows:

$$S_i = d_{ij}E_j \hspace{4cm} Eq.1$$

Here, S_i, E_j and d_{ij} represents induced stain, introduced electric field and piezo-coefficient respectively. The piezoelectric coefficient represents 3rd rank tensor, although it is stated in compact matrix notation in Equation (1) by describing the mechanical strain as a 1-Dmatrix with members $I = 1, 2,...6$. In actuator devices, the opposite piezoelectric effect is used [28].

Piezoelectricity can be improved via the process of doping and alloying as observed in PZT ceramics. Doping and substitution of impurities will alter its properties and structure depending upon the type of dopant and substituent. A soft ferroelectric material is formed when a PZT material is doped with a more positive donor ion than its host and on the other a hard ferroelectric behavior is observed when doped with a acceptor ion less positive than its host ion. Donor doping can lower the value of coercive field, increase dielectric constant and improve electromechanical coupling and acceptor doping gives a large coercive field, lower dielectric constant. [27, 29].

5.   Different lead-free piezoceramics and their applications

PZT ceramics have prevailed the ceramics industry for a long period of time. PZT ceramics have very fascinating properties, but this comes at a price such as even though PZT are good piezoelectric materials, they have electric field induced strain, low elastic moduli and low cohesive stress. The success of PZT ceramics and frequent application results in release of lead in environment during sintering and calcination processes. This toxicity of PZT ceramics is a major drawback and question mark for the safety of the environment. In 2003, PZT materials was rendered as unsafe for the environment which lead to a new field and search for lead free piezoceramics that can give equivalent or better piezoelectric properties as PZT materials [30]. Suitable lead-free alternative of PZT is still developing that can give comparable behavior as PZT.

Piezoelectric materials are most frequently utilized in sensors, actuators and resonators. Each application requires specific properties of the piezoelectric material such as application as an actuator demands for a high piezoelectric coefficient, similarly resonator application depends on coupling coefficient [31]. A tremendous amount of research has been performed in realizing actual application of lead free piezoceramics, the important contenders in this are KNN ceramics, BNT-ceramics $(Sr, Ca)_2NaNb_5O_{15}$ (SCNN), (BCZT), which have shown state of the art applications in as lead-free piezoceramics [31].

Lead free piezoceramics consist of three categories that are perovskites, bismuth layered ceramics and tungsten bronze ceramics. Perovskites based ceramics are ceramics that have the formula ABO_3 that include , KNN,BNT, $BaTiO_3$ (BT) [32] and $BiFeO_3$ ceramics [33]. We will discuss these lead free piezoceramics and their respective applications.

5.1   KNN based ceramics

KNN ceramics have been explored quite vigorously due to its high-performance qualities such as high piezoelectric coefficient and critical temperature [34-36]. The ferroelectricity in KNN ceramics is attributed to morphotropic phase boundary. Ferroelectricity and piezoelectricity peak in the composition $(K_{0.5}Na_{0.5})$ NbO_3. The first breakthrough in alkali niobate ceramics was made in 2004 by Saito and company in lithium, thallium, antimony modified KNN ceramics [37]. The value of piezoelectric coefficient found by Saito was 416 pC/N that was similar to conventional lead-based ceramics. KNN ceramics are very compatible with nickel electrode in comparison to PZT ceramics, which is crucial for industrial applications. KNN ceramics is combination of sodium niobate and potassium niobate and forms a solid solution at the morphotropic phase boundary (MPB), where potassium niobate exhibit ferroelectricity and sodium niobate are antiferroelectric.

As mentioned earlier actuators demand large electric field activated strain and critical temperature. The value of these parameters decides the range of actuator application. KNN based ceramics was employed in making a print head in 2012. Multiple pressure chamber is formed by gluing together two poled piezoceramics plates. The piezoceramic material serve as an actuator and also as wall of the pressure chamber that was formed by two poled piezoceramic plates. Electrodes are placed inside the pressure chamber and piezoceramics produce shear motion as a result of orthogonal electric field to polarization direction that produces stress on the ink present in the channels, this pressure on the ink result in ejection of droplets of ink through the nozzle. There are 60 nozzles and each nozzle has a density of 150 dots/inch in this KNN ceramics-based lead-free actuator. Shear mode of $BiFeO_3$-doped (KNL)(NTS) was also used to make inkjet printhead, but the driving voltage was two times higher than the PZT-based printhead [31] $0.96KNN-0.04CZO_3$+ $0.03ZrO_2$+0.05MCO is a 12 layered piezoceramic actuator that has the maximum reported induced strain. Although its half of the PZT based actuator, adding more piezoelectric layers along with limiting the thickness will give the same effect [37].

Sensors detect pressure or acoustic impulses via the transformation of mechanical signal into electrical signals. Sensors require high piezoelectric coefficient d, voltage coefficient g, in non-resonant conditions. In resonant conditions, sensors require high coupling coefficient k, and mechanical quality factor

A composite KNN-KTN($KTiNbO_5$) based piezoceramics with high planar coupling coefficient is used in knocking sensor [38]. Knocking sensor is implemented for the detection of combustion process and to control ignition time in automotive engines. This KNN-based sensors serves same output voltage in comparison to lead based sensor and gives superior performance in terms of durability [31].

There have been many applications with KNN-ceramics that are important to boost up the lead-free ceramic industry. A KNN composition such as $(K_{0.5-x}Li_x)Na_{0.5}(Nb_{1-y}Sb_y)O_3$ (KLNNS) with x=0-4% , y=0-8% was used for buzzer prototype [40].

5.2   Bismuth sodium titanate based piezoceramics and their applications

$Bi_{1/2}Na_{1/2}TiO_3$ is a ferroelectric material discovered in the 1960s and has a complex perovskite structure with elastic and electric fields due to variance between ionic radius of bismuth and sodium [41]. There is off centering in the octahedral site due to lone pair of bismuth. BNT has a small piezoelectric coefficient 73-95pC/N, low figures of merit, large coercive field but have high polarization and relative permittivity [42]. Similar to KNN based piezoceramics bismuth based ceramics (BNT, $Bi_{1/2}K_{1/2}TiO_3$ (BKT) have garnered a lot of attraction as contender of lead free piezoelectric ceramic material [43]. BNT-BKT solid solution and or $BaTiO_3$ (BT) mixed BNT-BKT gives improved piezoelectric properties in comparison to PZT ceramics [44]. BKT-BT show a piezo-coefficient of 60 pC/N, coupling coefficient 37%, and critical temperature of 290°C. BNT-BKT ceramics have larger curie temperature and piezoelectric coefficient in comparison to BNT-BT ceramics. The phase diagram of BNT-BT ceramic is given in Fig. 3 [30].

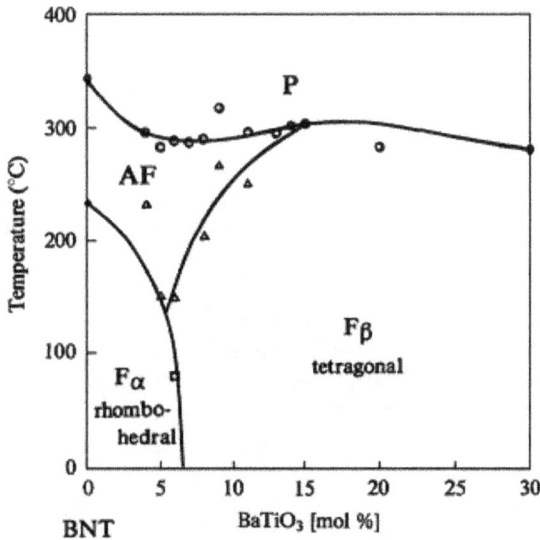

Figure 3. Phase diagram of BNT-BT piezoceramic [30].

Actuators application require an achievable strain. Fig. 4, compares the normalized strain of different lead free piezoceramics with lead based PZT ceramics

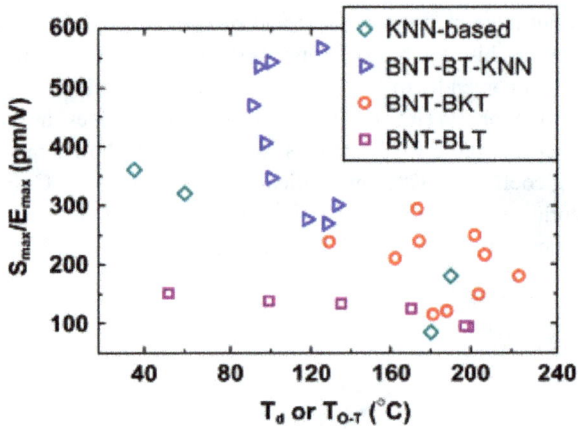

Figure 4. Normalized strain as a function of respective strain temperature of lead free piezoceramics [30].

BNT-BKT ceramics provide a lower strain of 0.18%, BNT-BT-KNN on the other hand gives a higher strain of 0.4%. The benefit of BNT-BKT lead free piezoceramics is its massive curie temperature. BNT ceramics have large coercive field and thus are hard to polarize, therefore BNT is modified in many ways in terms of composition such as $Bi_{0.5}Na_{0.5}TiO_3$-$(Bi_{0.5}Na_{0.5})TiO_3$ -$BaTiO_3$ (BNT-BKTBT), $(Bi_{0.5}Na_{0.5})TiO_3$-$BaTiO_3$ and $Bi_{0.5}Na_{0.5}TiO_3$- $Bi_{0.5}K_{0.5} TiO_3$ [45, 46].

Lithium doped (BNT-BKT-BT) is a prototype multilayer actuator with improved piezoelectric properties. Through gradient doping the temperature stability of the actuators are improved along with electric-strain [34]. $[(1–x–y) (BNT)_x -(BT)_y -(KNN)]$ for x = 0.06 and y = 0.02 system gives a strain that is 50% more than the soft PZT system. The lead free system exhibit a stain superior to PZT and thus can replace PZT material in actuators [47]. Fig. 5, gives a comparison of strain and piezoelectric coefficient for lead dependent and independent piezoelectric ceramics [48].

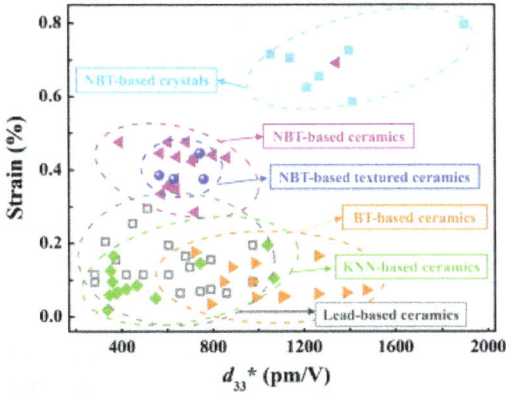

Figure 5. Comparison of strain and piezoelectric coefficient for lead free and lead based piezoceramics [48].

A piezoelectric immunosensor will work as transducer at resonance frequency that is dependent on transducer thickness. Such transducer can be used for the detection of bacteria and eukaryotic cells. A piezoceramic transducer was obtained by poling the composition $95(Bi_{0.5}Na_{0.5}) TiO_3 + 2.5(Bi_{0.5}K_{0.5}) TiO_3 + 2.5(BaTiO_3)$ under an electric field of 59 kVcm^{-1} and voltage of 14 kV for 1 hour and was embedded in silicone to avoid a breakdown. The full scheme of this immunosensor is given in Fig. 6 [49]

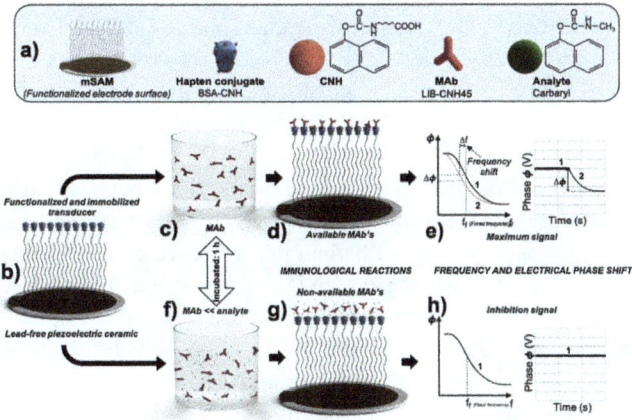

Figure 6. Scheme of lead free immunosensor principle and signal transduction energy [49].

Advanced Functional Piezoelectric Materials and Applications   Materials Research Forum LLC
Materials Research Foundations 131 (2022) 222-258   https://doi.org/10.21741/978164490209-8

5.3   BaTiO₃ (BT) based piezo-ceramics

In the 1940s and 1950s, $BaTiO_3$ was made as the main valuable piezoelectric [50]. Regardless of having a low piezoelectric constant , this material has a large permittivity, and thus is a valuable component for capacitors [51]. BaTiO3 (BT) is every now and again joined with other without lead compounds to produce a MPB (morphotropic phase boundary) with improved piezoelectric and dielectric attributes. Be that as it may, in light of the fact that BT has a low Curie temperature, there haven't been numerous developments lately for piezoelectric gadget applications [27].

5.3.1   BaTiO₃ ceramics phase boundary

The first oxide based piezoelectric material without lead first was $BaTiO_3$. There are three different phase change temperatures: 90°C for T_{R-O}, 120°C for Tc and 0°C for T_{O-T}. Albeit effective preparation techniques can extensively work on the piezoelectricity of pure BT ceramics. [52, 53], the traditional solid-state technique is still the greatest practicable instrument because of its economic efficacy and industrialization. Site engineering is widely known for its effectiveness in improving the electrical characteristics of piezoelectric ceramics [54]. It has previously been established that doping of Zr, Hf, Sn, or Ca to BT ceramics influences phase transitions [55]. Many phase boundaries of barium titanate can be constructed by shifting phase change temperature to room temperature by ion replacement. R-T, O-T, and R-O-T are the most commonly reported phase boundary types, related to KNN-based ((K, Na) NbO_3) ceramics. Commonly BT-based ceramics with orthogonal- tetragonal phase boundaries have less piezoelectricity performance than those rhombic-tetragonal and R-O-T phase barriers, and the R-O phase boundary was rarely explored due to its weak piezoelectricity. As a result, constructing a coexistence of R and T phases in BT-based ceramics can effectively improve piezoelectricity [50].

5.3.2   Factors in phase boundaries

Chemical compositions can have a big impact on the phase boundary of barium titanate-based ceramics, resulting in various electrical characteristics. The association amid configuration and phase boundary of barium titanate based ceramics make it insufficient since other parameters such as temperature [56], grain size and electric field can readily change the original phase structure. BT materials phase structure can also be affected by electric fields [57]. In 1999 it was established that assigned electric fields can persuade tetragonal-monoclinic a phase transition under 10 kV/cm, followed by monoclinic - rhombohedral phase transition under 30 kV/cm via an in-situ domain examination and Raman measurements of a barium titanate single crystal with constructive domain configuration. Furthermore, the monoclinic BT single crystal has the highest piezoelectric coefficient. The electric field activated monoclinic M_c phase in barium titanate single crystal was also discovered by high precision XRD in 2009 [58]. Ranjan et al. have examined the structural evolution of BT-based ceramics in the presence of electric fields. Through XRD structural investigation, three phases' coexistences (R + O + T) were postulated in the typical (BCTZ) ceramic at room temperature. According to the XRD data of unpoled and poled samples, at ambient temperature, an electric field can transition partial T phase to R + O, which is related with the gigantic piezoelectric response. Phase structure and associated electrical properties can be altered with introduction of temperatures and electric fields and grain size. The Curie temperature and c/a of $BaTiO_3$ ceramics, for example, fall as grain sizes decrease, suggesting that $BaTiO_3$ becomes less tetragonal as grain sizes decrease [59].

5.3.3 Sintering and curie temperature

It was also discovered that the piezoelectricity of $BaTiO_3$ -dependent ceramics is tightly connected with the calcination temperature as well as sintering temperatures. BCTZ ceramics can achieve an enhanced d33 of 650 pC/N via calcination on 1300°C and sintering on 1540°C [60]. High sintering temperatures of 1450°C are required to make high-performance BT-based ceramics, which limits their useful applications. Furthermore, lowering the sintering temperature in BT-based ceramics is essential to evade compositional fluctuation, and some sintering assistances (e.g., Li_2O, Ba $(Cu_{0.5}W_{0.5})$ O_3, MnO_2, CeO_2, CuO, Ga_2O_3ZnO, Li_2CO_3 etc.) [61-63] have been implemented to form liquid phases throughout the sintering process. Piezoelectricity of $BaTiO_3$- based ceramics can be enhanced with sintering aids which can significantly reduce the sintering temperature (TS 1400°C), improve sintering ability. CuO modified BCTS ceramics, for example, have a very high piezoelectric coefficient d33 of 683 pC/N [64]. Oxide based sintering supports can reduce sintering temperatures and thus increase piezoelectricity to some degree but their Curie temperature is always reduced. The T_c of BCTZ ceramics, for example, decreased linearly as the $Ba(Cu_{0.5}W_{0.5})$ concentration was increased. Ren et al. reported high-performance BZT-BCT ceramics with a low TC of 93°C [65]. The small Curie temperature is regarded to be the most significant obstruction to the practical deployment of $BaTiO_3$-based ceramics. Though, by doping with certain oxides, the Curie temperature can be raised. Xu et al. discovered that adding Dy_2O_3, Er_2O_3, and Y_2O_3 additives to BT-based ceramics can help enhance T_c. Inappropriately, no significant progress has been achieved in raising the Curie temperature. Consequently, BT still needs to find a stability between piezoelectricity and Curie temperature [66].

5.4   Bismuth based piezoceramics

Owed to their large Curie temperature (TC) and good electrical qualities, BiFeO3-based (BFO) ceramics are named as favorable materials for high-temperature applications [67]. Numerous examinations concerning the electrical properties of BFO piezoceramics have been investigated previously, however different issues have been met, like low resistivity, the creation impure phases and high leakage current [68]. The connection between configuration change and the phase boundary in BiFeO3-dependnet ceramics is momentarily looked into in this segment, and pragmatic procedures for property improvement are given.

5.4.1   Phase boundary in BFO-based ceramics

Bismuth ferrite-based ceramics have no definite phase boundary with ion replacement to advantage piezoelectricity enhancement. Electrical characteristics of BiFeO3-based ceramics can be amended by building a temperature-independent phase barrier for binary and ternary material systems. In this work, we look at the evolution of phase boundaries in BiFeO3-dependnet ceramics from the perspective of composition adjustment [69].

5.4.1.1   Ion substitution

Ion substitution is particularly useful for improving property augmentation in BFO-based ceramics, because of its enhanced resistivity, ion. Substitution at bismuth, iron oxygen are the three possibilities for ion substitution [70].

Electrical characteristics of BFO can be improved by when ion is substituted at Bi site, rather than constructing phase barriers, it is more beneficial to promote the ceramics with ion replacement by adjusting the processing procedures. Electrical characteristics cannot be considerably improved by substitution at Fe site or at both sites, since no phase barrier is formed [71].

5.4.1.2   Addition of ABO₃

Due to difficulties in formation of phase boundaries, it is problematic to improve the electrical characteristics of BiFeO3-ceramics by ion replacement (d_{33}=50 pC/N). As a result, research into BiFeO3- ceramics with the addition of oxide perovskites, expanded and their electrical properties has received a lot of attention. BiFeO$_3$-PbTiO$_3$ and BiFeO$_3$-BaTiO$_3$ are two common candidate materials [17]. Though the development of an R-T phase boundary in BiFeO$_3$-PbTiO$_3$ ceramics can result in rather strong piezoelectricity, this material has received less attention owed to the occurrence of the toxic lead element. BFO substituted in -SrTiO$_3$, CaTiO$_3$, NaNbO$_3$, BiFeO$_3$ and K$_{0.5}$TiO$_3$ are implemented to better the electrical characteristics of BFO ceramics [72, 73].

5.4.2 Temperature stability of strain properties

Strain investigations in lead-free piezo materials have received much attraction in current years, and BFO-based ceramics have gotten a lot of consideration because of their strong strain response and specific temperature dependency. By modifying the composition of BFO-BTO ceramics, the researchers were able to considerably improve the strain properties. It was discovered that adding $BaZrO_3$ to BFO-BTO ceramics can change their strain properties. By altering the $BaZrO_3$ concentration in BFO-BTO ceramics, a massively enhanced strain of 0.37 % sans negative strain is attainable, which is analogous to the condition in $BiNiTiO_3$-based (BNT) ceramics [47].

As a result, the strain augmentation in BFO-BTO dependent ceramics can be dedicated to the phase transition, which is comparable to MPB(II) in bismuth sodium titanate - dependent ceramics. More notably, phase structure in $BiFeO3$ dependent ceramics is simply affected by composition, but MPB(II) in bismuth sodium titanate -based ceramics is affected by both composition and temperature. As a result, in BFO-based ceramics, positive temperature dependency of strain characteristics may always be attained [47].

Temperature stability is owed to the simple electric field-instigated nonergodic relaxor - ferroelectric state phase transition, intrinsic lattice strain and thermally activated domain wall motion. $BiFeO_3$-$BaTiO_3$ based ceramics production can be expected as choices for piezoelectric actuators with large temperature, if we can beat the large leakage current density [74]. More research on the associations between configuration alteration, strain characteristics, and temperature stability in $BiFeO_3$-based ceramics should be done in the future.

5.4.3 Relationship between piezoelectricity and phase boundaries

The growth of piezoelectric characteristics in $BiFeO_3$-based ceramics was studied by looking at the collaboration between phase boundaries and electrical behavior. The following are some of the outcomes: Effective phase boundaries can amend piezoelectric characteristics and cannot be formed by ion substitution [75]. Due to the lack of actual MPB, its piezoelectric constant was limited to 50 pC/N even if composition adjustments and enhanced preparation techniques were widely used. Chemical alteration can successfully establish numerous kinds of phase boundaries (e.g., R-pseudocubic, R-T, R-monoclinic, etc.) in BFO-BTO systems, resulting in increased piezoelectricity (d_{33} = 100–402 pC/N)[76]. In addition to the creation of phase barriers, the composition of $BiFeO_3$-based ceramics must be optimized in order to find their piezoelectricity. $BiFeO_3$-$BaTiO_3$-based ceramics exhibit a significant strain response. The phase transition between the ferroelectric and relaxor states can be employed to endorse strain characteristics in $BiFeO_3$-$BaTiO_3$-based ceramics, just like in bismuth sodium titanate -based ceramics. Actual MPB

Advanced Functional Piezoelectric Materials and Applications   Materials Research Forum LLC
Materials Research Foundations 131 (2022) 222-258   https://doi.org/10.21741/978164490209-8

are the two types of phase barriers, on is the ferroelectric phase boundary for piezoelectric enhancement and second is the ferroelectric-relaxor phase transition for strain augmentation, [74, 76]. As a result, the strain behavior in BFO ceramics is remarkably similar to that of bismuth sodium titanate ceramics, and the strain improved as the temperature climbed, thus it is the most attractive contender for high-temperature lead-free actuators.

6.   Requirements for piezoceramic applications

Piezoelectric materials can be utilized in an assortment of electrochemical gadgets on account of their giving coupling among electrical and mechanical energy. This coupling was named a direct piezoelectric effect after its sighting by Curie siblings. With the discovery of piezoelectric ceramic materials significant progress in applications of piezoelectric materials was observed. The most commercially used piezoceramic materials are based upon lead-zirconate titanate that can also be abbreviated as (PZT) containing more than 60wt% lead [27, 77]. As a result of the toxic nature of lead, those piezoelectric materials are studied that have the same properties as that of PZT and can be a suitable replacement for PZT. Based on these new piezoceramic materials, their applications are going on the increase dominating the world market today. Depending upon functionalities and requirements for corresponding designs, piezoelectric materials were classified into four general categories: actuators, sensors, transducers, and resonators. Piezo materials can change mechanical energy into electrical energy thereby these can be used as sensors or even as energy harvesters. Likewise, these can also be used as an actuator when they change electrical energy into mechanical energy under the piezoelectric effect. There are some requirements and parameters that should follow by individual applications. The parameters affecting electromechanical effect are given as; piezoelectric coefficients describing mechanical (d), mechanical quality factor (Q_m), electrical (g) interactions and frequency constant (N_l), of piezoelectric ceramics, and piezoelectric coupling factors K_{31}, and K_t, K_{33}, $K_p$etc. Q_m and coupling coefficients (K) are critical parameters to explain the performance of resonators while piezoelectric coefficients (d and g) are important parameters for actuators. The ability of electromechanical devices to convert energy from one form to another is determined by effective electromechanical coupling coefficient K_{eff} by equation:

$$K_{eff}^2 = \frac{\text{mechanical energy Converted into electrical energy}}{\text{mechanical energy applied}} \qquad \text{Eq.2}$$

$$K_{eff}^2 \approx \frac{f_n^2 - f_m^2}{f_n^2}$$

$$\frac{K_p^2}{1 - K_p^2} = f\left(J_o, J_1, v\frac{f_p - f_s}{f_s}\right)$$

In the above equation, electrical energy can also be taken as input and then can be converted into mechanical energy at the output. The value of K^2 is always <1.0 because of the incomplete conversion of energy every time. In Equation-2 f_m and f_n are frequencies related to maximum and minimum impedance of the whole circuit whose measurement can be made using an appropriate bridge. Parallel and series resonant frequency of a whole circuit is represented using K_p where J_o, J_1 and v denotes Bessel functions of zero and first order and Poisson's ratio respectively. Mechanical loss of piezoceramics at resonance is defined by mechanical quality factor Q_m as:

$$Q_m = 2\pi \frac{\text{Amount of mechanically stored energy at resonance}}{\text{Dispersed energy per each resonant cycle}} \qquad \text{Eq.3}$$

Systematically, this Q_m is connected to sharpness of resonance frequency and can be obtained as:

$$Q_m = \frac{f_p^2}{2\pi f_r |Z|(C_1 + C_2)\left(f_p^2 - f_r^2\right)} \qquad \text{Eq.4}$$

Where Z, C_1 and C_2 elucidates minimum impedance and capacitance at resonance frequency f_r respectively. f_p is antiresonance frequency in hertz (HZ). The impedance of piezoelectric material is maximum at the resonance frequency and minimum at antiresonance frequency [78]. Last parameter frequency constant (N_l) can also be explained by the equation:

$$N_l = Lf_r = \frac{1}{2}\sqrt{\frac{Y}{\rho}} \qquad \text{Eq.5}$$

N_l is a product of resonance frequency f_r and linear dimensions of resonator. There can be various modes of it showing the direction of resonance frequency along any dimension when equivalent reactance is zero, Y is the young modulus, and ρ is density.

Advanced Functional Piezoelectric Materials and Applications   Materials Research Forum LLC
Materials Research Foundations 131 (2022) 222-258   https://doi.org/10.21741/978164490209-8

The equations of the above-explained parameters can be computed using the resonance behavior of piezoelectric materials when introduced to a varying electric field. Finally, different geometries can be used to find these piezoelectric coefficients. Under different doping conditions, requirements for particular applications can also be achieved as doping will produce new developments and properties in piezoelectric ceramics. For example, in the market for four main uses of piezoelectrics different substitutions in piezoelectric ceramics have been done that are given as: (1) for actuators to detect mechanical vibrations, (2) to control frequency, (3) generating charges at higher voltages, (4) to produce ultrasonic and acoustic vibrations [79]. All these uses demand different requirements such as the use of piezoelectrics for actuators requires low permittivity along with a high piezoelectric (g) coefficient. Time and temperature-related properties should be stable to use piezoelectrics for controlling the frequency to minimize losses and to achieve a high coupling coefficient. For the use of piezoelectric to produce charges at higher voltages combination of (g) coefficient with resistance is required to affect electromechanical properties by mechanical stress. To produce ultrasonic vibrations of useful amplitude; that piezoceramics material will be used that showed minimal losses when subjected to the necessary field. Here some applications of piezoceramics are explained in detail with their requirements.

6.1   Actuators

Piezoelectric materials have significant and extensive applications in today's actuator technologies. Piezoelectric actuators are acknowledged as ideal for many smart applications because of their fascinating properties such as high-power density and bandwidth, small size, and having the ability to act as both sensors and actuators. Actuators can uphold stresses in the range of tens of megapascals [80]. In many applications for the generation of large displacements without affecting the ability of a material to generate large forces actuators are utilized. For example, in several flex tensional underwater transducer designs, piezoelectric stack actuators are used that utilize large displacements of the stack to radiate large amplitude sound waves in presence of water mass. Now with the invention of multilayers actuators, its applications are going on an increase in many fields, especially in precision positioning, suspension system, and automotive fuel injection system [81]. In smart systems, multilayered actuators integrated with feedback control circuits advance and alter structures to the environment. For actuator devices, some properties that contribute to strong piezoelectric behavior are illustrated as higher environmental and chemical stability, high pressure per density ratio, and ability to operate at high temperatures and frequencies. The function given in Eq-(2) can be used as energy conversion in bandwidth, signal processing devices, in spacings of critical frequencies of resonators, and especially in actuators. Actuators are characterized by displacement and required large blocking force, higher electric field activated strain, and higher Curie temperature to be used in many applications [30]. Through these parameters actuator's operational range can be defined and adjusted by each device requirement like to derive motors and control fuel injection etc. An increase in displacement and force can be obtained by utilizing multilayered structures while an increase in strain is got by electromechanical coupling utilizing either electrostriction or piezoelectricity [31]. Using tensor representation of electric field activated strain (G_{ij}) is articulated as power series either in an electric field (E_k) and in polarization form (P_k) as:

$$G_{ij} = d_{ijk}E_k + M_{ijkl}E_kE_l + \cdots \quad (i,j,k,l) = \quad 1,2,3,4$$

Eq.6

$$G_{ij} = g_{ijk}P_k + Q_{ijkl}P_kP_l + \cdots \quad (i,j,k,l) = \quad 1,2,3,4$$

In tensor form, piezoelectric coefficients are denoted as d_{ijk} and g_{ijk} representing piezoelectric charge constant and piezoelectric voltage constant respectively. These piezoelectric coefficients are also named as third-order tensors of converse piezoelectric effect whose magnitude is very helpful in material selection and designing of multilayer actuators. d_{ijk} is termed as the proportion of electric charge produced per unit applied stress with unit of C/N. Whereas g_{ijk} is the ratio of field developed to mechanical stress applied having units of a volt. meter per newton Vm/N. E_k and P_k are electric field and polarization vector components whereas M_{ijkl} and Q_{ijkl} are electrostrictive coefficients. Direct and converse piezoelectric effects are basic piezoelectric coefficients. The first term in both equations of (6) describes the opposite piezoelectric effect and the 2nd term electrostriction. The only deliberation of longitudinal and transverse dislocations is sufficient for material development. In opposite piezoelectric effect when an electric field is subjected to piezo-material strain is produced in a material that is explained by simplifying Equation-6 as:

$$G_i = d_{ij}E_j$$

Eq.7

$$G_i = Q_{ij}P_j^2$$

In Equation 7 the second relation is a simplified version of electro strictive materials. For piezoelectric ceramic applications, coefficients d_{ij} are generally reported in matrix form as d_{31}, d_{33}, and d_{15} rendering to the direction of an electric field in any axis. Macroscopic symmetry of the material needs to be broken down to some extent in order to attain non-centrosymmetric nature in polycrystalline materials. Ferroelectric materials endure a phase transition at Curie temperature, in order to produce spontaneous polarization, ferroelectric. Then by applying a strong electric field spontaneous polarization tends to align along with metastable state; this process is named poling process. Now, these poled ferroelectric

Advanced Functional Piezoelectric Materials and Applications   Materials Research Forum LLC
Materials Research Foundations 131 (2022) 222-258   https://doi.org/10.21741/978164490209-8

materials are also piezoelectric because of the macroscopically non-centrosymmetric nature of the metastable state. So, poling process is a revolutionary finding in the ceramic industry by which isotropic ferroelectrics can be made anisotropic piezoelectric materials by applying specific electric fields at elevated temperatures. This poling direction is by convention in 3-direction along d_{33} demonstrating the direction of the applied field parallels to poling direction, creating strain also in the equivalent direction. In converse piezoelectric effect multilayer actuators are only related with d_{33}, especially in fuel injector actuators that require higher field strain correspondingly to $d_{33} \sim 550^{PC}/_{m^2}$ within a temperature range of -55-150°C [82]. Primarily, actuators were designed to obtain high strain at the low coaxial electric field. That drives to quantity $G_{3,max}/E_{3,max}$ having similarity with d_{33} coefficient only with the difference that this quantity possesses large-signal quantity because of maximum applied field. It is consequently named as huge signal d_{33} value. Using domain wall twisting and local switching of domain wall segments; domain walls in ferroelectric materials also contribute to a small signal piezoelectric coefficient. All these extrinsic contributions are revealed through hysteresis or by the frequency dependence of strain [83]. Nonlinear piezoelectric strain vary from electro strictive strain as it has a complex field dependency whereas strain in electrostriction has quadratic dependency of the field having the same sign for the positive or negative field. Dielectric displacement of actuators is critical in determining the required current for each actuator stroke in technological applications. The uniaxial force called blocking force is applied to compensate electric field-induced expansions in order to estimate the mechanical capabilities of actuators. When actuators are heated and made to run cold start, the temperature credence of available strain and dielectric displacement also becomes crucial. Finally, at a given field obtainable strain depends upon pressure, frequency, temperature, and electric field intervals.

6.2   Sensors

Piezoelectricity produces a voltage signal or a potential in retort to mechanical stress that consists of a crystal made up of piezoelectric material bounded with a surface of interest. Electrodes are connected across the ends of the crystal to detect the r potential. These can work at resonant or at off-resonant conditions by transforming exterior mechanical incentives into electrical signals. High piezoelectric coefficients (d and g) are required to operate sensors in non-resonant conditions while high coupling coefficients (k) and mechanical quality factor (Q_m) become important for those materials that sense at resonance conditions. The direct piezoelectric effect is used in many sensor applications to measure pressure or force. Depending upon the nature of pressure and force that has to be measured several piezoelectric materials are utilized in many sensor applications like in hydrophones, microphones, accelerometers [78], etc.

Piezoelectricity makes sense of the connection between mechanical factors like pressure, σ and strain, G and electrical factors. Piezoelectric charge and voltage coefficient are based on these independent variables and are crucial parameters for piezoelectric sensors. For direct piezoelectric effect these can be found when D and E are termed as independent variables as [77]:

$$d = \left(\frac{\partial D}{\partial \sigma}\right) E$$

Eq.8

$$g = -\left(\frac{\partial E}{\partial \sigma}\right) D$$

If we make changes in stress and electrical displacement along three orthogonal directions then the constitutive equation for piezoelectric material governing direct piezoelectric effect in sensors can be written as:

$$D_i = d_{ij}\sigma_j$$

Eq.9

$$D_3 = d_{33}\sigma_3$$

This equation tells that sensor are used to detect stresses σ_j by observing electric displacement. Stress will also be measured in three directions when an electric field is introduced parallel to three directions. Piezoelectric charge coefficient, d, is also known as small-signal piezoelectric coefficient, because of small external mechanical stress, the. d_{ij} is an imperative physical property that represents the intrinsic piezoelectric effect and is computed by quasi-static d_{33} meter. d_{33} is a longitudinal piezoelectric coefficient that is mostly preferred in many sensor applications. According to obtainable strain $G_{3,max}$, d_{33} is a complex function of the magnitude of an electrical-mechanical driving force, frequency, pressure, temperature, and biased electric field. Because of the most importance of d_{33} in piezoelectric sensors; this is often denoted as piezoelectric coefficient. In unipolar loading applications, a linear decrease in the piezoelectric coefficient has been noticed with increasing field [77]. In many materials value of d_{33} is specific such as greater than 200 $^{pm}/_V$ in PZT materials, and 2 $^{pm}/_V$ in quartz (SiO_2), 20 $^{pm}/_V$ in bismuth titanate based Aurivillius structures which is most important commercially used piezoelectric material used in timing devices and as acoustic wave filter [82], etc. From the above-given materials, only PZT shows an important input for domain wall displacements. Lead-free perovskite ferroelectric oxide $(K_{0.5}Na_{0.5})NbO_3$ (KNN) based compositions are mostly preferred to be used in knocking sensors to screen combustion in automobile engines. KNN based compositions are suited for sensing because of their high and temperature stable electromechanical coupling coefficients. Individual compositions were also developed that will use thermal exterior incentives for electrical power generation [84]. Lead-free sensors possessed the same level of voltages at the output and better sturdiness than those manufactured with lead-based ceramics and are currently available in the market. Finally, despite the magnitude of the piezoelectric coefficient its high electric resistivity, lowest dependence on temperature, and low permittivity are also basic requirements that should be fulfilled to make a successful piezoceramic sensor [31].

6.3   Transducers

Any device which interconverts different forms of energy is called transducer. The transducer which converts non-electrical physical quantity to the usable proportional electrical signal is termed as sensor. It is also known as "pick-up"[85]. The generated electrical signal may be in the form of frequency, voltage or current. The transducer transforms the electrical signal into another form then it is called actuator. The transducer can employ multiple effects to bring out such transformation. This process of conversion of signal from one form to another is called transduction. The non-electrical physical quantity may be mechanical, thermal, optical or thermal. The simplest transducer contains two important constituents sensing element and transduction element. The sensing element senses the rate of change of the non-electrical physical quantity while the transduction element converts the output signal of sensing element into the proportional electrical signal [86].

The transducers can be classified as follows:

Primary and secondary transducer

The primary transducer is directly connected to the input while the secondary transducer is indirectly connected to the input.

Analogue and digital transducer

Analogue transducers give output in analogue form that continues in time whereas the digital transducers give digital output signal (or pulses). LVDT (linear variable differential transformer) Strain gauge, thermistor etc. are the examples of analogue transducers while photo electric transducer gives the output in digital form [87].

Active and passive transducer

Active transducers utilize their own developed voltage or current to accord the output signal. These are also called "self-generating type transducers" as no external power is required as the obligatory energy for generating this output signal is extracted from the non-electrical physical quantity. The examples include photovoltaic cell, tacho-generator, thermocouple, piezoelectric transducers.

Passive transducers require external power source to transform the non-electrical quantity into the electrical signal. These are also called "externally powered transducers". Examples are R, L, C (Resistive, Inductive and capacitive) transducers [88].

6.3.1 Piezoelectric transducers

Piezoelectric transducers work on the principle of piezoelectricity. Naturally occurring piezoelectric crystals are Topaz, Quartz, Tourmaline group, Rochelle salt and certain organic substances (silk, wood, bone, enamel, rubber, hair etc.). Lead free piezoelectric crystals are also made artificially such as PVDF (Polyvinylidene difluoride), Potassium niobate, barium titanate, Lithium niobate etc. There are certain requirements which the piezoelectric crystals have to fulfil to be used for as transducers such as high output values, stable frequency response and negligible impact of external conditions (extreme temperature and humidity). Piezoelectric crystal is perfectly fit for all these requirements and hence every piezoelectric crystal has its own limitation. For example, Quartz (naturally occurring) has high stability but low output. Rochelle salt gives high output but is more sensitive towards external conditions [86].

The Piezoelectric quartz crystals are usually in the form of hexagonal crystals. These crystals have pyramids on which the coordinate axes are fixed. In the pyramids, the axes pass through the corners as well as the end points and are termed as optic axis (z-axis) and electric axis (x-axis) respectively. The mechanical axis (y-axis) passes through the middle of the opposite sides. Piezoelectric crystals are employed to measure small displacements (in the range of Angstrom) in scanning tunnelling microscopes (STM) and (AFM) [89].

Piezoelectric transducers use piezoelectric effect to convert changes in measurable physical quantities into electrical signal. A thin layer of coating material (silver) is coated on the faces of the piezoelectric material. The ions in the material are moved towards one of the ends when stress is applied. This modification in the charge distribution introduces charge polarity depending upon the compressive or tensile stress. The amount of charge displaced is measured by using a charge amplifier which is situated close to the piezoelectric sensor. High resistance is used to draw a very low current. The correction is affected by the value of the capacitance of the wire that joins the transducer and piezoelectric sensor. The electrical voltage thus produces can easily be gauged using voltmeter. The produced electric voltage will be the function of the applied force. Hence, these devices help in measuring the mechanical stress. They mainly produce electric polarization that is linearly proportional to applied stress. In other words, an emf is generated when an external force is applied. The force can be in the form of sound, vibration, pressure or acceleration etc.

Gas igniters, microphones, stress-strain gauges etc. mainly fall in the category of piezoelectric transducers. When an oscillating potential applied to the axis of a crystal, the dimensions of the crystal change. This effect is known as Piezo-electric effect. The reverse process is also true and is termed as Electro-resistive elements.

The amount of the generated voltage relies on the direction of the crystal. In longitudinal orientation, the charge produced can be calculated as

$$Q = F \times d \qquad\qquad \text{Eq.10}$$

While in transverse orientation, the charged is calculated as

$$Q = F \times d \times (b/a) \qquad\qquad \text{Eq.11}$$

Here, F is the applied force and d is the piezoelectric coefficient of the crystal. The differentiating terms between transverse and longitudinal orientation of the crystal here is the ratio (b/a) which shows that the transverse orientation gives more change when the ratio is greater than unity i.e., $(b/a) > 1$.

The best feature of piezoelectric transducers is the rapid variation of parameter and thus high-frequency response [90]. They are self-generating as there is no need of any external power source and are easy to handle due to their small displacements and long-range measurements. However, they provide small charge and low output. These devices are affected by environmental conditions like temperature, humidity etc. and are not suitable for static conditions.

Figure of merit is the product of voltage coefficients and piezoelectric charge and for piezo-materials, is important for applications in transducers. PZT ceramics were used as transducers due to their electromechanically and piezoelectric properties but now lead free ceramics such as (KNN) and (BNT) based ceramics are used as piezoelectric materials for transducers. These lead-free alternative offer high piezoelectric properties but provide low curie temperature in comparison to non-perovskite type lead free piezoelectric materials. Hard piezoelectric perovskites have higher quality factor, low dielectric loss in comparison to lead based piezoelectric perovskites. These piezoelectric perovskites based transducers have wide range of applications such as in ultrasonic diagnostic imaging, ultrasound therapy [91].

6.4   Resonators

Resonator is a device that generates waves with specified frequencies using resonance phenomenon. The resonators that produce electromagnetic waves are called electromagnetic resonators while the resonators that generate mechanical waves are termed as mechanical resonators (including acoustic resonators). Along with the generation of the waves, resonators can also be employed to select wave of specific frequency. The acoustic resonators are commonly used in musical instruments to produce specified sound wave. Whereas Quartz crystals produce electromagnetic waves i.e., radio waves in radio transmitters. For electromagnetic resonators, there are various types including optical cavity, cavity resonator, resonance circuits, dielectric resonators etc. [92].

Piezoelectric resonators are mechanical resonators that are used to produce signal of desired frequency in electronic circuits as well as filters. These resonators are based on the piezoelectricity which is coupled to mechanical and electrical process. The following equation expresses the electrical behavior of materials:

$$D = \epsilon E \ (1) \hspace{4cm} \text{Eq.12}$$

here E is the electric field strength, D represents the electrical displacement and ϵ shows the permittivity of the material, and. Similarly, the mechanical behavior of materials is described by the Hook's law:

$$S = sT \ (2) \hspace{4cm} \text{Eq.13}$$

where the symbol S represents the strain, s represents the compliance, the stress is symbolized by T. The above-mentioned equations are modified for the piezoelectric crystals as follows:

$$D = \epsilon^T E + dT \hspace{4cm} \text{Eq.14}$$

$$S = d^t E + s^E T \hspace{4cm} \text{Eq.15}$$

In these equations, d is the piezoelectric constant and ϵ^T represents the permittivity , s^E is the compliance at a constant electric field at a constant stress [93].

Advanced Functional Piezoelectric Materials and Applications   Materials Research Forum LLC
Materials Research Foundations 131 (2022) 222-258   https://doi.org/10.21741/978164490209-8

Resonator contains piezoelectric substrate hanging from the edge. The two surfaces of the substrate are covered with tinny layer of metal which acts as electrodes. The electronic signal lines are connected to these electrodes. The resonance mode is activated when the applied signal frequency become equal to that of natural frequency of the construction setup. The change in the position of the suspension elements and the electrode patterns strongly influence the activation of the targeted-mode. The modification of the strain field is in the direction of applied electric field to give rise to thickness extensional mode. This mode is frequently employed for resonance in piezoelectric devices. In order to substitute the electrical field parallel to the axis which is normal to the plane of the substrate, an alternative voltage is required between the two electrodes. Consequently, the application of the non-zero piezoelectric constant for piezoelectric substrate give rise to a parallel field in the construction setup by following this relation.

$$S_3 = d_{33}E_3 \hspace{5cm} \text{Eq.16}$$

The strain field develops in the other directions of the body of the resonator because there are other non-zero parameters along with the piezoelectric coefficient d_{33}.

The main strain field, produced due to d_{31}, is orthogonal to the applied electric field and is in the plane of the substrate.

$$S_1 = d_{31}E_3 \hspace{5cm} \text{Eq.17}$$

The clamping of the substrate influences the strain field (S_1) to give rise to various resonance modes. In particular, the clamping of the piezoelectric plate along the nodal points provokes the lateral mode excitation. These nodal points influence the length depending resonance mode simultaneously with the development of the thick film.

A partial differential equation called wave equation governs the displacement of each particle in the system [94]. Assumption such as small vibration amplitude are made in order to remove non linearities from equation and to obtain a closed form solution. Resonators require high quality factor and coupling coefficient f piezoelectric materials. The composition $(Bi_{1/2}Na_{1/2})$ TiO_3-$BaTiO_3$-$(Bi_{1/2}Na_{1/2})$ $(Mn_{1/3}Nb_{2/3})O_3$, is a contender for a lead-free resonator and is being used commercially. The perovskite oxides were assorted via ball milling and calcinated at 850°C for 2 h and then sintered at 1100–1200°C for 2 h [95]. Another type of resonator called Langevin type transducer was built by piecing together two leads free piezoceramics, the frequency of poles were adjusted to resonance frequency. The velocity achieved with lead free piezoceramic was higher than PZT ceramics with the same input [96]. Tetragonal-tungsten-bronze structured lead free piezoceramic (SCNN) and KNN based piezoceramics is another lead free piezoelectric ultrasonic motor which exhibits superior qualities to PZT based motor [97].

In the modern worlds, piezoelectric materials are predominant because of their capability of conversion between electrical energy and mechanical energy. Such materials are important in the fields of communication, medical diagnostics and imaging and industrial automation. Lead-free piezo-materials came under the focus after the restriction of the use of the toxic substances for electrical purposes. Lead-free piezoelectric materials provide environmental-friendly alteration of electrical energy into mechanical energy and vice versa. In the past decade, this field has been explored because of the toxicity free materials like lead-free (KNN)-based piezoelectric materials [98].

Conclusion

Lead free piezoelectric perovskites or piezoceramics have paramount significance in the field of perovskites and ceramic industry. The toxicity of PZT ceramics instigated the search for lead free alternatives of these ceramics. Ever since there have been continuous efforts in this regard. There are three categories in the lead free piezoceramics that are perovskites, bismuth based and tungsten bronze compositions. Solid solutions of different perovskites and their compositions served as lead free alternative.

Niobate and bismuth-based ceramics are such class of perovskites that served as lead free alternative piezoelectric perovskites and have variety of applications. Different compositions of KNN ceramics are used as knocking sensor, actuator and buzzer and are briefly discussed here. KNN ceramics have large electric field induced strain and piezoelectric coefficient. Bismuth based piezoceramics are an excellent lead alternative and have exciting applications, many compositions and solid solutions of bismuth-based perovskites are proposed over the years. Some applications of bismuth based piezoceramics as a multilayer actuator, immunosensor is discussed here. Different lead free piezo-ceramic materials were discussed in detail along with their applications. Barium titante was the first lead free alternative of PZT ceramics, even though it has low piezoelectric coefficient, many techniques such an ion substitution and sintering techniques have been projected to progress its piezoelectric properties, Strain energy, phase boundary, high sintering and low curie temperature of $BaTiO_3$ material were deliberated briefly. The piezoelectric properties of bismuth ferrite perovskites can be modified through substitution of different perovskites and this subject have been briefly touched here. Phase boundary, temperature stability of strain properties and association between piezoelectricity/strain and phase boundaries were also discussed in detail. The applications of piezoceramics as actuators, sensors require different requirements. Such as application as actuator require high piezoelectric coefficient and high-quality factor. Parameters such as coupling coefficient, piezoelectric coefficient, quality factor, strain, curie temperature are important in deciding the future for application of these lead-free piezoelectric perovskites. Different lead free piezoceramics as actuators, sensors and resonators that are in commercial use are deliberated briefly.

References

[1] A. Verma, A. Kumar, Bulk modulus of cubic perovskites, J. Alloy. Comp. 541 (2012) 210-214. https://doi.org/10.1016/j.jallcom.2012.07.027

[2] D.P. Agrawal, Different Types of Transducers, in Embedded Sensor Systems, Springer. (2017) 65-104. https://doi.org/10.1007/978-981-10-3038-3_3

[3] E. Aksel, J.L. Jones, Advances in lead-free piezoelectric materials for sensors and actuators, Sensors. 10 (2010) 1935-1954. https://doi.org/10.3390/s100301935

[4] E. Aksel, J.S. Forrester, J.C. Nino, K. Page, D.P. Shoemaker, J.L. Jones, Local atomic structure deviation from average structure of $Na_{0.5}Bi_{0.5}TiO_3$: Combined x-ray and neutron total scattering study, Phys. Rev. B. 87 (2013) 104110- 104113. https://doi.org/10.1103/PhysRevB.87.104113

[5] A. Arnau, Piezoelectric transducers and applications, second de., Springer., Berlin, 2004, pp. 97-116 https://doi.org/10.1007/978-3-662-05361-4

[6] P. Baettig, C.F. Schelle, R. LeSar, U.V. Waghmare, N.A. Spaldin, Theoretical prediction of new high-performance lead-free piezoelectrics, Chem. Mater. 17 (2005)1376-1380. https://doi.org/10.1021/cm0480418

[7] N. Balke, D.C. Lupascu, T. Granzow, J. Rödel, Fatigue of lead zirconate titanate ceramics II: sesquipolar loading, J.Amer. Ceram.Soc. 90(4) (2007)1088-1093. https://doi.org/10.1111/j.1551-2916.2007.01521.x

[8] D. Berlincourt, H. Jaffe, Elastic and piezoelectric coefficients of single-crystal barium titanate, Phys.Rev.111 (1958)140-143. https://doi.org/10.1103/PhysRev.111.143

[9] S.Bhattacharjee, D. Pandey, Effect of stress induced monoclinic to tetragonal phase transformation in the multiferroic $(1-x)BiFeO_3$-$_xPbTiO_3$ system on the width of the morphotropic phase boundary and the tetragonality, J.App.Phys. 110 (2011) 084100-084105. https://doi.org/10.1063/1.3647755

[10] R.J. Bobber, New types of transducers, in Underwater Acoustics and Signal Processing, Springer. (1981) 243-261. https://doi.org/10.1007/978-94-009-8447-9_20

[11] K. Brajesh, M. Abebe, R. Ranjan, Structural transformations in morphotropic-phase-boundary composition of the lead-free piezoelectric system Ba(Ti$_{0.8}$Zr$_{0.2}$)O$_3$–(Ba$_{0.7}$Ca$_{0.3}$)TiO$_3$, Phys. Rev.B. 94 (2016) 104100-104108. https://doi.org/10.1103/PhysRevB.94.104108

[12] X. Chao, Z. Wang, Y. Tian, Y. Zhou, Z. Yang, Ba(Cu$_{0.5}$W$_{0.5}$)O$_3$-induced sinterability, electrical and mechanical properties of (Ba$_{0.85}$Ca$_{0.15}$Ti$_{0.90}$Zr$_{0.10}$)O$_3$ ceramics sintered at low temperature, MRS. Bulletin. 66 (2015) 16-25. https://doi.org/10.1016/j.materresbull.2015.02.022

[13] F. Chen, Q. Zhang, J. Li, Y. Qi, C. Lu, X. Chen, X. Ren, Y. Zhao, Sol-gel derived multiferroic BiFeO3 ceramics with large polarization and weak ferromagnetism, Appl. Phys. Lett. 89 (2006) 092905-092910. https://doi.org/10.1063/1.2335367

[14] T. Chen, T. Zhang, G. Wang, J. Zhou, J. Zhang, Y. Liu, Effect of CuO on the microstructure and electrical properties of Ba$_{0.85}$Ca$_{0.15}$Ti$_{0.90}$Zr$_{0.10}$O$_3$ piezoceramics, J.Mater.Sci. 47 (2012) 4612-4619. https://doi.org/10.1007/s10853-012-6326-1

[15] Y. Cui, X. Liu, M. Jiang, Y. Hu, Q. Su, H. Wang, Lead-free (Ba$_{0.7}$Ca$_{0.3}$)TiO$_3$-Ba(Zr$_{0.2}$Ti$_{0.8}$)O$_3$-xwt% CuO ceramics with high piezoelectric coefficient by low-temperature sintering, J. Mater. Sci.: Mater. Electron. 23 (2012) 1342-1345. https://doi.org/10.1007/s10854-011-0596-2

[16] Y. Cui, C. Yuan, X. Liu, X.Zhao, X.Shan, Lead-free (Ba0.85Ca0.15)(Ti0.9Zr0.1)O3-Y2O3 ceramics with large piezoelectric coefficient obtained by low-temperature sintering, J. Mater. Sci.: Mater. Electron. 24 (2013) 654-657. https://doi.org/10.1007/s10854-012-0785-7

[17] Y. Cui, X. Liu, M. Jiang, X. Zhao, X. Shan, W. Li, C. Yuan, C.Zhou, Lead-free (Ba0.85Ca0.15)(Ti0.9Zr0.1)O3-CeO2 ceramics with high piezoelectric coefficient obtained by low-temperature sintering, Ceram. Intern. 38 (2012) 4761-4764. https://doi.org/10.1016/j.ceramint.2012.02.063

[18] M. Davis, Picturing the elephant: Giant piezoelectric activity and the monoclinic phases of relaxor-ferroelectric single crystals, J.Electroceram. 19 (2007) 25-47. https://doi.org/10.1007/s10832-007-9046-1

[19] O. Deubzer, Y. Baron, N. Nissen, K.-D. Lang, Status of the RoHS directive and exemptions. in 2016 Electronics Goes Green 2016 (EGG), IEEE.(2016) https://doi.org/10.1109/EGG.2016.7829868

[20] Y. Doshida, S. Kishimoto, T. Irieda, H. Tamura, Y. Tomikawa, S. Hirose, Double-mode miniature cantilever-type ultrasonic motor using lead-free array-type multilayer piezoelectric ceramics, Jpn. J.App. Phys. 47 (2008) 4240-4242. https://doi.org/10.1143/JJAP.47.4242

[21] Y. Doshida, S. Kishimoto, K. Ishii, H. Kishi, H. Tamura, Y. Tomikawa, S. Hirose, Miniature cantilever-type ultrasonic motor using Pb-free multilayer piezoelectric ceramics, Jpn. J.Appl.Phys. 46 (2007) 4920-4921. https://doi.org/10.1143/JJAP.46.4921

[22] P. Fan, K. Liu, W. Ma, H. Tan, Q. Zhang, L. Zhang, C. Zhou, D. Salamon, S.-T. Zhang, Y. Zhang, Progress and perspective of high strain NBT-based lead-free piezoceramics and multilayer actuators, J. Materiom. 7 (2021) 508-544. https://doi.org/10.1016/j.jmat.2020.11.009

[23] D. Fernández-Benavides, L. Cervera-Chiner, Y. Jiménez, O.A. de Fuentes, A. Montoya, J. Muñoz-Saldaña, A novel bismuth-based lead-free piezoelectric transducer immunosensor for carbaryl quantification, ens. Actuators B Chem. 285 (2019) 423-430. https://doi.org/10.1016/j.snb.2019.01.081

[24] T. Fujii, S. Watanabe, M. Suzuki, T. Fujiu, Application of lead zirconate titanate thin film displacement sensors for the atomic force microscope, J. Vac. Sci. Technol. B. 13 (1995)1119-1122. https://doi.org/10.1116/1.587914

[25] G.H. Haertling, Ferroelectric ceramics: history and technology, J.Amer. Ceram. Soc. 82 (1999) 797-818. https://doi.org/10.1111/j.1151-2916.1999.tb01840.x

[26] A.A. Heitmann, G.A. Rossetti Jr, Thermodynamics of ferroelectric solid solutions with morphotropic phase boundaries, J J.Amer. Ceram. Soc. 97 (2014) 1661-1685. https://doi.org/10.1111/jace.12979

[27] P. Herzig, J. Zemann, AB$_3$ nets built from corner-connected octahedra: geometries, electrostatic lattice energies, and stereochemical discussion, Z. Kristallogr. Krist. 205 (1993)85-97. https://doi.org/10.1524/zkri.1993.205.Part-1.85

[28] C.-H. Hong, H.-S. S.-S. Lee, K. Wang, H-Z. Yao, J-F. Li, J.-H. Gwon, N.V. Quyet, J-K. Jung, W. Jo, Ring-type rotary ultrasonic motor using lead-free ceramics, J. Sens. Sci. Tech. 24 (2015) 228 - 231 https://doi.org/10.5369/JSST.2015.24.4.228

[29] W. Jo, J.E. Daniels, J.L. Jones, X. Tan, P.A. Thomas, D. Damjanovic, J. Rödel, Evolving morphotropic phase boundary in lead-free $(Bi_{1/2}Na_{1/2})TiO_3$-$BaTiO_3$ piezoceramics, J.App.Phys.109 (2011) 014100-014110. https://doi.org/10.1063/1.3530737

[30] M. Johnsson, P. Lemmens, Perovskites and thin films-crystallography and chemistry, J. Phys. Cond. Matter. 20 (2008) 263990-264001. https://doi.org/10.1088/0953-8984/20/26/264001

[31] D. Jones, S. Prasad, J. Wallace, Piezoelectric materials and their applications. in Key Engineering Materials, Trans. Tech. Publ.122 (1996) 71-144 https://doi.org/10.4028/www.scientific.net/KEM.122-124.71

[32] T. Kainz, M. Naderer, D. Schütz, O. Fruhwirth, F.-A. Mautner, K. Reichmann, Solid state synthesis and sintering of solid solutions of BNT-Xbkt, J.Euro.Ceram.Soc. 34 (2014) 3685-3697. https://doi.org/10.1016/j.jeurceramsoc.2014.04.040

[33] A.K. Kalyani, K. Brajesh, A. Senyshyn,R. Ranjan, Orthorhombic-tetragonal phase coexistence and enhanced piezo-response at room temperature in Zr, Sn, and Hf modified $BaTiO_3$, App.Phys.Lett. 104 (2014) 252900-252906. https://doi.org/10.1063/1.4885516

[34] X-Y.Kang, Z.-H. Zhao, Y.-K. Lv, Y.Dai, BNT-based multi-layer ceramic actuator with enhanced temperature stability, J.Alloy.Comp. 771 (2019) 541-546. https://doi.org/10.1016/j.jallcom.2018.08.311

[35] M. Karpelson, G.-Y. Wei, R.J. Wood, Driving high voltage piezoelectric actuators in microrobotic applications, Sen. Actuat. A. Phys. 176 (2012) 78-89. https://doi.org/10.1016/j.sna.2011.11.035

[36] C.H. Kim, G. Qi, K. Dahlberg, W. Li, Strontium-doped perovskites rival platinum catalysts for treating NOx in simulated diesel exhaust, Science. 327 (2010) 1624-1627. https://doi.org/10.1126/science.1184087

[37] R.M. Langdon, Resonator sensors-a review,J.Phys.E. Sci.Instr. 18(2) (1985) 100- 103. https://doi.org/10.1088/0022-3735/18/2/002

[38] P.S. Lavers, The electronic structure of oxide perovskites and related materials, Doctor of Philosphy Theiss. (2015)

[39] H.J. Lee, S. Zhang, Y. Bar-Cohen, S. Sherrit, High temperature, high power piezoelectric composite transducers, Sens. 14 (2014)14526-14552. https://doi.org/10.3390/s140814526

[40] Q. Li, J. Wei, J. Cheng, J.Chen, High temperature dielectric, ferroelectric and piezoelectric properties of Mn-modified $BiFeO_3$-$BaTiO_3$ lead-free ceramics, J.Mater.Sci. 52 (2017) 229-237. https://doi.org/10.1007/s10853-016-0325-6

[41] W. Li, Z. Xu, R. Chu, P. Fu, G. Zang, Improved piezoelectric property and bright up conversion luminescence in Er doped $(Ba_{0.99}Ca_{0.01})(Ti_{0.98}Zr_{0.02})O_3$ ceramics, J. Alloy. Comp. 583 (2014) 305-308. https://doi.org/10.1016/j.jallcom.2013.08.103

[42] W. Liu, X. Ren, Large piezoelectric effect in Pb-free ceramics, Phys. Rev. Lett. 103 (2009) 257600- 257602. https://doi.org/10.1103/PhysRevLett.103.257602

[43] X. Liu, X.Tan, Giant strains in non-textured $(Bi_{1/2}Na_{1/2})TiO_3$ based lead free ceramics, Adv. Mater. 28(3) (2016) 574-578. https://doi.org/10.1002/adma.201503768

[44] J. Lv, X. Lou, J. Wu, Defect dipole-induced poling characteristics and ferroelectricity of quenched bismuth ferrite-based ceramics, J. Mater. Chem. C. 4 (2016) 6140-6151. https://doi.org/10.1039/C6TC01629D

[45] J. Lv, J. Wu, W. Wu, Enhanced Electrical Properties of Quenched $_{(1-x)}Bi_{1-y}Sm_yFeO_3$-$_xBiScO_3$ Lead-Free Ceramics. The Journal of Physical Chemistry C, 119 (2015) 21105-21115. https://doi.org/10.1021/acs.jpcc.5b07249

[46] J. Ma, X. Liu, W. Li, High piezoelectric coefficient and temperature stability of Ga_2O_3-doped $(Ba_{0.99}Ca_{0.01})(Zr_{0.02}Ti_{0.98})O_3$ lead-free ceramics by low-temperature sintering, J. Alloy.Comp. 581 (2013) 642-645. https://doi.org/10.1016/j.jallcom.2013.07.131

[47] R. Machado, V.B. dos Santos, D.A. Ochoa, E. Cerdeiras,L. Mestres, J.E. García, Elastic, dielectric and electromechanical properties of $(Bi_{0.5}Na_{0.5})TiO_3$-$BaTiO_3$ piezoceramics at the morphotropic phase boundary region, J. Alloy. Comp. 690 (2017) 568-574. https://doi.org/10.1016/j.jallcom.2016.08.116

[48] B. Malič, M. Otoničar, K. Radan, J. Koruza, A. Heterogeneity challenges in multiple-element-modified Lead-free piezoelectric ceramics, Mater. 12 (2019) 4040-4049. https://doi.org/10.3390/ma12244049

[49] T. Matsuoka, H. Kozuka, K. Kitamura, H. Yamada, T. Kurahashi, M. Yamazaki, K. Ohbayashi, KNN-NTK composite lead-free piezoelectric ceramic, J. App. Phys. 116 (2014) 154100-154104. https://doi.org/10.1063/1.4898586

[50] H.D. Megaw, Crystal structure of double oxides of the perovskite type, Procee. Phys. Soc. 58 (1946) 128-133. https://doi.org/10.1088/0959-5309/58/2/301

[51] A. Navrotsky, D.J. Weidner, Perovskite: a structure of great interest to geophysics and materials science, Geophys. 45 (1989) https://doi.org/10.1029/GM045

[52] R. Nunamaker, Frequency control devices for mobile communications. in 25th Annual Symposium on Frequency Control, IEEE. (1971) https://doi.org/10.1109/FREQ.1971.199836

[53] G. Piazza, Piezoelectric aluminum nitride vibrating RF MEMS for radio front-end technology, University of California, Berkeley (2005) https://doi.org/10.1109/MWSYM.2006.249702

[54] A. Popovič, L. Bencze, J. Koruza, B. Malič, Vapour pressure and mixing thermodynamic properties of the KNbO3-NaNbO3 system, RSC Adva. 5(93) (2015) 76249-76256. https://doi.org/10.1039/C5RA11874C

[55] C. Randall, A. Kelnberger, G. Yang, R. Eitel, T. Shrout, High strain piezoelectric multilayer actuators-a material science and engineering challenge, J. Electroceram. 4 (2005) 177-191. https://doi.org/10.1007/s10832-005-0956-5

[56] W.M. Roberts, The synthesis and characterisation of lead-free piezoelectric ceramics, University of Birmingham.(2012)

[57] J. Rödel, W. Jo, K.T. Seifert, E.M. Anton, T. Granzow, D. Damjanovic, Perspective on the development of lead free piezoceramics, J. Amer. Ceram. Soc. 92 (2009)1153-1177. https://doi.org/10.1111/j.1551-2916.2009.03061.x

[58] T. Rojac, M. Kosec, B. Budic, N. Setter, D. Damjanovic, Strong ferroelectric domain-wall pinning in $BiFeO_3$ ceramics, J. App. Phys. 108 (2010)074100- 074107. https://doi.org/10.1063/1.3490249

[59] J.F. Rosenbaum, Bulk acoustic wave theory and devices, Artech House Acoustics Library (1988)

[60] Y. Saito, H. Takao, T. Tani, T.Nonoyama, K. Takatori, T. Homma, T.Nagaya, M. Nakamura, Lead-free piezoceramics, Nature. 432 (2004) 84-87. https://doi.org/10.1038/nature03028

[61] A. Sasaki, T. Chiba, Y. Mamiya, E. Otsuki, Dielectric and piezoelectric properties of $(Bi_{0.5}Na_{0.5})TiO_3$-$(Bi_{0.5}K_{0.5})TiO_3$ systems, JPN. J. App. Phys. 38 (1999) 5560-5564.

[62] K. Sen, K.Singh, A. Gautam, M. Singh, Study of Dielectric and Ferroelectric Properties of Multiferroic $BiCo_xFe_{1-x}O_3$ Ceramic, Integr. Ferroelectr. 120 (2010) 122-130. https://doi.org/10.1080/10584587.2010.504126

[63] H. Sengul, Life cycle analysis of quantum dot semiconductor materials, University of Illinois at Chicago. (2009)

[64] N. Setter, ABC of piezoelectric materials. Piezoelectric materials in devices, Swiss Institute Technol. (2002) 1-518

[65] Z. -Y. Shen, J.-F. Li, Enhancement of piezoelectric constant d33 in $BaTiO_3$ ceramics due to nano-domain structure, J. Ceram. Soci. Jpn. 118 (2010) 940-943. https://doi.org/10.2109/jcersj2.118.940

[66] K. Shibata, R. Wang, T. Tou, J. Koruza, Applications of lead-free piezoelectric materials, MRS. Bulletin. 43 (2018)612-616. https://doi.org/10.1557/mrs.2018.180

[67] V. Shuvaeva, D. Zekria, A. Glazer, Q. Jiang, S. Weber,P. , Bhattacharya, P.Thomas, Local structure of the lead-free relaxor ferroelectric $(K_xNa_{1-x})_{0.5}Bi_{0.5}TiO_3$, Phys.Rev. B. 71 92005) 174114.

[68] V. Shvartsman, W. Kleemann, R. Haumont, J. Kreisel, Large bulk polarization and regular domain structure in ceramic $BiFeO_3$, App. Phys. Lett. 90 (2007) 172110-172115. https://doi.org/10.1063/1.2731312

[69] I. Sinclair, Sensors and transducers, third ed. Elsevier, Boston 2000, pp. 220-256

[70] G. Smolensky, V. Isupov, A. Agranovskaya, N. Krainik, New materials of AIIBIVOVI type, Trans. Sov. Phys. Solid State. 2 (1961)2651-2654.

[71] H.Sun, S. Duan, X. Liu, D. Wang, H. Sui, Lead-free $Ba_{0.98}Ca_{0.02}Zr_{0.02}Ti_{0.98}O_3$ ceramics with enhanced electrical performance by modifying MnO_2 doping content and sintering temperature, J. Alloy. Comp. 670 (2016) 262-267. https://doi.org/10.1016/j.jallcom.2016.02.008

[72] H. Takahashi, Y. Numamoto, J.T. Tani, S. Tsurekawa, Piezoelectric properties of BaTiO3 ceramics with high performance fabricated by microwave sintering, Jpn. J. App. Phys. 45 (2006) 7400- 7405. https://doi.org/10.1143/JJAP.45.7405

[73] T. Takenaka, K. -I.M.K.-I. Maruyama, K.S.K. Sakata, $(Bi_{1/2}Na_{1/2})TiO_3$-$BaTiO_3$ system for lead-free piezoelectric ceramics, Jpn. J. App. Phys. 30 (1991) 2230-2236. https://doi.org/10.1143/JJAP.30.2236

[74] H.-C. Thong, C. Zhao, Z. Zhou, C.-F. Wu, Y.-X. Liu, Z.-Z. Du, J.-F. Li, W. Gong, K.Wang, Technology transfer of lead-free (K, Na)NbO3-based piezoelectric ceramics, Mater.Today. 29 (2019) 37-48. https://doi.org/10.1016/j.mattod.2019.04.016

[75] K. Uchino, Glory of piezoelectric perovskites. Science and Technology of Advanced Materials, T&F. (2015) 1-6 https://doi.org/10.1088/1468-6996/16/4/046001

[76] A. Verma, A.Kumar, Bulk modulus of cubic perovskites, J. Alloy. Comp. 541 (2012) 210-214. https://doi.org/10.1016/j.jallcom.2012.07.027

[77] D. Viehland, Effect of uniaxial stress upon the electromechanical properties of various piezoelectric ceramics and single crystals, J. Amer.Ceram. Soc. 89 (2006) 775-785. https://doi.org/10.1111/j.1551-2916.2005.00879.x

[78] S. Wada, S. Suzuki, T. Noma, T.Suzuki, M. Osada, M.Kakihana, S-E. Park, L.E. Cross, T.R. Shrout, Enhanced piezoelectric property of barium titanate single crystals with engineered domain configurations, Jpn. J. App. Phys. 38 (1999) 5500-5505. https://doi.org/10.1143/JJAP.38.5500

[79] D. Wang, Z. Jiang, B. Yang, S. Zhang, M. Zhang, F. Guo, W. Cao, Phase transition behavior and high piezoelectric properties in lead-free $BaTiO_3$-$CaTiO_3$-$BaHfO_3$ ceramics, J. Mater. Sci. 49 (2014) 62-69. https://doi.org/10.1007/s10853-013-7650-9

[80. P. Wang, Y. Li, Y.Lu, Enhanced piezoelectric properties of $(Ba_{0.85}Ca_{0.15})(Ti_{0.9}Zr_{0.1})O_3$ lead-free ceramics by optimizing calcination and sintering temperature, J. Euro. Ceram. Soci. 31 (2011) 2005-2012. https://doi.org/10.1016/j.jeurceramsoc.2011.04.023

[81] R.E. Watson, Analytic Hartree-Fock solutions for O, Phys. Rev. 111 (1958) 1100-1108. https://doi.org/10.1103/PhysRev.111.1108

[82] D.I. Woodward, I.M. Reaney, R.E. Eitel, C.A. Randall, Crystal and domain structure of the $BiFeO_3$-$PbTiO_3$ solid solution, J. Appl. Phys. 94 (2003)3313-3318. https://doi.org/10.1063/1.1595726

[83] B. Wu, H. Wu, J. Wu, D. Xiao, Zhu, J.,Pennycook, S.J., Giant piezoelectricity and high Curie temperature in nanostructured alkali niobate lead-free piezoceramics through phase coexistence, J.Amer. Chem. Soc. 138 (2016) 15459-15464. https://doi.org/10.1021/jacs.6b09024

[84] J. Wu, Advances in lead-free piezoelectric materials, Springer. 98(2018) 552-624 https://doi.org/10.1016/j.pmatsci.2018.06.002

[85] J. Wu, Perovskite lead-free piezoelectric ceramics, J. App.Phys. 127 (2020)190900-190901. https://doi.org/10.1063/5.0006261

[86] L. Wu, D. Xiao, J. Wu, Y. Sun, D. Lin, J. Zhu, P. Yu, Y. Zhuang, Q. Wei, Good temperature stability of $K_{0.5}Na_{0.5}NbO_3$ based lead-free ceramics and their applications in buzzers, J. Euro. Ceram. Soc. 28 (2008) 2963-2968. https://doi.org/10.1016/j.jeurceramsoc.2008.04.033

[87] K. Xu, J. Li, X. Lv, J. Wu, X. Zhang, D. Xiao, J. Zhu, Superior piezoelectric properties in potassium-sodium niobate lead free ceramics, Adva. Mater. 28 (201) 8519-8523. https://doi.org/10.1002/adma.201601859

[88] G. Yuan, S.W. Or, Y. Wang, Z. Liu, J. Liu, JPreparation and multi-properties of insulated single-phase $BiFeO_3$ ceramics, Solid State Commun. 138 (2006) 76-81. https://doi.org/10.1016/j.ssc.2006.02.005

[89] S.-T. Zhang, A.B. Kounga, E. Aulbach, H. Ehrenberg, J. Rödel, Giant strain in lead-free piezoceramics $Bi_{0.5}Na_{0.5}TiO_3$-$BaTiO_3$-$K_{0.5}Na_{0.5}NbO_3$ system, App. Phys. Lett. 91 (2007) 112900-112906. https://doi.org/10.1063/1.2783200

[90] Q. Zhao, J. Zhao, X. Tan, Classification, preparation process and its equipment and applications of piezoelectric ceramic, Mater. Phys. Chem. 1 (2018) https://doi.org/10.18282/mpc.v1i1.560

[91] X.-G. Zhao, D. Yang, J.-C. Ren, Y. Sun, Z. Xiao, L. Zhang, Rational design of halide double perovskites for optoelectronic applications, Joule. 2 (2018)1662-1673. https://doi.org/10.1016/j.joule.2018.06.017

[92] Z. Zhao, V. Buscaglia, M. Viviani, M.T. Buscaglia, L. Mitoseriu, A. Testino, M. Nygren, M. Johnsson, P. Nanni, Grain-size effects on the ferroelectric behavior of dense nanocrystalline BaTiO3 ceramics, Phys. Rev. B. 70 (2004) 024100- 024107. https://doi.org/10.1103/PhysRevB.70.024107

[93] T. Zheng, J. Wu, D. Xiao, J. Zhu, Recent development in lead-free perovskite piezoelectric bulk materials, Prog. Mater. Sci. PROG. 298 (2018) 552-624. https://doi.org/10.1016/j.pmatsci.2018.06.002

[94] T. Zheng, H. Wu, Y. Yuan, X. Lv, Q. Li, T. Men, C. Zhao, D. Xiao, J. Wu, K. Wang, The structural origin of enhanced piezoelectric performance and stability in lead free ceramics, Energy Environ. Sci.10 (2017) 528-537. https://doi.org/10.1039/C6EE03597C

[95] P.-F. Zhou, B.-P. Zhang, L. Zhao, X.-K. Zhao, L.-F. Zhu, L. Cheng, J.-F. Li, High piezoelectricity due to multiphase coexistence in low-temperature sintered (Ba, Ca)(Ti, Sn)O3-CuOx ceramics., App. Phys. Lett. 103 (2013) 172900-172904. https://doi.org/10.1063/1.4826933

[96] X. Zhu, Piezoelectric ceramic materials: processing, properties, characterization, and applications, UK ed., Nova Science Publishers Inc., New York , 2010, pp.1-63

[97] X.N. Zhu, W. Zhang, X.M. Chen, Enhanced dielectric and ferroelectric characteristics in Ca-modified BaTiO3 ceramic, AIP. Adv. 3 (2013) 082120-082125. https://doi.org/10.1063/1.4819451

[98] J, Zylberberg, A. Belik, E. Takayama-Muromachi, Z.-G. Ye, Bismuth aluminate: a new high-Tc lead-free piezo-/ferroelectric, Chem. Mater. 19 (2007)6385-6390. https://doi.org/10.1021/cm071830f

Advanced Functional Piezoelectric Materials and Applications   Materials Research Forum LLC
Materials Research Foundations 131 (2022) 259-280   https://doi.org/10.21741/978164490209-9

Chapter 9

Piezoelectric Materials for Sensor Applications

Dnyandeo Pawar*[1,2], Rajesh Kanawade[3], Dattatray Late[4], DeLiang Zhu[1], PeiJiang Cao*[1]

[1]College of Materials Science and Engineering, Shenzhen University; Shenzhen Key Laboratory of Special Functional Materials, Shenzhen Engineering Laboratory for Advanced Technology of Ceramics and Guangdong Research Center for Interfacial Engineering of Functional Materials, Shenzhen 518055, PR China

[2]Key Laboratory of Optoelectronic Devices and Systems of Ministry of Education and Guangdong Province, College of Optoelectronic Engineering, Shenzhen University, Shenzhen 518060, PR China

[3]Physical and Materials Chemistry Division, CSIR- National Chemical Laboratory (NCL), Pashan, Pune- 411008, India

[4]Centre for Nanoscience and Nanotechnology, Amity University Maharashtra, Mumbai-Pune Expressway, Bhatan 410206, India

pjcao@szu.edu.cn, pawar.dnyandeo@gmail.com

Abstract

Today, the piezoelectric materials are the state-of-art materials due to their mesmerizing property of tuning mechanical strain or vibrational energy into an electrical energy and vice versa, and therefore, it is considered very promising for future technological applications. In this chapter, the details including fundamental mechanism of piezoelectricity, fabrication methods, newly designed piezoelectric materials, and their applications in various fields are deeply reviewed. The conclusion and the future outlook are discussed.

Keywords

Piezoelectricity, Energy-Harvesting, Nanogenerator, Piezoelectric-Coefficient, Electro-Mechanical Coupling

Contents

1.   Introduction

In recent years, various semiconductor materials and polymers have drawn the attention of many researchers due to their fascinating properties and therefore, considerable research has been devoted toward the fabrication and sensor development for various applications such as food packaging [1], gas/volatile compounds sensing [2–6], energy storage [7], photonics [8–10], healthcare [11,12], pesticides detection [13,14], temperature [15], wind energy [16], and electronics [17]. When specific material or composite is under applied certain mechanical stress or high voltage, it exhibits piezoelectric property which could be utilized for energy scavenging.

Piezoelectricity is a phenomenon of having coupling between electromechanical in which the mechanical energy converts into an electrical charge in the piezoelectric materials and vice versa [18]. Because of this unique property, the piezoelectric materials could be widely employed in both sensors and actuators field. The electrical output can be generated from the vibrations related to the object i.e. may be a human body movements, machine, or environmental parameter such as solar energy, heat or wind. Therefore, there is a keen interest to develop the new piezoelectric materials and miniaturized devices which could harvest these parameters.

Numerous polymeric materials such as polymers, ceramics, single crystals, biological materials have been considered for diversified applications [19]. Various morphological structures in the form of nano thin films and layer-by-layer structures are fabricated by using numerous synthesis methods. These structures have shown enhanced in physical-chemical properties such as high piezoelectric coefficient, long-term durability, high stretching ability and therefore, very promising for wireless technologies, wearable electronics, and biological applications [20].

The most common piezoelectric polymeric materials i.e. PZT, ceramics such as PVDF, BTO, inorganic semiconductors such as ZnO, InN are extensively been used in numerous applications [19–22]. The material like PVDF has a good mechanical stability and flexibility than that of PZT. But the piezoelectric coefficient of PVDF is lesser than PZT. The inorganic materials are having high piezoelectric coefficient but they have associated certain drawbacks like brittleness and less deformation ability [18,19,23]. It is also observed that lead-based materials exhibit piezoelectric properties and therefore, widely selected for numerous applications. However, lead has a negative impact on the environment and therefore, researchers now focused on the development of lead-free based piezoelectric materials [23]. The properties like curie temperature, piezoelectric coefficients, stress coefficients, dielectric constants, Youngs modulus, and poisson ratio decides the performance of a piezoelectric sensor. Generally, the device is designed by considering the specific property, for e.g. designing a strain sensor is based on the piezoelectric coefficient, and the capacitive sensor is relied on dielectric constant property of the material. Designing a novel piezoelectric material is very critical task and therefore, various parameters, synthesis method, and design of the device fabrication are need to be focused.

In this chapter, firstly, the concept of piezoelectric mechanism and various fabrication methods are introduced. Later, the various properties of piezoelectric materials, and their use in wearable and biomedical fields, energy harvesting and applications in different fields with their merits and limitations are reviewed. Lastly, the conclusion and future outlook is discussed.

2.   Piezoelectric mechanism

Piezoelectric material generates an electricity when it encounters with any external mechanical stress. Generally, certain parameters are involved in the piezoelectric mechanism. In direct piezoelectricity, the electrical displacement (D) is produced due to the applied mechanical stress (T) and in the inverse piezoelectric effect, the electric field (E) is produced because of strain (S) (shown in Fig. 1(a, b)) [18,24]. The expression is shown below:

$$D = dT + \varepsilon E \ (Direct \ effect)$$
$$S = sT + dE \ (Inverse \ effect)$$

Advanced Functional Piezoelectric Materials and Applications — Materials Research Forum LLC
Materials Research Foundations 131 (2022) 259-280 — https://doi.org/10.21741/978164490209-9

where, ε is dielectric permittivity, d is piezoelectric coefficient, and s is mechanical compliance.

Mathematically, the polarization is a vector that fits to the first rank tensor, the stress and strain are the second rank tensor and hence, the piezoelectric coefficients are the third rank tensor. The piezoelectric coefficient d directs the total of electric displacement is made in the material due to applied mechanical stress, which is represented by the following equation:

$$D = dT$$

The piezoelectric coefficient is related to D and T and therefore, can be denoted by two subscripts.

Now, the D is accumulated charge density (Q) per unit area (A), and T is mechanical force (F) per unit area (A). Therefore, the piezoelectric coefficient (d) is expressed as follows:

$$d = D / T = Q / F$$

There is another parameter called it as piezoelectric voltage coefficient (g). It is the amount of electric field generated to the applied mechanical stress.

$$E = -gT$$

The g and d parameters are related by the following equation:

$$g = d / e = d / (e_0 \, e_r)$$

where e_0 and e_r are permittivity in vacuum and medium.

Both parameters (d, g) describe the strength of the piezoelectric effect of the material. In sensor design point of view, the contribution of g is more significant than d. The g shows the electric permittivity of the material and hence, for a given material, smaller the permittivity, lesser it takes to be charged by applied stimuli, and therefore, it strongly decides the sensitivity. The product between d and g shows the energy density stored inside the material. The higher the amount of energy store per unit volume, higher will be the energy density and thus, suggesting its application in the field of energy harvesting.

For direct photoelectric effect, the parameter called it as a coupling coefficient (K) is also important, which indicates the relation between amount of mechanical/electrical energy into electrical/mechanical energy. So, K is directly connected to variance among the resonance and antiresonance frequency at a specified vibration mode.

$$K^2 = (Mechnical\ energy\ converted\ into\ electrical\ energy\ ) \\ /(Input\ mechnical\ energy)$$

Advanced Functional Piezoelectric Materials and Applications   Materials Research Forum LLC
Materials Research Foundations 131 (2022) 259-280   https://doi.org/10.21741/978164490209-9

Generally, when the deformation occurs inside the material, the corresponding position of positive charges and negative charges get displace relative to each other and thus, changes the electric polarization of the material. For a symmetric material, the polarization is zero. When the crystal compressed horizontally, the polarization vector push towards downward and vice versa [22].

A piezoelectric generator consists of 2 main operating modes (Fig. 1(b)) as follows:

1) Mode 33: In this mode, the applied stress is along the polarization. It is displayed in Fig. 1(c).

2) Mode 31: In this configuration, the stress is perpendicularly to the polarization as shown in Fig. 1(b).

The d33 mode provides a better electromechanical coupling than d31 as experienced by most of the piezoelectric materials. Generally, in most of the materials, the dipoles are randomly oriented in the crystal. For an effective polarization to occur, the process called as "poling" is applied on the material. It is categorized in two methods namely electrode poling and corona poling. In the electrode poling, a very high voltage (5–100 MV/m) is applied on the piezoelectric material via electrodes attached to the two sides of the material. In corona poling, a needle operated at large potential of (8 to 20 kilovolt) is placed above the grid, which is maintained at low potential of (0.2 to 3 kilovolt). The piezoelectric object placed above an electrode which is operated with a hot substrate. The piezoelectric material is exposed to dry air or inert gas environment [20].

As per the direct piezoelectric phenomenon, the charge get generated with respect to used force and hence, this mechanism can be effectively used for energy harvesting. The mechanism is explained as shown in Fig. 2. When polarization is produced inside the material, which leads to formation of aligned diploes and generate opposite charge carriers at each side of the material surface as shown in Fig. 2(a). The current will not flow under this condition as no stress is applied on the material. Now, when any stress or compression is applied to the material, the polarization level gets changed and therefore, with increase in applied stress or compression the polarization of the material decreases as depicted in Fig. 2(b). The electrical current is produced due to the free charge carriers at the interface. The reverse phenomenon is happening when the material is under applied tension. In fig. 2(c), the polarization increases as the material goes under tension and hence, the electrical current flows opposite to balance the surface charge. Generally, the material with high ratio of d/e is the ideal for energy harvesting.

3.   Types of piezoelectric materials

The piezoelectric materials are classified into many categories depending upon their dielectric and piezoelectric properties. The most common materials are polymers due to their softness and flexibility. In energy harvesting, the stretchable piezoelectric materials and their composites are very effective due to their flexible nature, ease in use and integration with printable electronics circuitry. Table 1, shows various polymeric materials along with their piezoelectric coefficients. Generally, the materials are categorized as follows:

1. Piezoelectric polymers and composites

   (e.g. PVDF, PVDF-MgO, PVDF-ZnO, PVDF-Al$_2$O$_3$)

2. Inorganic piezoelectric nanomaterials

3. (e.g. PZT, BTO, PDMS-MWCNT, PZT-PDMS)

4. 3-D microstructured piezoelectric nanomaterials

   (e.g. 3D cellular-structured PZT-PDMS, microstructured P(VDF-TrFE)

5. Multilayered piezoelectric nanomaterials

   (e.g. rGO/PVDF composite, BT/PVDF composite)

4.   Fabrication methods

As the technology is progressing, new sophisticated deposition techniques are also designed and developed for the fabrication of nanoscale and submicron thin films. The thin films are suitable for flexible electronics applications and therefore, very attractive in wearable devices. The piezoelectric thin films of small size, lightweight, printable and ease in preparation can be prepared by using various physical and chemical methods. The main methods such as radio frequency magnetron sputtering, pulsed laser deposition, chemical vapor deposition, chemical bath deposition, sol–gel method, electro-spraying technique, atomic layer deposition, and laser pyrolysis etc. have been used in preparation of thin films. Figure 3 and 4 show some typical fabrication methods of thin films of nanomaterials and polymers. Many methods have been used for nanostructured and piezoelectric thin film preparation. The detail description of these methods is explained elsewhere [21,25].

The magnetron sputtering is extensively used in preparation of piezoelectric thin films due to its ability of controlling uniformity and deposit large area of the film. However, the film deposition cost is very high due to its expensive substrate. The certain piezoelectric materials have been deposited by using this technique and have shown potential applications [26,27].

The pulsed laser deposition has been used in depositing the piezoelectric thin films. However, achieving a large surface occupied film is difficult with this method. Some of the reports based on this technique are described elsewhere [28,29].

Among all the techniques, the sol-gel method is a simple method to prepare piezoelectric thin films. The multilayer thin films can be prepared easily by using this method. However, maintaining the film surface morphology is a challenge. This method is effectively used in piezoelectric film preparations [30,31].

5.   Applications of piezoelectric materials

5.1   Applications in wearable and implanted biomedical devices

At present, harvesting energy from the human body motion is of great interest to various researchers. It not only provides efficient continuous energy but also replace the unnecessarily use of an external battery. The flexible device can be effortlessly integrated with fabric and change the mechanical energy into an electrical energy. These kinds of devices are very favorable due to its comfort. The few examples are discussed in detail.

The highly stretchable, reliable, and efficient self-powered energy harvesters for wearable and implantable electronic structures are shown in Fig. 5. Different materials such as organic piezoelectric materials, piezoelectric composite, piezoelectric macrofoam, different stretchable structural designs were used to fabricate highly efficient energy generators, which generated the power from the human motion.

A $BaTiO_3$ nanowire-PVC fabric material structure was united with cotton threads and harvested biomechanical energy into an electrical energy attaching the structure onto the human arm [32]. It exhibited a Voc: 1.9 volt, Isc: 24 nanoampere, and power: 10.02 nanowatt on 80 MΩ applied load. This material can be simply combined with cloth and could produce electrical energy. A woven nanogenerator based on ZnO and Pd was fabricated and achieved an Voc (3 mV) and Isc of (17 pA) [33]. The device stayed sensitive to wind and sound. Wearable nanogenerators based on soft and flexible PZT textile was reported by Wu et al. [34]. The group was working through the elongating and freeing the chemical fabric. The nanogenerator generated output voltage of 6 volt and 45 nanoampere as an output current, which was sufficient to power a LCD. A power-generating textile sensor array based on PVDF was demonstrated [35]. The maximum power and power density of this device was 850 µW and 105 mW/m². Due to its woven textile structure, it can be applied to a large surface area. A very flexible piezoelectric nanogenerator using Al-foil and ZnO was also fabricated [36]. The film can detect small skin motion owing to its very minimum resistance to motion. The active sensor shown maximum V_{OUT} 0.2 V and

I_{OUT} 2 nA during flashing motion. The PVDF constructed self-powered system was fabricated for energy harvesting [37]. The system was harnessing machine-driven energy from humanoid gesture and converting to an electrical energy through rectifier for charging a patterned electrochromic supercapacitor, which was supplying continuous energy to wearable electronics. In non-stop palm influence for initial 50 s, the system detected a voltage of 0.071 V and further decreased to 0.051 V in the discharge mode for 50 s once palm influence stopped. Shin et al. [38] designed and fabricated a graphene, $BaTiO_3$ and PVDF based self-powered energy generator, which harvested energy from human activity of finger-pressing method. This shown voltage of 112 V and high durability over 1800 cycles, which is sufficient to power up 15 LEDs. The energy is also harvested from the parts such as heart and diaphragm and was performed by using PZT based devices [39]. The single-layer PZT thin film shown the output voltage higher than 3.7 V, which enough to function a cardiac pacemaker. A highly flexible PZT based device on a plastic substrate was explored using a laser-lift off method [40]. The PZT layers were moved on the PE base. It harvested an output voltage of ~200 V and ~150 $\mu A \cdot cm^{-2}$ from the small mechanical distortions. The nanogenerator of size 3.5 cm x 3.5 cm was capable of producing short-circuit current of ~ 8 μA, and can light 100 LEDs, under irregular human finger motions. Table 2 shows the properties of some stretchable piezoelectric polymeric materials.

The piezoelectric generator made of PVDF was demonstrated by producing an electrical energy from active energy of ascending aorta by Zhang et al. [41]. The piezoelectric device was fixed and wrapped to the ascending aorta (Fig. 6(a, b)). Two electrodes were connected to the DAQ system to get an electrical output. The graph of output voltage and current during the counter-pulsation period is revealed in Fig. 6(c, d). The harvester showed a power density, V_{OUT}, and I_{OUT} 10.3 volt, 170 nW/cm^3, and 400 nA, respectively. The device exhibited a supreme output voltage 1.5 volt and current 300 nanoampere, below the heart frequency and blood pressure of 120 beats per minute and 160/105 mmHg, respectively. The current was decreased with rise in external resistance (Fig. 6(e)). The highest power (681 nW) was gained at 33 MΩ resistor (Fig. 6(f)).

5.2   Piezoelectric materials for energy applications

Numerous piezoelectric materials and its composites have been utilized for energy applications. Few examples are described as follows. The sonic wave applied ZnO based energy harvester was fabricated [42]. The ZnO nanorods were placed in between two Au layered polyester fabric bases. The layer of polyethylene was placed in between ZnO and Au electrode substrate from the top. This flexible generator was able to produce V_{OC}: 8 volt, and I: 2.5 μA, under applied 100 dB sonic wave at 100 Hz. Another study reported a

Advanced Functional Piezoelectric Materials and Applications   Materials Research Forum LLC
Materials Research Foundations 131 (2022) 259-280   https://doi.org/10.21741/978164490209-9

nanogenerator based on cellulose microfiber, MWCNTs, and PDMS structure [43]. Under repeating human hand punching, it generated an open circuit voltage: 30 volt, power density: 9 $\mu W/cm^3$ and Isc: 500 nA. This was sufficient to light many LEDs, LCD screen or a wristwatch. The vitamin B_2 assisted PVDF was explored for energy harvesting and achieved a V_{OUT}: 61.5 V, I_{OUT}: 12.2 μA, and energy conversion efficiency: 62%, respectively [44]. The energy harvester was capable to illuminate 100 LEDs and a CD.

Some natural materials are also utilized for energy harvesting. The cellulose fibrous structure of an untreated onion was used for energy harvesting applications by Maiti et al. [45], as shown in Fig. 7. The graphic of production steps of a device as depicted in Fig. 7(a). The image of the onion skin, flexibility and ready nanogenerator are presented in Fig. 7(b-f). The nanogenerator achieved remarkable results of V_{OC}: 18 V, power density: 1.7 $\mu W/cm^2$, and converting efficiency: 61.7%, respectively. Below a pressure of 34 kPa at 3Hz, the device powered 30 green LEDs. This work shown a green technology-based application.

The porous cellulose nanofibril (CNFs)/poly(dimethylsiloxane) (PDMS) aerogel material was used to fabricate a high-performance nanogenerator [46]. The device was produced a rate of energy flow of 6.3 mW/cm^3 with an open-circuit voltage 60.2 V. The device shown high stability of 1200 cycles per day. The fish swim bladder (FSB) was utilized to fabricate a nanogenerator [47]. A simple human finger tapping was sufficient to create a V_{OUT}: 10 V and I_{SC}: 51 nA. This glowed more than fifty commercial blue LEDs. The power generation is also taken from tire rolling. The piezoelectric perovskite material ($ZnSO_3$) and polymer PDMS lacking any electric poling treatment is described [48]. This material was attached to a road and power generation of 20 V and current density of 1 $\mu W/cm^2$ was reported. This performance was stable over 50 cycles of tire rolling.

The practical application of the energy harvesting from the unlike automobiles was demonstrated [49] as shown in Fig. 8. The schematic of dissimilar vehicles such as cycle, motorbike and car are depicted in Fig. 8(a-c). The $BaTiO_3$/PVDF dispersed diffusing the Ag-NWs composite film utilized to construct a piezoelectric device. The electrical output generated to each vehicle, when they pass on the harvester are presented in Fig. 8(d-f). The results shown that the device was capable to produce sufficient energy from the motion of the vehicles. Additionally, the device studied under forceful force of 3 N at 5 Hz, and observed that the nanogenerator with Ag-NWs/BTO/PVDF composite film displayed a V_{OUT}: ~ 14 V and I_{SC}:~ 0.96 μA as that of BTO/PVDF (V_{OUT}/ I_{SC} ~ 11 volt/0.78 μA). Table 3 displayed the sensing characteristics of few energy harvesters.

5.3   Piezoelectric materials in tissue engineering

Generally, in a piezoelectric polymer the piezoelectric effect can be observed. With a mechanical stimulus, an electrical signal get produces and thus, impacts the cell response. The piezoelectricity is present in several parts of human body for example bone, skin, muscles, DNA tendon, cartilage, collagen, and cell membranes, and therefore, employing the piezoelectric materials in tissue engineering is always beneficial and advantageous.

In tissue engineering, the smart scaffolds are very much important to regenerate the specific tissues. Various electrically active synthetic and natural polymeric materials were used in tissue and biomedical engineering.   The natural polymers are preferable as they are obtained from natural sources but most of them exhibit poor electrical properties. Therefore, to support the tissue and cells, several polymers have been developed. The conductive polymer such as polypyrrole [50] is widely used in cell growth. However, due to its inability of biodegradation, it may create certain problems. Therefore, polymers like polyaniline are preferred for cell growth due to its biocompatibility and support [51]. Certain conducting polymeric composites like polypyrrole/poly(L-lactic acid), poly(L-lactic   acid)/carbon   nanotubes,   polypyrrole/chitosan,   and polyaniline/polycaprolactone/gelatin are applied in tissue engineering [52]. Mechanical stimulation of bone cells study based on PVDF was conducted by Frias et al. [53]. It observed that the mechanical stimulation induces the newly bone creation in vivo study and rises metabolic action, when the stimulation was applied with 5 V ac current at 1 and 3 Hz for 15 min. The polyurethane/polyvinylidene fluoride (PU/PVDF) scaffolds were used in wound healing applications [54]. The frameworks were deformed of 8% at frequency of 0.5 Hz (24 h duration), and observed more rapid wound healing due to improved relocation, adhesion and secretion.

5.4   Piezoelectric materials in other applications

Numerous piezoelectric polymers show attractive properties such as high flexibility, durability, biocompatibility etc. In this part, the energy harvesting based on various piezoelectric nanostructured materials and polymers for different applications is reviewed.

Piezoelectric materials were tested for gas sensing and achieved good results. The piezoelectric CuO/ZnO nanoarray based nanogenerator [55] was demonstrated for H_2S gas sensing at RT. At 800 ppm H_2S, the sensitivity increased to 629.8, which was significant as compared with only ZnO. A piezoelectric nanogenerator made by ZnO was also utilized for H_2S gas sensing [56]. The sensor response to O_2, hydrogen sulfide and water vapor was verified. The sensitivity of the device at 1000 ppm to H_2S was observed 127.3%. A self-powered room temperature formaldehyde sensor based on MXene/Co_3O_4 driven by ZnO/MXene-based piezoelectric nanogenerator was investigated [57]. At constant force of

Advanced Functional Piezoelectric Materials and Applications   Materials Research Forum LLC
Materials Research Foundations 131 (2022) 259-280   https://doi.org/10.21741/978164490209-9

10 newton and frequency 6 Hz, the V_{OUT} of nanogenerator was reached to 750 mV. It observed that the output voltage of piezoelectric nanogenerator was increased with increase in formaldehyde concentration. A gas flow sensing based on piezoelectric ZnO films deposited on Phynox (Elgiloy) substrate was performed by Joshi et al. [58]. This material was tested for gas flow rate in the range of 2 to 18 L min^{-1} and achieved the sensitivity of ~18 mV/(Lmin^{-1}) and quick response time of 20 ms. The piezoelectric material also tested in wireless data transmission. A self-powered ZnO nanowire was explored for long-distance data transmission [59]. When it strained up to 0.12% (strain rate of 3.56% S^{-1}), the V_{OUT}: 10 volt and power density: 10 mW/cm^3 was achieved. The wireless signal sent out by this device could be sensed by a radio up to a distance of 5 to 10 m, displaying its possible use in wireless transmission-based applications. Flexible BaTi$_{(1-x)}$Zr$_x$O$_3$)/PVDF device for fluid velocity measurement was studied by Alluri et al. [60]. When water was flowing through the pipe, the sensor shown an output peak power of 0.2-15.8 nanowatt for water speeds in the range of 31.43-125.7 m/s, Also, it was observed that at 11 N pressure, the nanogenerator generated an output voltage and shot-circuit current of ~11.9 V and ~1.35 μA. A pH sensor based on piezoelectric ZnO/PVDF was tested [61]. Under uniaxial compression, the device shown a highest V_{OUT}: 6.9 V, Isc: 0.96 μA, and power: 6.624 μW, respectively. It was sufficient to light 5 green LEDs without an external storage. The sensor was exposed to different pH range of 6.02 to 12.2 and observed that as the pH value increased the corresponding output voltage also decreased. The piezoelectric materials are also examined as a temperature sensor. The rare earth ions Er^{3+} doped in the piezoelectric material ZnS was investigated as a temperature sensor [62]. The Er^{3+} acted as a luminescent center and therefore, this material was displaying different colors transition as orange–yellow–green. As the excitation wavelength changed, the corresponding mode of emission wavelength was also changed. The sensor shown good temperature response in between 300 K to 500 K. The ZnO nanowire-based energy harvester and single walled CNT based FET as Hg^{2+} ion sensor was performed by Lee et al. [63]. A LED was used to serve as an indicator. A single device with an area of ~0.8 cm^2 touched a Voc, and Isc of around 350 mV and ~125 nA cm^{-2}, respectively. The performance further improved by stacking 10 layers and reached V_{OUT} and Isc reach up to ~2.1 volt and ~105 nA.

Piezoelectric PZT-5H was selected as a vibrator material and tested for road piezoelectricity generating test [64]. The wheel impact produced an electric energy of 0.23 mJ and output voltage generated around 65.2 V. With this an electric capacity of 0.8 kW/h was formed, which was sufficient to light signal traffic lights. However, further structural optimization may enhance the efficient of the piezoelectricity generating test. The piezoelectric materials are also tested in railroad track monitoring [65]. The design consisted of a piezoelectric material which was placed at the bottom of the rail. During on-field testing, a maximum output power of 53 microwatt with a 387 kilo-ohm load resistor was obtained. A power output of 1.1 mW was predicted with a higher axle-load vehicle run on the line. The detailed information of application of piezoelectric material in the railway field was summarized by Bosso et al. [66]. The possible application of a underwater thrust and power production by using a self-powered piezolectric composites was demonstrated [67]. In this work, a fish-like bimorph designs with and without a passive caudal fin (tail) were designed and analyzed. At thrust of 19 mN and 6 Hz, produced power of 120 mW with a 10 g piezoelastic fish. Various piezoeelctric nanostructured materials incluidng ZnO, LiNbO$_3$, PZT, GaN and their composites are utilized for energy harvesting and their device performnace is tabulated in Table 4.

Piezoelectric based human motion remotely control device was proposed by Deng et al. [68]. The cowpea-structured PVDF/ZnO nanofibers (CPZNs) was used to fabricate the piezoelectric sensor. Fig. 9 shows the schematic of a material placed at an innermost sideways of a human finger side and also on robot fingers for directing the same bending actions via the electronic way and respective bendy motions. The structure of PVDF/ZnO nanofiber as revealed in Fig. 9(b,c,d). The flexibility is obtained through a flexible PVDF/ZnO nanofibers and MXene (Ti$_3$C$_2$) electrode sandwich structure (shown in Fig. 9(e)). The SEM and TEM images confirmed the design of the material (shown in Fig. 9(f). The finite element method is conducted to study the electrical signal variation under bending motions as display in Fig. 9(h). The application of controlling robot hand via human hand through the device is shown in Fig. 9(i). After the hand signaled a 'Two', the robot hand also completed the same action. It observed that under wide ranging angle from 44° to 122°, the self-powered piezoelectric sensor showed remarkable bending responsivity (4.4 milli-volt per degree) and quick speed (76 milli-seconds). It also exhibited better stability during 5000 cycles.

The innovative technique of altering energy from the steps into energy was studied by Puscasu et al. [69]. With a thickness of 8.5 mm active layer, the device was capable of producing energy density is 0.49 J/m^2 per step. The piezoelectric material was also explored for Wind energy harvesting applications. A piezoelectric flag was fabricated by using a 28 μm thick PVDF sandwiched by layers of Ag (electrode) and PET (protection) [70]. This device was capable to generate an electrical power at low wind speed. For a flag length of 60 mm and 100 mm, the device showed a peak electrical output power (1–5 mW/cm^3) during wind speed varied from 5 to 9 meter per second. The piezoelectirc mataral is also used as an acceleration and pressure sensor. The β-PVDF (hβ-PVDF) utilized acceleration sensor was demonstrated by Jin et al. [71]. The sensor shown a sensitivity of 2.405 nAs^{2}m^{-1} in 5 m/s^2 to 30 m/s^2. This sensor can monitor the safty condition in vehicle safty system. The shock wave piezoelectric pressure sensor was developed by Wang et al. [72]. The ceramics (PZT (Pb(Zr_xTi_{1-x})O_3)) was choosen for force detection. The natural frequency of the sensor was 200 kHz. The sensitivity of standard sensor and the developed sensor was 0.025 pC/g and 0.359. The strain sensor by using a piezoelectric composite of milled PZT ceramic powder mixed with a polymer binder (water-based acrylic) was demonstrated [73]. The response was tested under strain of ±200 με and observed that the sensitivity was flat within ±2 dB in the frequancy range of 5-500 Hz. This type of sensor can be used in structural vibration monitoing.

Conclusion and outlook

In this chapter, numerous piezoelectric materials and their applications in various fields are discussed. The principle mechanism of piezoelectricity and fabrication methods are reviewed. Finally, the applications of various self-powered piezoelectric materials for many sensing applications are reviewed with their advantages and limitations.

In recent years, numerous inorganic and inorganic piezoelectric polymers of different structures have been studied extensively. However, the brittleness, toxicity and piezoelectric performance have certain limitations of few piezoelectric materials. Therefore, the materials with porous structure, polymer matrix, surface functionalization, and biocompatibility can be considered to increase the performance of the device. At present, lead-free piezoelectric materials are in high demand due to their serious impact on human and environmental concern and therefore, development of lead-free piezoelectric materials are highly preferable. Along with these, many natural biological materials are also being explored and show remarkable piezoelectric properties. The new materials with enhanced piezoelectric properties will have great potential applications.

At present, development of self-powered piezoelectric materials is a keen interest for researchers. The high energy conversion efficiency, high durability, lack of depolarization and higher flexibility are the key features of the piezoelectric materials. The mimic in size and integration ability makes it ideal to deploy at preferred location. Due to high flexibility, advances in materials, and device integration, the piezoelectric materials are effectively used in wearable and biomedical applications. There are some reports on implantation of piezoelectric nanogenerators for health monitoring. However, in vivo studies are essential to show their potential use in clinical applications. Also, the flexible and less bulky design is preferable in implantation. Various piezoelectric based sensors are deployed to harvest energy from rolling tire, vehicle or railway passing. However, the amount of energy can be increased by using multiple sensor array and continuous loading frequency. In the energy harvesting, the high-performance piezoelectric based nanogenerator are fabricated. In the future, the demands in wearing devices in the form of smart wrist watches and health bands for health monitoring will increase, and so, indicates the necessity of production of piezoelectricity-based smart devices. Therefore, considering the mandate of sensors, the high piezoelectric coefficient-based devices are highly desirable. With technology and advanced piezoelectric materials, a reliable, sufficiently low cost, highly miniaturized, deployable and self-powered piezoelectric-based energy harvesters can fulfill future demands.

References

[1]   H. Wang, J. Qian, F. Ding, Emerging Chitosan-Based Films for Food Packaging Applications, J. Agric. Food Chem. 66 (2018) 395–413. https://doi.org/10.1021/acs.jafc.7b04528

[2]   D. Pawar, B.V.B. Rao, S.N. Kale, Fe3O4-decorated graphene assembled porous carbon nanocomposite for ammonia sensing: study using an optical fiber Fabry–Perot interferometer, Analyst. 143 (2018) 1890–1898. https://doi.org/10.1039/C7AN01891F

[3]   R. Kitture, D. Pawar, C.N. Rao, R.K. Choubey, S.N. Kale, Nanocomposite modified optical fiber: A room temperature, selective H2S gas sensor: Studies using ZnO-PMMA, J. Alloys Compd. 695 (2017) 2091–2096. https://doi.org/https://doi.org/10.1016/j.jallcom.2016.11.048

[4]   D. Pawar, R. Kanawade, A. Kumar, C.N. Rao, P. Cao, S. Gaware, D. Late, S.N. Kale, S.T. Navale, W.J. Liu, D.L. Zhu, Y.M. Lu, R.K. Sinha, High-performance dual cavity-interferometric volatile gas sensor utilizing Graphene/PMMA nanocomposite, Sensors Actuators B Chem. 312 (2020) 127921. https://doi.org/https://doi.org/10.1016/j.snb.2020.127921

[5]   R. Kanawade, A. Kumar, D. Pawar, K. Vairagi, D. Late, S. Sarkar, R.K. Sinha, S. Mondal, Negative axicon tip-based fiber optic interferometer cavity sensor for volatile gas sensing, Opt. Express. 27 (2019) 7277–7290. https://doi.org/10.1364/OE.27.007277

[6]   A. Kumar, D. Pawar, K. Vairagi, S. Mondal, R. Kanawade, Polyvinyl alcohol filled negative axicon tip based highly sensitive fiber optic sensor for acetone sensing, Mater. Today Proc. 28 (2020) 1816–1819. https://doi.org/https://doi.org/10.1016/j.matpr.2020.05.220

[7]   S.A. Paniagua, Y. Kim, K. Henry, R. Kumar, J.W. Perry, S.R. Marder, Surface-Initiated Polymerization from Barium Titanate Nanoparticles for Hybrid Dielectric Capacitors, ACS Appl. Mater. Interfaces. 6 (2014) 3477–3482. https://doi.org/10.1021/am4056276

[8]   C.N. Rao, D. Pawar, U.T. Nakate, R. Aepuru, X. Gui, R. V Mangalaraja, S.N. Kale, E. Suh, W. Liu, D. Zhu, Y. Lu, P. Cao, Electric field controlled near-infrared high-speed electro-optic switching modulator integrated with 2D MgO, Opt. Lett. 45 (2020) 4611–4614. https://doi.org/10.1364/OL.393796

[9]   S. Dutta, E.A. Goldschmidt, S. Barik, U. Saha, E. Waks, Integrated Photonic Platform for Rare-Earth Ions in Thin Film Lithium Niobate, Nano Lett. 20 (2020) 741–747. https://doi.org/10.1021/acs.nanolett.9b04679

[10]   Q. Su, M. Fang, D. Zhu, W. Xu, S. Han, M. Fang, W. Liu, P. Cao, Y. Lu, D. Pawar, Ultrahigh-responsivity deep-UV photodetector based on heterogeneously integrated AZO/a-Ga2O3 vertical structure, J. Alloys Compd. 889 (2021) 161599. https://doi.org/https://doi.org/10.1016/j.jallcom.2021.161599

[11]   N. Karim, S. Afroj, K. Lloyd, L.C. Oaten, D. V Andreeva, C. Carr, A.D. Farmery, I.-D. Kim, K.S. Novoselov, Sustainable Personal Protective Clothing for Healthcare Applications: A Review, ACS Nano. 14 (2020) 12313–12340. https://doi.org/10.1021/acsnano.0c05537

[12]   S. Karagoz, N.B. Kiremitler, G. Sarp, S. Pekdemir, S. Salem, A.G. Goksu, M.S. Onses, I. Sozdutmaz, E. Sahmetlioglu, E.S. Ozkara, A. Ceylan, E. Yilmaz, Antibacterial, Antiviral, and Self-Cleaning Mats with Sensing Capabilities Based on Electrospun Nanofibers Decorated with ZnO Nanorods and Ag Nanoparticles for Protective Clothing Applications, ACS Appl. Mater. Interfaces. 13 (2021) 5678–5690. https://doi.org/10.1021/acsami.0c15606

[13]   Y.-H. Zhang, H.-H. Ren, L.-P. Yu, Development of molecularly imprinted photonic polymers for sensing of sulfonamides in egg white, Anal. Methods. 10 (2018) 101–108. https://doi.org/10.1039/C7AY02283B

[14]   B. Zhang, B. Li, Z. Wang, Creation of Carbazole-Based Fluorescent Porous Polymers for Recognition and Detection of Various Pesticides in Water, ACS Sensors. 5 (2020) 162–170. https://doi.org/10.1021/acssensors.9b01954

[15]   D. Pawar, A. Kumar, R. Kanawade, S. Mondal, R.K. Sinha, Negative axicon tip micro-cavity with a polymer incorporated optical fiber temperature sensor, OSA Contin. 2 (2019) 2353–2361. https://doi.org/10.1364/OSAC.2.002353

[16]   M.M. Alam, S.K. Ghosh, A. Sultana, D. Mandal, An Effective Wind Energy Harvester of Paper Ash-Mediated Rapidly Synthesized ZnO Nanoparticle-Interfaced Electrospun PVDF Fiber, ACS Sustain. Chem. Eng. 6 (2018) 292–299. https://doi.org/10.1021/acssuschemeng.7b02441

[17]   Z. Wang, H. Cui, S. Li, X. Feng, J. Aghassi-Hagmann, S. Azizian, P.A. Levkin, Facile Approach to Conductive Polymer Microelectrodes for Flexible Electronics, ACS Appl. Mater. Interfaces. 13 (2021) 21661–21668. https://doi.org/10.1021/acsami.0c22519

[18]   N. Sezer, M. Koç, A comprehensive review on the state-of-the-art of piezoelectric energy harvesting, Nano Energy. 80 (2021) 105567. https://doi.org/https://doi.org/10.1016/j.nanoen.2020.105567

[19]   H. Zhou, Y. Zhang, Y. Qiu, H. Wu, W. Qin, Y. Liao, Q. Yu, H. Cheng, Stretchable piezoelectric energy harvesters and self-powered sensors for wearable and implantable devices, Biosens. Bioelectron. 168 (2020) 112569. https://doi.org/https://doi.org/10.1016/j.bios.2020.112569

[20]   S. Das Mahapatra, P.C. Mohapatra, A.I. Aria, G. Christie, Y.K. Mishra, S. Hofmann, V.K. Thakur, Piezoelectric Materials for Energy Harvesting and Sensing Applications: Roadmap for Future Smart Materials, Adv. Sci. (2021) 2100864. https://doi.org/https://doi.org/10.1002/advs.202100864

[21]   A.M. Roji M, J. G, A.B. Raj T, A retrospect on the role of piezoelectric nanogenerators in the development of the green world, RSC Adv. 7 (2017) 33642–33670. https://doi.org/10.1039/C7RA05256A

[22]   C. Wan, C.R. Bowen, Multiscale-structuring of polyvinylidene fluoride for energy harvesting: the impact of molecular-, micro- and macro-structure, J. Mater. Chem. A. 5 (2017) 3091–3128. https://doi.org/10.1039/C6TA09590A

[23]   H. Wei, H. Wang, Y. Xia, D. Cui, Y. Shi, M. Dong, C. Liu, T. Ding, J. Zhang, Y. Ma, N. Wang, Z. Wang, Y. Sun, R. Wei, Z. Guo, An overview of lead-free piezoelectric materials and devices, J. Mater. Chem. C. 6 (2018) 12446–12467. https://doi.org/10.1039/C8TC04515A

[24]   Z. Yang, S. Zhou, J. Zu, D. Inman, High-Performance Piezoelectric Energy Harvesters and Their Applications, Joule. 2 (2018) 642–697. https://doi.org/https://doi.org/10.1016/j.joule.2018.03.011

[25]   A.U.U. Olayinka Oluwatosin Abegunde, Esther Titilayo Akinlabi, Oluseyi Philip Oladijo, Stephen Akinlabi, Overview of thin film deposition techniques, AIMS Mater. Sci. 6 (2019) 174–199. https://doi.org/10.3934/matersci.2019.2.174

[26]   X. He, Q. Wen, Z. Lu, Z. Shang, Z. Wen, A micro-electromechanical systems based vibration energy harvester with aluminum nitride piezoelectric thin film

deposited by pulsed direct-current magnetron sputtering, Appl. Energy. 228 (2018) 881–890. https://doi.org/https://doi.org/10.1016/j.apenergy.2018.07.001

[27]   M. Akhtari Zavareh, B. Abd Razak, M.H. Bin Wahab, B.T. Goh, R. Mahmoodian, K. Wasa, Fabrication of Pb(Zr,Ti)O3 thin films utilizing unconventional powder magnetron sputtering (PMS), Ceram. Int. 46 (2020) 1281–1296. https://doi.org/https://doi.org/10.1016/j.ceramint.2019.09.013

[28]   A. Schatz, D. Pantel, T. Hanemann, Towards low-temperature deposition of piezoelectric Pb(Zr,Ti)O3: Influence of pressure and temperature on the properties of pulsed laser deposited Pb(Zr,Ti)O3, Thin Solid Films. 636 (2017) 680–687. https://doi.org/https://doi.org/10.1016/j.tsf.2017.06.045

[29]   S. Jiao, Y. Zhang, Z. Duan, T. Wang, Y. Tang, X. Zhao, D. Sun, W. Shi, F. Wang, Influence of oxygen pressure on the electrical properties of Mn-doped Bi0.5Na0.5TiO3BaTiO3 thin films by pulsed laser deposition, Ceram. Int. 45 (2019) 13518–13522. https://doi.org/https://doi.org/10.1016/j.ceramint.2019.04.056

[30]   Z. Zhang, Y. Long, L. Nie, S. Yao, Molecularly imprinted thin film self-assembled on piezoelectric quartz crystal surface by the sol–gel process for protein recognition, Biosens. Bioelectron. 21 (2006) 1244–1251. https://doi.org/https://doi.org/10.1016/j.bios.2005.05.009

[31]   C.-C. Tsai, S.-Y. Chu, C.-S. Hong, Y.-C. Chien, C.-C. Lin, Effects of annealing temperature and pressure of vacuum infiltration on the electrical properties of Pb(Zr0.52Ti0.48)O3 thick films prepared via a modified sol–gel method, Thin Solid Films. 706 (2020) 138071. https://doi.org/https://doi.org/10.1016/j.tsf.2020.138071

[32]   M. Zhang, T. Gao, J. Wang, J. Liao, Y. Qiu, Q. Yang, H. Xue, Z. Shi, Y. Zhao, Z. Xiong, L. Chen, A hybrid fibers based wearable fabric piezoelectric nanogenerator for energy harvesting application, Nano Energy. 13 (2015) 298–305. https://doi.org/https://doi.org/10.1016/j.nanoen.2015.02.034

[33]   S. Bai, L. Zhang, Q. Xu, Y. Zheng, Y. Qin, Z.L. Wang, Two dimensional woven nanogenerator, Nano Energy. 2 (2013) 749–753. https://doi.org/https://doi.org/10.1016/j.nanoen.2013.01.001

[34]   W. Wu, S. Bai, M. Yuan, Y. Qin, Z.L. Wang, T. Jing, Lead Zirconate Titanate Nanowire Textile Nanogenerator for Wearable Energy-Harvesting and Self-Powered Devices, ACS Nano. 6 (2012) 6231–6235. https://doi.org/10.1021/nn3016585

[35]   Y. Ahn, S. Song, K.-S. Yun, Woven flexible textile structure for wearable power-generating tactile sensor array, Smart Mater. Struct. 24 (2015) 75002. https://doi.org/10.1088/0964-1726/24/7/075002

[36]   S. Lee, S.-H. Bae, L. Lin, Y. Yang, C. Park, S.-W. Kim, S.N. Cha, H. Kim, Y.J. Park, Z.L. Wang, Super-Flexible Nanogenerator for Energy Harvesting from Gentle Wind and as an Active Deformation Sensor, Adv. Funct. Mater. 23 (2013) 2445–2449. https://doi.org/https://doi.org/10.1002/adfm.201202867

[37]   Z. He, B. Gao, T. Li, J. Liao, B. Liu, X. Liu, C. Wang, Z. Feng, Z. Gu, Piezoelectric-Driven Self-Powered Patterned Electrochromic Supercapacitor for Human Motion Energy Harvesting, ACS Sustain. Chem. Eng. 7 (2019) 1745–1752. https://doi.org/10.1021/acssuschemeng.8b05606

[38]   K. Shi, B. Sun, X. Huang, P. Jiang, Synergistic effect of graphene nanosheet and BaTiO3 nanoparticles on performance enhancement of electrospun PVDF nanofiber mat for flexible piezoelectric nanogenerators, Nano Energy. 52 (2018) 153–162. https://doi.org/https://doi.org/10.1016/j.nanoen.2018.07.053

[39]   C. Dagdeviren, B.D. Yang, Y. Su, P.L. Tran, P. Joe, E. Anderson, J. Xia, V. Doraiswamy, B. Dehdashti, X. Feng, B. Lu, R. Poston, Z. Khalpey, R. Ghaffari, Y. Huang, M.J. Slepian, J.A. Rogers, Conformal piezoelectric energy harvesting and storage from motions of the heart, lung, and diaphragm, Proc. Natl. Acad. Sci. 111 (2014) 1927 LP – 1932. https://doi.org/10.1073/pnas.1317233111

[40]   K.-I. Park, J.H. Son, G.-T. Hwang, C.K. Jeong, J. Ryu, M. Koo, I. Choi, S.H. Lee, M. Byun, Z.L. Wang, K.J. Lee, Highly-Efficient, Flexible Piezoelectric PZT Thin Film Nanogenerator on Plastic Substrates, Adv. Mater. 26 (2014) 2514–2520. https://doi.org/https://doi.org/10.1002/adma.201305659

[41]   H. Zhang, X.-S. Zhang, X. Cheng, Y. Liu, M. Han, X. Xue, S. Wang, F. Yang, S. A S, H. Zhang, Z. Xu, A flexible and implantable piezoelectric generator harvesting energy from the pulsation of ascending aorta: in vitro and in vivo studies, Nano Energy. 12 (2015) 296–304. https://doi.org/https://doi.org/10.1016/j.nanoen.2014.12.038

[42]   H. Kim, S.M. Kim, H. Son, H. Kim, B. Park, J. Ku, J.I. Sohn, K. Im, J.E. Jang, J.-J. Park, O. Kim, S. Cha, Y.J. Park, Enhancement of piezoelectricity via electrostatic effects on a textile platform, Energy Environ. Sci. 5 (2012) 8932–8936. https://doi.org/10.1039/C2EE22744D

[43]   M.M. Alam, D. Mandal, Native Cellulose Microfiber-Based Hybrid Piezoelectric Generator for Mechanical Energy Harvesting Utility, ACS Appl. Mater. Interfaces. 8 (2016) 1555–1558. https://doi.org/10.1021/acsami.5b08168

[44]   S.K. Karan, S. Maiti, A.K. Agrawal, A.K. Das, A. Maitra, S. Paria, A. Bera, R. Bera, L. Halder, A.K. Mishra, J.K. Kim, B.B. Khatua, Designing high energy conversion efficient bio-inspired vitamin assisted single-structured based self-powered piezoelectric/wind/acoustic multi-energy harvester with remarkable power density, Nano Energy. 59 (2019) 169–183. https://doi.org/https://doi.org/10.1016/j.nanoen.2019.02.031

[45]   S. Maiti, S. Kumar Karan, J. Lee, A. Kumar Mishra, B. Bhusan Khatua, J. Kon Kim, Bio-waste onion skin as an innovative nature-driven piezoelectric material with high energy conversion efficiency, Nano Energy. 42 (2017) 282–293. https://doi.org/https://doi.org/10.1016/j.nanoen.2017.10.041

[46]   Q. Zheng, H. Zhang, H. Mi, Z. Cai, Z. Ma, S. Gong, High-performance flexible piezoelectric nanogenerators consisting of porous cellulose nanofibril (CNF)/poly(dimethylsiloxane) (PDMS) aerogel films, Nano Energy. 26 (2016) 504–512. https://doi.org/https://doi.org/10.1016/j.nanoen.2016.06.009

[47]   S.K. Ghosh, D. Mandal, Efficient natural piezoelectric nanogenerator: Electricity generation from fish swim bladder, Nano Energy. 28 (2016) 356–365. https://doi.org/https://doi.org/10.1016/j.nanoen.2016.08.030

[48]   K.Y. Lee, D. Kim, J.-H. Lee, T.Y. Kim, M.K. Gupta, S.-W. Kim, Unidirectional High-Power Generation via Stress-Induced Dipole Alignment from ZnSnO3 Nanocubes/Polymer Hybrid Piezoelectric Nanogenerator, Adv. Funct. Mater. 24 (2014) 37–43. https://doi.org/https://doi.org/10.1002/adfm.201301379

[49]   B. Dudem, D.H. Kim, L.K. Bharat, J.S. Yu, Highly-flexible piezoelectric nanogenerators with silver nanowires and barium titanate embedded composite films for mechanical energy harvesting, Appl. Energy. 230 (2018) 865–874. https://doi.org/https://doi.org/10.1016/j.apenergy.2018.09.009

[50]   L. Ghasemi-Mobarakeh, M.P. Prabhakaran, M. Morshed, M.H. Nasr-Esfahani, H. Baharvand, S. Kiani, S.S. Al-Deyab, S. Ramakrishna, Application of conductive polymers, scaffolds and electrical stimulation for nerve tissue engineering, J. Tissue Eng. Regen. Med. 5 (2011) e17–e35. https://doi.org/https://doi.org/10.1002/term.383

[51]   P.R. Bidez, S. Li, A.G. MacDiarmid, E.C. Venancio, Y. Wei, P.I. Lelkes, Polyaniline, an electroactive polymer, supports adhesion and proliferation of cardiac

myoblasts, J. Biomater. Sci. Polym. Ed. 17 (2006) 199–212.
https://doi.org/10.1163/156856206774879180

[52] C. Ribeiro, V. Sencadas, D.M. Correia, S. Lanceros-Méndez, Piezoelectric polymers as biomaterials for tissue engineering applications, Colloids Surfaces B Biointerfaces. 136 (2015) 46–55.
https://doi.org/https://doi.org/10.1016/j.colsurfb.2015.08.043

[53] C. Frias, J. Reis, F. Capela e Silva, J. Potes, J. Simões, A.T. Marques, Polymeric piezoelectric actuator substrate for osteoblast mechanical stimulation, J. Biomech. 43 (2010) 1061–1066. https://doi.org/https://doi.org/10.1016/j.jbiomech.2009.12.010

[54] H.-F. Guo, Z.-S. Li, S.-W. Dong, W.-J. Chen, L. Deng, Y.-F. Wang, D.-J. Ying, Piezoelectric PU/PVDF electrospun scaffolds for wound healing applications, Colloids Surfaces B Biointerfaces. 96 (2012) 29–36.
https://doi.org/https://doi.org/10.1016/j.colsurfb.2012.03.014

[55] Y. Nie, P. Deng, Y. Zhao, P. Wang, L. Xing, Y. Zhang, X. Xue, The conversion of PN-junction influencing the piezoelectric output of a CuO/ZnO nanoarray nanogenerator and its application as a room-temperature self-powered active H2S sensor, Nanotechnology. 25 (2014) 265501. https://doi.org/10.1088/0957-4484/25/26/265501

[56] X. Xue, Y. Nie, B. He, L. Xing, Y. Zhang, Z.L. Wang, Surface free-carrier screening effect on the output of a ZnO nanowire nanogenerator and its potential as a self-powered active gas sensor, Nanotechnology. 24 (2013) 225501.
https://doi.org/10.1088/0957-4484/24/22/225501

[57] D. Zhang, Q. Mi, D. Wang, T. Li, MXene/Co3O4 composite based formaldehyde sensor driven by ZnO/MXene nanowire arrays piezoelectric nanogenerator, Sensors Actuators B Chem. 339 (2021) 129923.
https://doi.org/https://doi.org/10.1016/j.snb.2021.129923

[58] S. Joshi, M. Parmar, K. Rajanna, A novel gas flow sensing application using piezoelectric ZnO thin films deposited on Phynox alloy, Sensors Actuators A Phys. 187 (2012) 194–200. https://doi.org/https://doi.org/10.1016/j.sna.2012.08.032

[59] Y. Hu, Y. Zhang, C. Xu, L. Lin, R.L. Snyder, Z.L. Wang, Self-Powered System with Wireless Data Transmission, Nano Lett. 11 (2011) 2572–2577.
https://doi.org/10.1021/nl201505c

[60] N.R. Alluri, B. Saravanakumar, S.-J. Kim, Flexible, Hybrid Piezoelectric Film (BaTi(1–x)ZrxO3)/PVDF Nanogenerator as a Self-Powered Fluid Velocity Sensor,

ACS Appl. Mater. Interfaces. 7 (2015) 9831–9840.
https://doi.org/10.1021/acsami.5b01760

[61]   B. Saravanakumar, S. Soyoon, S.-J. Kim, Self-Powered pH Sensor Based on a
Flexible Organic–Inorganic Hybrid Composite Nanogenerator, ACS Appl. Mater.
Interfaces. 6 (2014) 13716–13723. https://doi.org/10.1021/am5031648

[62]   W. Yan, G. Bai, R. Ye, X. Yang, H. Xie, S. Xu, Dual-mode luminescence tuning
of Er3+ doped Zinc Sulfide piezoelectric microcrystals for multi-dimensional anti-
counterfeiting and temperature sensing, Opt. Commun. 475 (2020) 126262.
https://doi.org/https://doi.org/10.1016/j.optcom.2020.126262

[63]   M. Lee, J. Bae, J. Lee, C.-S. Lee, S. Hong, Z.L. Wang, Self-powered
environmental sensor system driven by nanogenerators, Energy Environ. Sci. 4 (2011)
3359–3363. https://doi.org/10.1039/C1EE01558C

[64]   R. Li, Y. Yu, B. Zhou, Q. Guo, M. Li, J. Pei, Harvesting energy from pavement
based on piezoelectric effects: Fabrication and electric properties of piezoelectric
vibrator, J. Renew. Sustain. Energy. 10 (2018) 54701.
https://doi.org/10.1063/1.5002731

[65]   C.A. Nelson, S.R. Platt, D. Albrecht, V. Kamarajugadda, M. Fateh, Power
harvesting for railroad track health monitoring using piezoelectric and inductive
devices, in: Proc.SPIE, 2008. https://doi.org/10.1117/12.775884

[66]   N. Bosso, M. Magelli, N. Zampieri, Application of low-power energy harvesting
solutions in the railway field: a review, Veh. Syst. Dyn. 59 (2021) 841–871.
https://doi.org/10.1080/00423114.2020.1726973

[67]   A. Erturk, G. Delporte, Underwater thrust and power generation using flexible
piezoelectric composites: an experimental investigation toward self-powered
swimmer-sensor platforms, Smart Mater. Struct. 20 (2011) 125013.
https://doi.org/10.1088/0964-1726/20/12/125013

[68]   W. Deng, T. Yang, L. Jin, C. Yan, H. Huang, X. Chu, Z. Wang, D. Xiong, G.
Tian, Y. Gao, H. Zhang, W. Yang, Cowpea-structured PVDF/ZnO nanofibers based
flexible self-powered piezoelectric bending motion sensor towards remote control of
gestures, Nano Energy. 55 (2019) 516–525.
https://doi.org/https://doi.org/10.1016/j.nanoen.2018.10.049

[69]   O. Puscasu, N. Counsell, M.R. Herfatmanesh, R. Peace, J. Patsavellas, R. Day,
Powering Lights with Piezoelectric Energy-Harvesting Floors, Energy Technol. 6
(2018) 906–916. https://doi.org/https://doi.org/10.1002/ente.201700629

[70]   S. Orrego, K. Shoele, A. Ruas, K. Doran, B. Caggiano, R. Mittal, S.H. Kang, Harvesting ambient wind energy with an inverted piezoelectric flag, Appl. Energy. 194 (2017) 212–222. https://doi.org/https://doi.org/10.1016/j.apenergy.2017.03.016

[71]   L. Jin, S. Ma, W. Deng, C. Yan, T. Yang, X. Chu, G. Tian, D. Xiong, J. Lu, W. Yang, Polarization-free high-crystallization β-PVDF piezoelectric nanogenerator toward self-powered 3D acceleration sensor, Nano Energy. 50 (2018) 632–638. https://doi.org/https://doi.org/10.1016/j.nanoen.2018.05.068

[72]   G. Wang, Y. Li, H. Cui, X. Yang, C. Yang, N. Chen, Acceleration self-compensation mechanism and experimental research on shock wave piezoelectric pressure sensor, Mech. Syst. Signal Process. 150 (2021) 107303. https://doi.org/https://doi.org/10.1016/j.ymssp.2020.107303

[73]   I. Payo, J.M. Hale, Dynamic characterization of piezoelectric paint sensors under biaxial strain, Sensors Actuators A Phys. 163 (2010) 150–158. https://doi.org/https://doi.org/10.1016/j.sna.2010.08.005

Keyword Index

About the Editors

Dr. Inamuddin is working as Assistant Professor at the Department of Applied Chemistry, Aligarh Muslim University, Aligarh, India. He obtained Master of Science degree in Organic Chemistry from Chaudhary Charan Singh (CCS) University, Meerut, India, in 2002. He received his Master of Philosophy and Doctor of Philosophy degrees in Applied Chemistry from Aligarh Muslim University (AMU), India, in 2004 and 2007, respectively. He has extensive research experience in multidisciplinary fields of Analytical Chemistry, Materials Chemistry, and Electrochemistry and, more specifically, Renewable Energy and Environment. He has worked on different research projects as project fellow and senior research fellow funded by University Grants Commission (UGC), Government of India, and Council of Scientific and Industrial Research (CSIR), Government of India. He has received Fast Track Young Scientist Award from the Department of Science and Technology, India, to work in the area of bending actuators and artificial muscles. He has also received the Sir Syed Young Researcher of the Year Award 2020 from Aligarh Muslim University. He has completed four major research projects sanctioned by University Grant Commission, Department of Science and Technology, Council of Scientific and Industrial Research, and Council of Science and Technology, India. He has published 200 research articles in international journals of repute and nineteen book chapters in knowledge-based book editions published by renowned international publishers. He has published 150 edited books with Springer (U.K.), Elsevier, Nova Science Publishers, Inc. (U.S.A.), CRC Press Taylor & Francis Asia Pacific, Trans Tech Publications Ltd. (Switzerland), IntechOpen Limited (U.K.), Wiley-Scrivener, (U.S.A.) and Materials Research Forum LLC (U.S.A). He is a member of various journals' editorial boards. He is also serving as Associate Editor for journals (Environmental Chemistry Letter, Applied Water Science and Euro-Mediterranean Journal for Environmental Integration, Springer-Nature), Frontiers Section Editor (Current Analytical Chemistry, Bentham Science Publishers), Editorial Board Member (Scientific Reports-Nature), Editor (Eurasian Journal of Analytical Chemistry), and Review Editor (Frontiers in Chemistry, Frontiers, U.K.). He is also guest-editing various special thematic special issues to the journals of Elsevier, Bentham Science Publishers, and John Wiley & Sons, Inc. He has attended as well as chaired sessions in various international and national conferences. He has worked as a Postdoctoral Fellow, leading a research team at the Creative Research Initiative Center for Bio-Artificial Muscle, Hanyang University, South Korea, in the field of renewable energy, especially biofuel cells. He has also worked as a Postdoctoral Fellow at the Center of Research Excellence in Renewable Energy, King Fahd University of Petroleum and Minerals, Saudi Arabia, in the field of polymer electrolyte membrane fuel cells and computational fluid dynamics of polymer electrolyte membrane fuel cells. He is a life member of the Journal of the Indian Chemical Society. His research interest includes ion exchange materials, a sensor for heavy metal ions, biofuel cells, supercapacitors and bending actuators.

Tariq Altalhi, PhD, is working as Associate Professor in the Department of Chemistry at Taif University, Saudi Arabia. He received his doctorate degree from University of Adelaide, Australia in the year 2014 with Dean's Commendation for Doctoral Thesis Excellence. He has worked as head of Chemistry Department at Taif university and Vice Dean of Science college. In 2015, one of his works was nominated for Green Tech awards from Germany, Europe's largest environmental and business prize, amongst top 10 entries. He has co-edited various scientific books. His group is involved in fundamental multidisciplinary research in nanomaterial synthesis and engineering, characterization, and their application in molecular separation, desalination, membrane systems, drug delivery, and biosensing. In addition, he has established key contacts with major industries in Kingdom of Saudi Arabia.

Dr. Mohammad Luqman has 12+ years of post-PhD experience in Teaching, Research, and Administration. Currently, he is serving as an Assistant Professor of Chemical Engineering in Taibah University, Saudi Arabia. Before joining there, he served as an Assistant Professor in College of Applied Science at A'Sharqiyah University, Oman, and in College of Engineering at King Saud University, Saudi Arabia. He served as a Research Engineer in SAMSUNG Cheil Industries, South Korea. Moreover, he served as a post-doctoral fellow at Artificial Muscle Research Center, Konkuk University, South Korea, in the field of Ionic Polymer Metal Composites for the development of Artificial Muscles, Robotic Actuators and Dynamic Sensors. He earned his PhD degree in the field of Ionomers (Ion-containing Polymers), from Chosun University, South Korea. He successfully served as an Editor to three books, published by world renowned publishers. He published numerous high-quality papers, and book chapters. He is serving as an Editor and editorial/review board members to many International SCI and Non-SCI journals. He has attracted a few important research grants from industry and academia. His research interests include but not limited to Development of Ionomer/Polyelectrolyte/non-ionic Polymer Nanocomposites/Blends for Smart and Industrial/Engineering Applications.

Dr. Hamida-Tun-Nisa Chisti is currently working as an Associate Professor in the Department of Chemistry, National Institute of Technology Srinagar, India. She has received her B.Sc. Hons., Masters and Ph.D. in Chemistry from Aligarh Muslim University, Aligarh, and later has joined her services as Lecturer in 2008 at NIT Srinagar. Her research focus is in the fields of Inorganic Chemistry, Environmental Chemistry, and Material Chemistry. She has published several research articles in international journals of repute. She has authored 4 books and 6 book chapters. She is a Member Royal Society of Chemistry (MRSC), ACS and also a life member of Asian Polymer Association, Indian Council of Chemists, Life member of the International Association of Engineers (**IAENG**), Hong Kong, Life member of Eco Ethics International Union (**EEIU**), Germany and many more. She is also serving as a reviewer for many journals.

www.ingramcontent.com/pod-product-compliance
Lightning Source LLC
Chambersburg PA
CBHW071332210326
41597CB00015B/1428